INVESTIGATING MICROBIOLOGY

A Laboratory Manual for General Microbiology

Philip E. Stukus, Ph.D.

INVESTIGATING MICROBIOLOGY

A Laboratory Manual for General Microbiology

Philip E. Stukus, Ph.D.
Denison University

SAUNDERS COLLEGE PUBLISHING
Harcourt Brace College Publishers

FORT WORTH PHILADELPHIA SAN DIEGO NEW YORK ORLANDO AUSTIN
SAN ANTONIO TORONTO MONTREAL LONDON SYDNEY TOKYO

Text Typeface: Palatino 11/13
Compositor: York Graphic Services
Executive Editor: Edith Beard Brady
Developmental Editor: Cathleen E. Petree
Project Editor: Vanessa Ray
Production Manager: Arnold Lynch, Jr.
Art Director: Caroline McGowan
Text Designer: Ruth Hoover
Cover Designer: Chazz Bjanes
Photo Researcher: Amy Ellis Dunleavy
Text Artwork: Rolin Graphics

Cover Image: *E. coli* Bacteria. (NIBC/Science Photo Library, Courtesy of Photo Researchers, Inc.)

Printed in the United States of America

Investigating Microbiology: A Laboratory Manual for General Microbiology

ISBN 0-03-018394-4

Library of Congress Catalog Number: 96-71993
7890123456 071 10 987654321

To my beloved and courageous brother, Tim, whose dedication, perseverance, and love provided the greatest inspiration in completing this manual.

PREFACE

Approach

This laboratory manual, *Investigating Microbiology: A Laboratory Manual for General Microbiology,* includes a number of traditional laboratories as well as more experimental, investigative open-ended laboratories. In some cases, students will play an active role in the design of experimental investigations, and will need to make decisions concerning the protocol and execution of experiments. Many of the core laboratories involve the characterization of organisms isolated from nature. A semester-length project provides the opportunity to perform research on the isolation of a specific type of microorganism. The product of the research is a formal manuscript.

In one of his speeches, Louis Pasteur stated: "Life is a Germ and a Germ is Life." Microorganisms (germs) are essential to life as we know it on the planet Earth, and it behooves us to gain a greater understanding of the vital roles that they play.

Overview of Laboratories

Unit I includes the basic microbiological techniques of macroscopic and microscopic examination, and staining. The mastery of basic techniques is vital to the process of good microbiology.

Unit II provides information on various ways in which cultures can be maintained. Also included is a laboratory involving the preparation of media.

Unit III describes three investigative projects. Laboratory 13 involves the isolation and microbiological characterization of an environment of your choice. Laboratory 14 fosters teamwork by requiring the execution of a group-designed experiment. The strategy for a semester-length project is given in Laboratory 15.

Unit IV focuses on the growth and metabolism of microorganisms. In Laboratory 16 the nutrition and growth patterns of three organisms are monitored on a variety of media. Sharing of class results is key for this experiment. The construction of a Winogradsky column (Laboratory 17) is another type of investigative laboratory, requiring creativity and diligent, careful observations over many weeks. Anaerobic cultivation techniques are learned (Laboratory 18) and applied to the isolation of sulfate-reducing bacteria in Laboratory 19. The growth of a strict autotroph, *Thiobacillus thioxidans,* is monitored in Laboratory 20.

Unit V includes most of the standard biochemical tests and miniaturized procedures used for the identification of microorganisms isolated from nature in Laboratory 13. The tests are grouped according to the type of metabolic reaction: extracellular degradation, fermentation of carbohydrates, measurement of other products of metabolism, and biochemical reactions important in types of respiration. Sections are also included for tests significant in the identification of pathogens, and for rapid miniaturized tests.

Unit VI explores the effects that chemical and physical agents have on bacteria. Some of the organisms in Unit V can be carried forth in the laboratories in Unit VI.

Unit VII laboratories deal with the isolation of microorganisms from air, water, soil, and food. Core laboratories present basic sampling strategies. Other applied labs provide opportunities for more investigative explorations.

Unit VIII includes laboratories dealing with the isolation and identification of algae, fungi, and protozoa from pond water and other sources.

Unit IX laboratories provide techniques used in the isolation and propagation of bacterial viruses.

Unit X includes laboratories that contrast bacterial variation and mutation. Potential carcinogenicity of chemicals is determined in the Ames test (Laboratory 63), and an examination of genetic exchanges (transformation and conjugation) is presented in Laboratories 64 and 65.

Unit XI incorporates laboratories that provide an overview of medical microbiology and immunological applications.

General Guidelines

The laboratory study of microorganisms is tedious, time-consuming, and dependent upon the perfection of fundamental laboratory technique. Sloppiness cannot be tolerated. In the laboratory, millions of microorganisms will be encountered on a daily basis. Since living organisms will be manipulated, care must be exercised not to release these organisms into the environment. On occasion, work will involve known pathogens (disease-causing microorganisms). However, it must be remembered that any organism is capable of eliciting an adverse reaction if it enters an appropriate environment in a susceptible host. You must be diligent at all times to protect yourself and others from potentially dangerous exposure to these agents. **Please carefully read the safety regulations found in the Student Safety Contract and sign to acknowledge that you have read them.** Safety and biohazard issues are identified in boxes at the beginning of each laboratory. In addition, safety and biohazard symbols appear in the text to indicate when a hazard exists and when special precautions need to be taken.

Manual Organization

There are 74 experiments in the manual. Each of the experiments has stated objectives to be mastered in that laboratory. Basic background information is provided. It is expected that students will consult their textbooks for more detailed information on the topic. A list of all material is provided, and carefully outlined procedural instructions are presented. A report section appears at the end of each laboratory and should be carefully completed at the conclusion of each laboratory. Thought-provoking review questions are incorporated. These provide an opportunity to reflect on the significance of the work accomplished, as well as to extend results to the interpretation of other problems. In some cases, consultation with other references may be necessary in answering the questions. Appendices provide information on the preparation of all stains, reagents, and media used in the experiments. A listing of all organisms used in the manual is also provided. Other appendices illustrate techniques in spectrophotometer use and in pipetting and dilution techniques. Information on identification schemes and semester-length project guidelines are found in other appendices.

Instructor's Manual

An instructor's manual accompanies this laboratory manual, providing a complete listing of all media, reagents, cultures, and supplies needed for each experiment. Helpful suggestions for the successful implementation of each laboratory are offered. Answers to all of the review questions are also found in the instructor's manual.

Acknowledgments

I extend my deep appreciation to the many reviewers whose comments were extremely valuable in the preparation of this manual:

Rodney Anderson, Ohio Northern University
James Barbaree, Auburn University
Joan Braddock, University of Alaska at Fairbanks
William Coleman, University of Hartford
Jan Decker, University of Arizona
Diane Gambill, Southern Methodist University
John Lennox, Pennsylvania State University at Altoona
Lynne Lucher, University of Alaska at Anchorage
Gordon D. Schrank, St. Cloud State University
Sara Silverstone, SUNY at Brockport
Linda Simpson, University of North Carolina at Charlotte
Brenda Speer, University of Maine
Pam Tabery, North Hampton County Area Community College
Susan Williams, Oregon State University

Finally, special thanks goes out to my wife, Kathy, for her patience, understanding, and support during the writing of this manual.

Philip E. Stukus, Ph.D.
Denison University
January, 1997

Saunders College Publishing may provide complimentary instructional aids and supplements or supplement packages to those adopters qualified under our adoption policy. Please contact your sales representative for more information. If as an adopter or potential user you receive supplements you do not need, please return them to your sales representative or send them to:

Attn: Returns Department
Troy Warehouse
465 South Lincoln Drive
Troy, MO 63379

STUDENT SAFETY CONTRACT

Safety Rules*

There are a number of safety rules that need to be followed carefully while you are working in the laboratory. Read the following rules and acknowledge your reading and understanding by signing your name at the end of the list.

1. No food or drink is allowed in the laboratory.
2. A lab coat must be worn at all times.
3. When you are boiling any material or pipetting acids or strong bases, you must wear protective goggles or safety glasses.
4. Report all spills at once to your instructor or lab assistant. Cover spills with paper towels that have been soaked with a disinfectant such as sodium hypochlorite, 70% ethanol, or Lysol. Keep the paper towels in place for at least 15 min before discarding in a biohazard bag.
5. Mouth pipetting is strictly forbidden. Use mechanical pipetting devices or rubber bulbs and discard disposable pipettes in a tray of disinfectant. Glass pipettes should be placed in pipette jars containing a disinfectant.
6. The lab bench area should be cleaned with disinfectant before you begin work and again before you leave the laboratory.
7. Hands should be thoroughly washed with disinfectant soap before laboratory work and after the day's work has been completed.
8. Lab benches should be kept clear of coats, books, or other personal items. Keep only those items on the lab bench that are needed to perform experiments.
9. Discard old cultures in designated decontamination areas. Remove all tape and markings from tubes before discarding.
10. Cultures, slides, or other material should never be taken from the laboratory unless permission from the instructor is given.
11. If access to the laboratory is allowed after class hours, make sure to go with a lab partner or a friend; do not work in the lab alone.
12. Shoes must be worn at all times.
13. Tie back long hair to avoid contact with the open flame of burners.
14. Place uncontaminated broken glassware in a separate disposal container. Any broken glassware that is contaminated with microorganisms must be placed in a biohazard container to be decontaminated by autoclaving.
15. Do not attempt to operate equipment that is not essential to the experiment being performed.
16. Igniting paper or burning tubing in the lab are grounds for dismissal from the laboratory.

*Suggested from the work, and with permission, of John M. Lammert, Gustavus Adolphus College.

17. Do not wet adhesive labels by mouth. If such labels are used, moisten them with a wet paper towel.
18. Locate the nearest eyewash station and become familiar with its operation, in the event that any chemical splashes in your eyes.
19. A first aid kit is available to treat burns or cuts. Report any accident to your instructor or lab assistant.
20. Locate the fire extinguisher. To extinguish a small fire caused by ignition of a small amount of paper or alcohol, smother with a towel, beaker, or glass plate. If a fire gets out of control, activate the fire alarm, turn off all gas, and leave the laboratory via the nearest exit.
21. Blood and blood products have been implicated in the transmission of human immunodeficiency virus (HIV), the causative agent of AIDS. In the laboratory experiment involving human blood typing, students should wear gloves, and work only with their own blood. Dispose of all slides in the disinfectant container. Place all blood-contaminated material (cotton, Band-Aids, towels) in the biohazard bag.
22. If you smell gas in the laboratory, check for a leak in burner tubing or an open stopcock.
23. On occasion, some of your friends may visit the lab with you during unscheduled laboratory hours. They are welcome, but they must be made aware of all laboratory rules and are expected to abide by them.

Safety and biohazard icons are placed in the manual to alert you to steps that pose a health or safety risk to you. These steps should be carried out with extreme caution.

I, the undersigned, have read the above rules, and agree to observe and abide by them.

______________________________ ______________________________

Student Signature Date

General Suggestions

To ensure an understanding of the lab concepts, precautions, and preliminary preparation, it is essential that students carefully read the laboratories before coming to lab.

The first exposure to a discipline like microbiology bombards one with a dizzying array of unfamiliar terms. It is critical that microbiological jargon be understood so that effective communication is possible.

Make sure to locate the eyewash station and the fire extinguisher. Identify the location of the incubators, refrigerators, balances, and autoclave that will be used during the course of the semester.

You should have routine access to the following items:

(a) Bunsen burner
(b) Glass slides
(c) Coverslips
(d) Lens paper
(e) Bibulous paper
(f) Inoculating loop and needle
(g) Forceps
(h) Culture tube rack
(i) Marking pens
(j) Slide storage box
(k) Clothespins

Carefully complete the report sections for each of the laboratories. Maintaining complete data records is fundamental to the understanding and interpretation of scientific phenomena. Enjoy your explorations in the wondrous world of microbiology.

CONTENTS

UNIT I

Basic Microbiological Techniques

This first unit on basic microbiological techniques is extremely important in that it provides exposure to the basic manipulative techniques that are used repeatedly during the semester. Mastery of these techniques is essential for successful completion of work that lies ahead. Exposure is also provided to basic microscopy and staining techniques that help to visualize and differentiate the various morphological bacterial types. The Gram stain is central to microbiology since it allows for the differentiation of two major groups of bacteria.

LABORATORY

1 Macroscopic Observation of Microorganisms

Safety Issues

- Colonies on the Rodac plates comprise millions of microorganisms. Be careful not to touch any open plate as you may contaminate yourself with the bacteria.
- Tubes of inoculated broth must always be kept upright in a rack or support. Screw-capped tubes may not always be tight and other types of closures may not have leak-proof barriers. If any culture is spilled, notify your lab instructor immediately.

OBJECTIVES

1. Differentiate microorganisms based upon differences in colony morphology.
2. Describe the nature of bacterial growth on a solid surface and in a liquid.

Materials (per 20 students)

1. Exposed nutrient agar plates (standard or Rodac) showing bacterial colonies (20)
2. Nutrient broth cultures (10)

Each student has been provided with a plastic plate called a Petri plate which contains solid material that promotes the growth of bacteria. We speak of this material as a **medium** (plural, **media**) used for the propagation of bacteria. Media can exist in three forms: solid, liquid, or semi-solid. A solid medium is used to propagate organisms leading to the formation of bacterial colonies. Semi-solid media have lower concentrations of agar and are used for determination of bacterial motility. In solid or semi-solid media, solidifying agents are used. In most cases the agent used is **agar**, a complex carbohydrate produced by red algae, consisting of galacturonic acid and galactose. In a solid medium, the concentration of agar used is usually 15 g per liter or 1.5% (w/v). Agar is used because it will remain solid under most routine laboratory incubation temperatures. Agar must be heated to boiling in order to dissolve and will resolidify at 42 to 45°C.

Examine the surface of the agar medium for any macroscopic growth. Each identifiable unit is referred to as a **colony** and represents a population of bacteria or mold which has arisen from a single cell or small cluster of cells deposited on the plate. Microbes were either impacted or spread onto the solid surface. Since the medium contains nutrients in addition to the agar, the microbial cells divide and give rise to macroscopic colonies. A study of the types of nutrients used to cultivate organisms will be performed

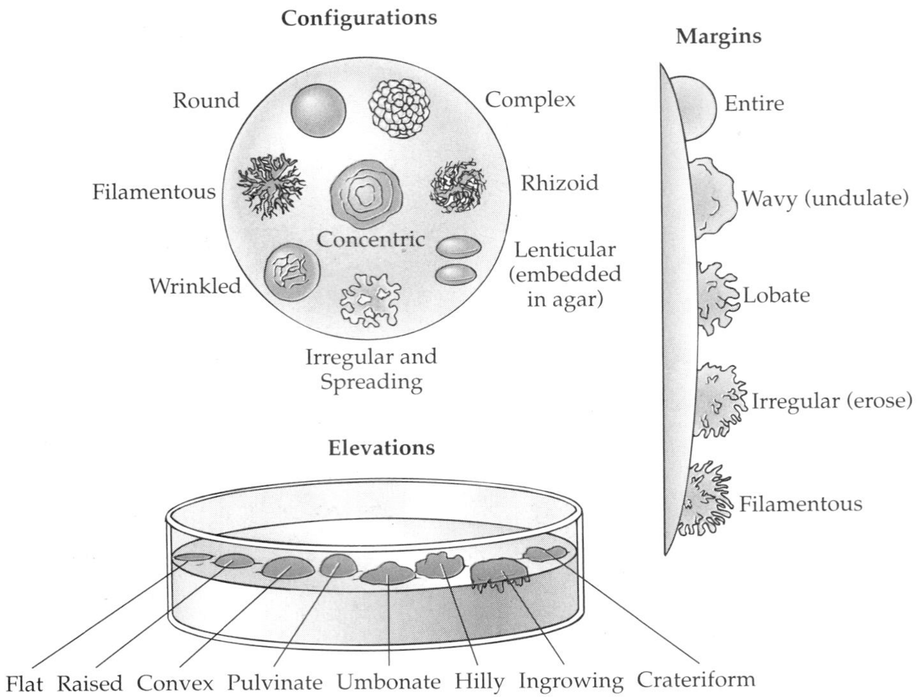

FIGURE 1.1 Morphologies of bacterial colonies.

later, but for now be aware that the medium used contains a readily utilizable source of protein for the bacteria.

Do not open plate. Touching any of the colonies will expose you to millions of organisms.

Carefully examine the nutrient agar plate (Color Plate 1) containing the bacterial colonies. How many different colony types can be observed? Develop criteria to differentiate the various colonial morphologies. List them below.

Each microorganism will give rise to a unique type of colony. Differences in the types of colonies provide information that can be used to help classify and identify bacteria. Care must be exercised in noting the type of medium being used and the environmental conditions employed, as colonial morphologies vary under different conditions. After developing the list of criteria, compare it with the colony characteristics described in Figure 1.1.

With a laboratory marking pen, place a mark on the bottom of the agar plate to identify three colonies for further study. Describe the colonial morphologies of these three colonies.

Make sure to keep broth tubes in an upright position to prevent spillage.

Examine the inoculated tubes of nutrient broth that have been prepared. Compare them with tubes that have not been inoculated with microorganisms. There are a number of observations that should be made routinely when observing the growth of bacteria in liquid media. Note whether the growth, as evidenced by a cloudiness or turbidity, is distributed throughout the tube or seems to be concentrated at the top or bottom of the tube. A culture showing a just visible turbidity usually contains approximately 10^6 microorganisms/ml. What factors may influence where microorganisms grow in the tube?

Figure 1.2 illustrates the growth of microorganisms on the surface of a liquid medium. A number of patterns are possible: some organisms form a thin layer referred to as a **pellicle**, whereas others form a much thicker layer

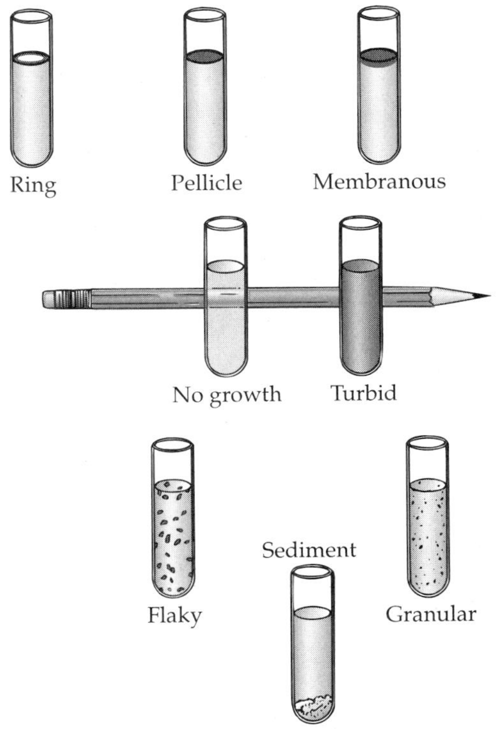

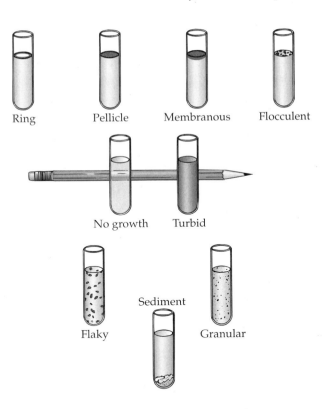

FIGURE 1.2 Bacterial growth in liquid-nutrient media.

known as **membrane**. A **flocculent** pattern of growth distributed throughout the medium represents large clusters of organisms.

The growth below the surface may appear to be uniformly **turbid** (cloudy), **flaky**, or **granular**.

If a sediment is present, gently agitate the tube and determine the type of sediment. Does it appear flaky, granular, or rise as a confluent mass from the bottom of the tube?

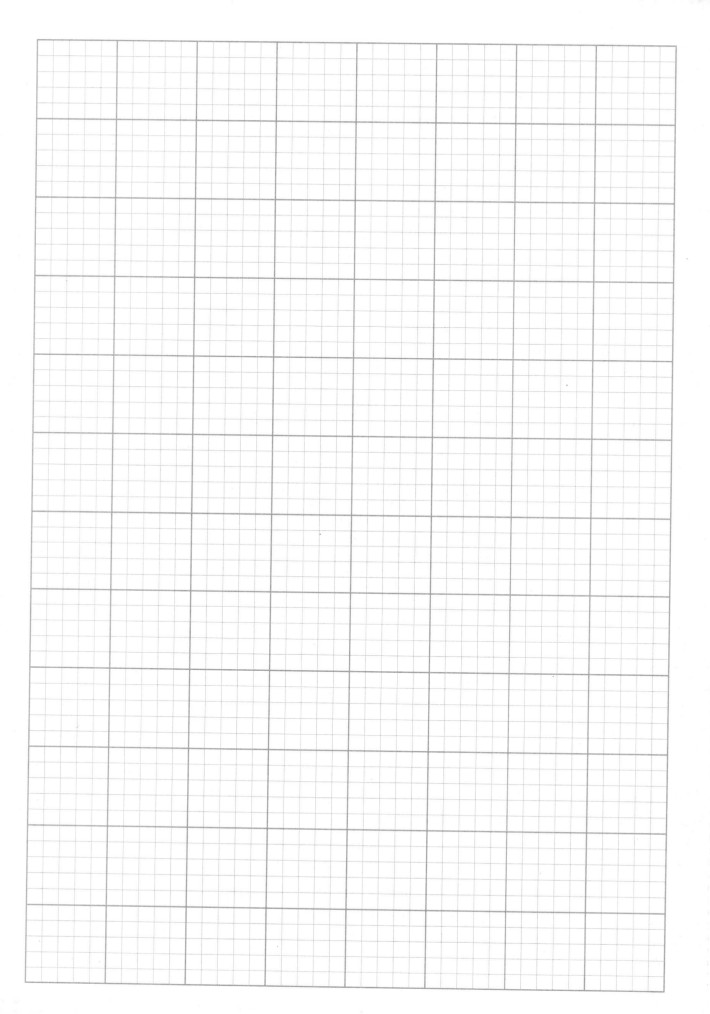

Laboratory Report

Name ______ Date ______ Section ______

1

Macroscopic Observations

1. List the criteria developed to help differentiate bacterial colonies on the surface of the agar plate.

______

______

______

______

______

2. Describe the colonial morphologies of the three selected colonies:

OBSERVATIONS OF COLONY CHARACTERISTICS

	Colony 1	Colony 2	Colony 3
Size			
Shape			
Color			
Margin			
Texture			

3. **OBSERVATIONS OF GROWTH IN NUTRIENT BROTH TUBES**

	Culture A	Culture B	Culture C
Surface Growth			
Sediment			
Overall			

How do you account for the viscosity of any sediment that may be present?

Review Questions

1. Why do some microorganisms give rise to colonies that spread over the surface of a nutrient agar plate?

2. Why do some bacteria grow near the surface of a liquid culture medium whereas others accumulate in the bottom of the tube?

3. Why can't one be certain that each colony on an agar plate has arisen from a single microorganism that has been deposited on the plate?

LABORATORY

2 Manipulation of Cultures

Safety Issues

- Caution must be exercised when using a lighted burner. Keep books and clothing a safe distance away from any flame. Long hair should be tied back to prevent accidental exposure to the flame. Never attempt to reach for any item that may bring you into contact with the burner.
- All culture tubes should be kept upright in a rack to prevent spillage.
- Allow any heated inoculating loop time to cool before removing growth from a colony or liquid from a broth tube. Contact with a hot loop may cause splattering of material and the aerosol produced may be harmful.

OBJECTIVES

1. Learn aseptic culture methods.
2. Learn techniques for transfer of microorganisms from an agar plate to broth and slants.
3. Separate in pure culture all microorganisms from a mixed culture by mastering the streak-plating technique.

ASEPTIC TECHNIQUE

Before learning about the manipulation of cultures, one must first become acquainted with the tools of the microbiologist. **Inoculating loops** or **needles** are used to transfer microorganisms from solid or liquid surfaces to other media. These devices are constructed of either nichrome or platinum (if the laboratory is well-funded!). Since microorganisms are ubiquitous, the loops and needles are contaminated with microorganisms from the environment. Before needles or loops can be used, they must first be sterilized (ensuring that all microorganisms have been destroyed) by passing them through or into the hottest part of the flame of the Bunsen burner, as illustrated in Figure 2.1. Heat the wire first and then slowly draw the loop into the flame. The wire portion must always be sterilized in addition to the loop. This type of sterilization is sterilization by incineration. (Exercise caution that hair or clothing do not pass near the open flame of the Bunsen burner.) The entire length of the wire should turn red hot in the flame. One is not forging steel, so do not hold the wire in the flame for a prolonged period. You must always sterilize wires before and after transferring cultures. This is done to ensure that organisms are not carried over from a previous transfer or acquired as environmental contaminants. When working with patho-

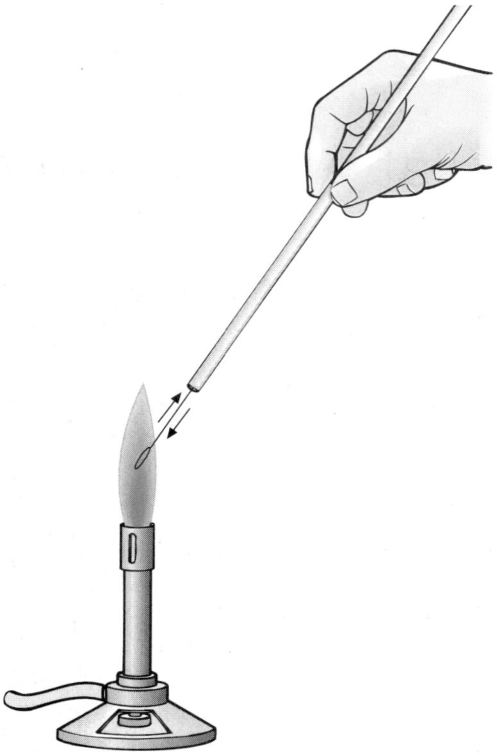

FIGURE 2.1 Technique for loop sterilization.

genic microorganisms, alternative methods of loop sterilization, such as electric incineration, may be used. There is a danger that aerosols can be released from a loop when it is heated in the flame of a Bunsen burner. When transferring microorganisms from one medium to another, the process is referred to as **subculturing**. All transfers must take place using **aseptic technique**. (Asepsis [*a*, neg. + *Gr. sepesthai*, to decay]: freedom from infection or the prevention of contact with microorganisms.) A culture containing a single type of microorganism is called a **pure culture**. In many situations, mixtures of microorganisms are used and such cultures are known as **mixed cultures**. Since microorganisms are present in all environments, precautions must be taken to prevent contamination of cultures.

Humans are good sources of microorganisms: We expel them from the respiratory tract during speaking, coughing, or sneezing; and we liberate numerous organisms as dead flecks of skin are shed from body surfaces! Pigpen, the Charles Schultz character in the Peanuts cartoon strip, is surrounded by a cloud of dust just as humans are surrounded by an invisible cloud of microorganisms.

Materials (per 20 students)

1. Tubes of nutrient broth (20)
2. Nutrient agar slants (20)
3. Petri dishes with nutrient, tryptic soy, or plate count agar (80)
4. Tubes containing a mixed population of bacteria (10)
5. Rodac or Petri plates from Laboratory 1.

Procedure

Using the techniques illustrated in Figure 2.2, carry out the following transfers:

A. Transfer from Plate to Broth

Allow inoculating loop to cool before removing growth from a colony. Using a hot loop may cause splattering of organisms.

1. Sterilize the inoculating loop by passing it though the flame of the Bunsen burner (Figure 2.2a).
2. Using the loop, remove a small amount of growth from a colony on a plate from Laboratory 1 (Figure 2.2b).

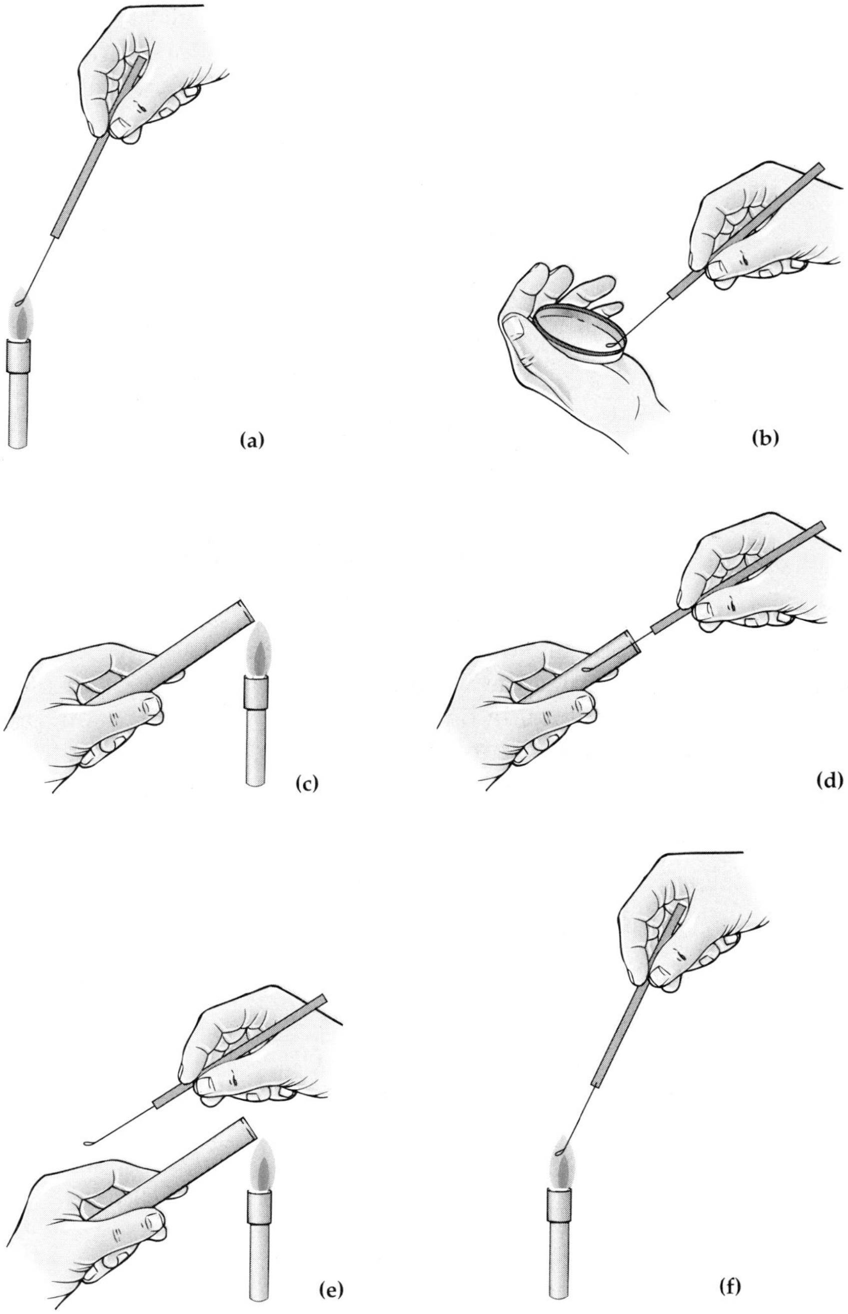

FIGURE 2.2 Technique of transfer from agar plate to broth.

3. Remove the cap of the sterile nutrient broth tube and gently flame the lip of the tube (Figure 2.2c).
4. Transfer the inoculum on the loop to the broth by immersing the inoculating loop in the broth (Figure 2.2d). Twirl or shake the loop to remove organisms.
5. Reflame the lip of the tube and replace the cap (Figure 2.2e).
6. Flame the loop to kill any organisms remaining before placing the loop on the lab table (Figure 2.2f).
7. Incubate at 37°C for 24 to 48 hours.

B. Transfer from Plate to Slant

1. Repeat steps 1 and 2 from part A.
2. Remove cap of the nutrient agar slant and gently flame the lip of the tube.
3. Transfer the inoculum on the loop of the surface of the agar slant. Two types of inoculations are usually performed. The first is a straight-line inoculation, where the inoculating loop is touched to the bottom of the slant in a straight line (Figure 2.3a). In a second method, deposit the inoculum on the slant by moving the loop in a zigzag fashion over the surface of the slant (Figure 2.3b). Care should be taken not to gouge the agar surface with the inoculating loop.
4. Incubate the slants at 37°C for 24 to 48 hours.
5. After incubation and observation place the slants in your desk drawer.
6. This culture will be used in the performance of the Gram stain in Laboratory 8.
7. Twenty-four hours prior to Laboratory 8, transfer some of the growth from the reserved slant to a fresh nutrient agar slant. Incubate at 37°C.

PURE CULTURE TECHNIQUES

In most environments one is likely to encounter a mixture of microbial populations. For purposes of identification and further study, the microbiologist separates the various species into what are termed **pure cultures**. These cultures contain only one type of microorganism, providing an opportunity for a careful study of colonial and cellular morphology and for distinguishing biochemical characteristics. In this procedure, determine the number of dif-

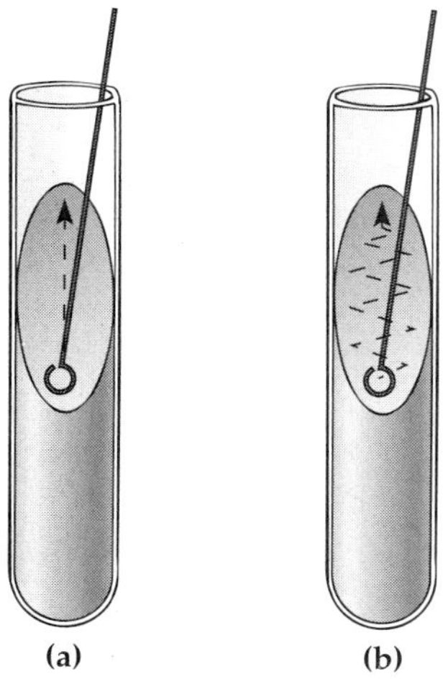

FIGURE 2.3 Inoculation of agar slants.

ferent types of microorganisms present in a tube of nutrient broth. In order to do this, use the **streak-plate** method. Use the inoculating loop to dilute out organisms on the surface of a nutrient agar plate.

When opening and inverting the plate, be careful not to touch the bacterial growth on the agar surface.

1. Obtain a tube of nutrient broth containing the mixed bacterial population and three tryptic soy agar or nutrient agar Petri plates.
2. Open the agar plate, being careful to protect the agar surface from contamination. This can be accomplished in a number of ways. In one technique, the plate is opened, and the lid is placed face down on the lab bench. The bottom portion of the plate containing the agar is then inverted and placed on top of the lid (Figure 2.4). At no time should the plate with the agar surface be placed face up on the bench. Exposure to the air might cause significant contamination.
3. Using the aseptic technique and an inoculating loop, remove a loopful of culture from the broth tube containing the mixed population of bacteria.
4. Lift the bottom of the plate and hold it in a slightly inverted position; now deposit a loopful of the culture in one corner of the plate. Make sure to grasp the end of the inoculating loop between your thumb and forefinger and, with a whiplike motion, spread the drop over the surface of the agar. The lines of movement over the surface should extend from one edge of the plate to the other and should not overlap (Figure 2.5a).

Remember, the purpose of performing this technique is to separate the organisms that are in the broth. The technique illustrated in Figure 2.5a may not achieve proper separation of organisms. Using an alternative method (Figure 2.5b), streak about one half of the plate as illustrated. Then remove the loop from the agar surface and pass it in the flame of the Bunsen burner to kill any microorganisms remaining on the loop. After allowing the loop to cool sufficiently, pick up the last line of inoculation and continue to spread

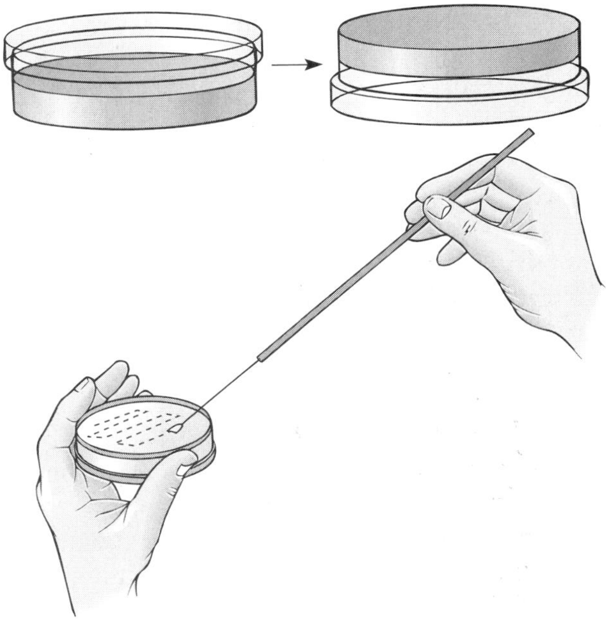

FIGURE 2.4 One technique for opening Petri plate.

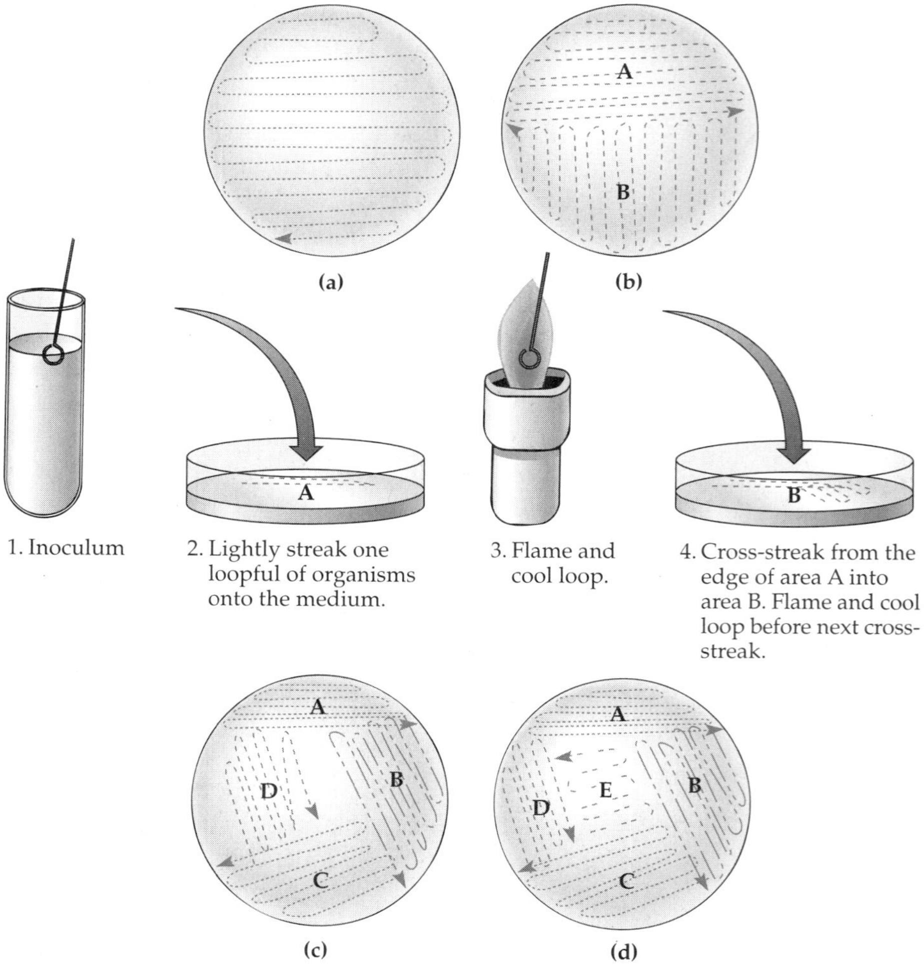

FIGURE 2.5 Various streaking techniques to obtain isolated colonies.

out the organisms, but this time streak at a 90° angle to the first streaks (Figure 2.5b). Be sure to overlap the last few lines from the original inoculum so that a further dilution of the organisms can be accomplished. Other techniques are the four- or five-way streaks (Figure 2.5c and d). The intent of both of these techniques is to achieve the smallest population of organisms in the center of the plate. The plate is inoculated by depositing a drop of culture on the surface and dragging it across the surface to spread out the liquid as illustrated in a. The loop is flamed to kill any remaining organisms and, after allowing to cool, is used to further separate deposited organisms, as illustrated in A–E.

After the plate has been allowed to incubate, note whether there are fewer organisms in each progressively streaked area. The fewest number of organisms should appear in the center of the plate. An alternative way of manipulating the Petri plate is to grasp the plate in one hand and, while slightly opening the lid, proceed to streak using one of the described techniques (Figure 2.6). In a subsequent exercise we will use the spread-plate technique to isolate microorganisms on the agar surface.

Invert the plates and incubate for 48 hours at room temperature, then examine to see if there are isolated colonies on the plates. Verify that the colonies selected actually represent pure cultures by carefully noting the

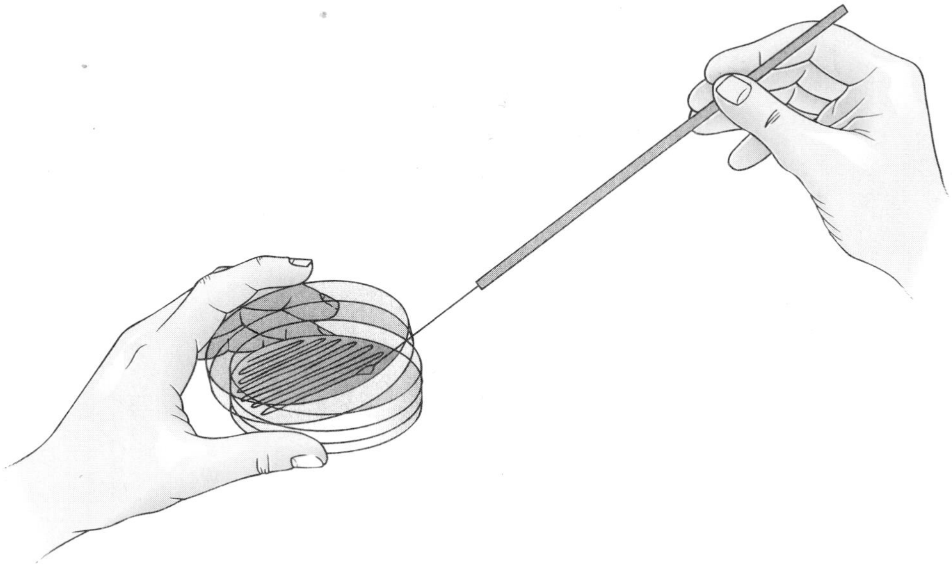

FIGURE 2.6 Alternative plate-opening technique.

characteristics of the isolated colonies. If they are pure cultures, rejoice and record the number of different organisms in the unknown suspension, and describe each of their colonial morphologies. If success is not achieved, not all is lost. Obtain some additional Petri plates and a fresh mixed culture and repeat the exercise until successful. This procedure will be used repeatedly during the course of the semester so it is critical that the technique be mastered.

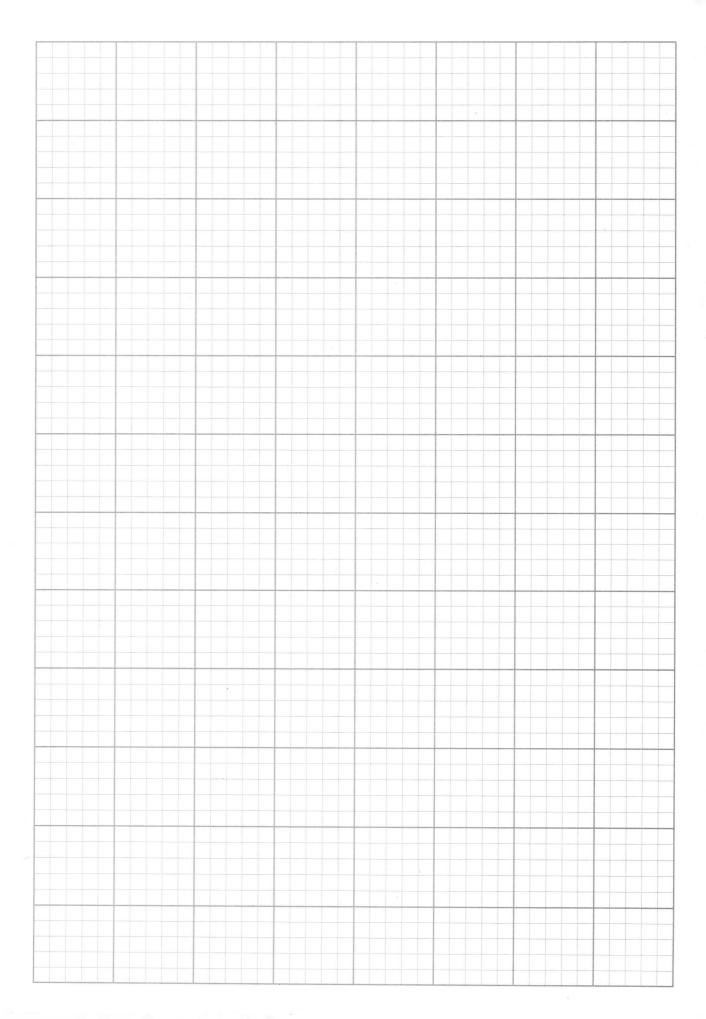

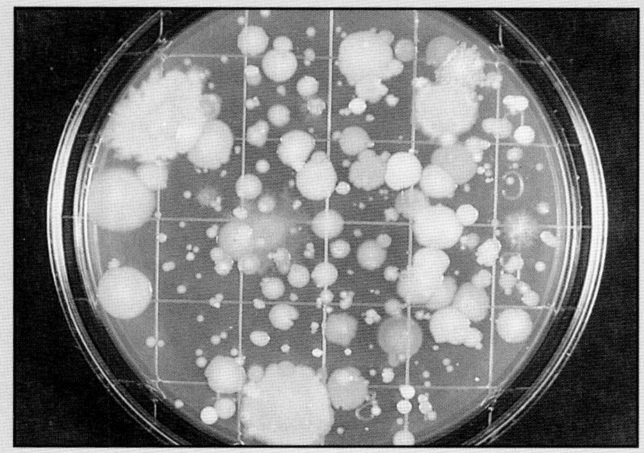

Plate 1
Bacterial colonies on agar plate
(Courtesy, Philip Stukus).

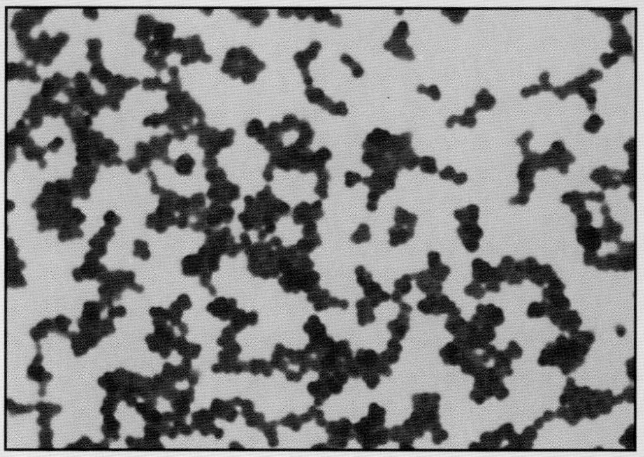

Plate 2
Cellular morphologies of bacteria using a simple stain *(Courtesy, Philip Stukus).*

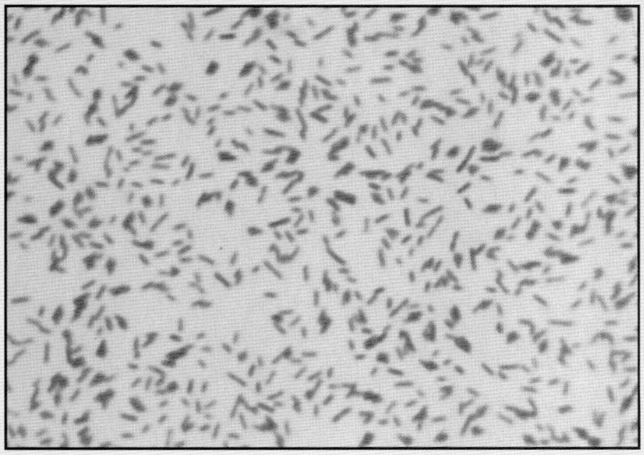

Plate 3
Cellular morphologies of bacteria using a simple stain *(Courtesy, Philip Stukus).*

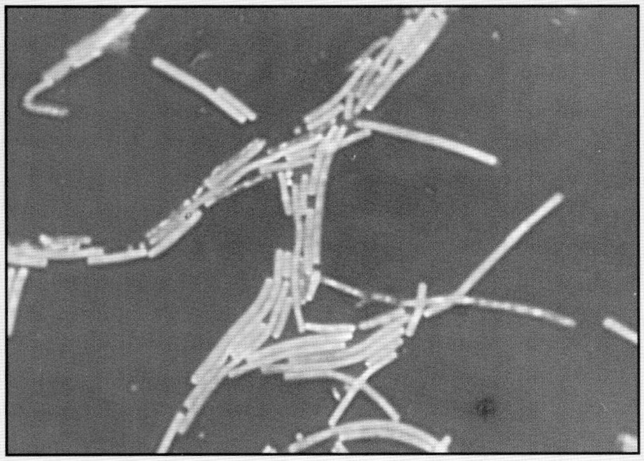

Plate 4a
Negative stain: bacilli (*Lineola longa* at 1,000 x) *(Courtsey, Steven W. Woeste and Paul H. Demchick).*

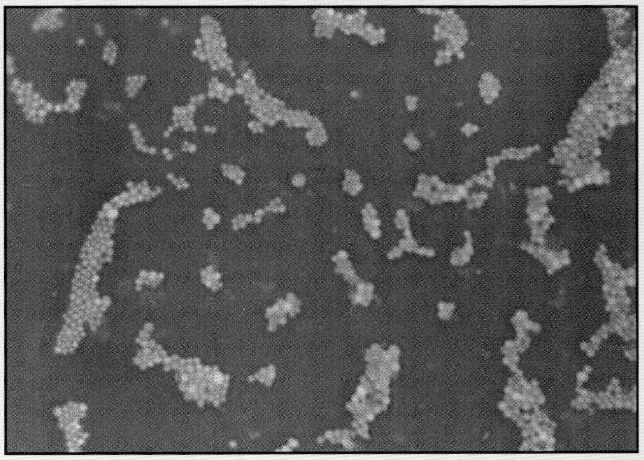

Plate 4b
Positive stain: cocci (*Staphylococcus aureus* at 1,000 x) *(Courtsey, Steven W. Woeste and Paul H. Demchick).*

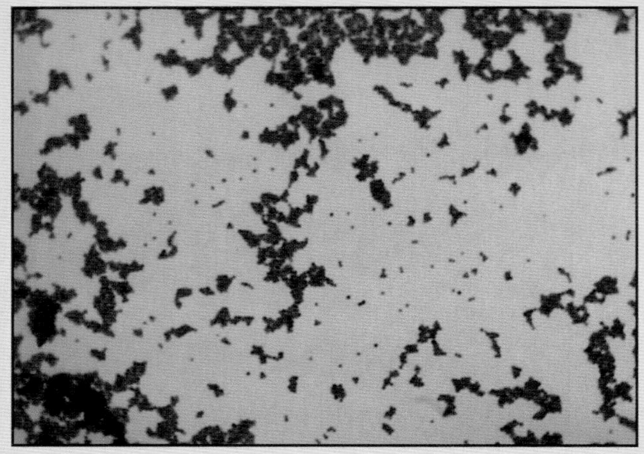

Plate 5a
Gram-positive cocci *(Courtesy, Phillip Stukus).*

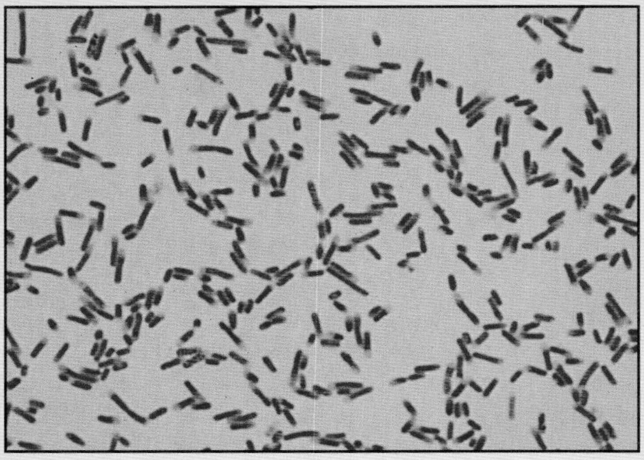

Plate 5b
Gram-negative rods *(© G.W. Willis, Oschner Medical Institution/ Biological Photo Services).*

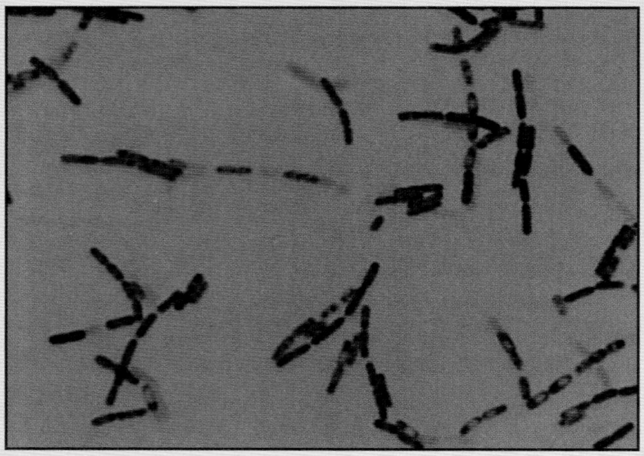

Plate 5c
Gram-positive rods *(Courtesy, Philip Stukus).*

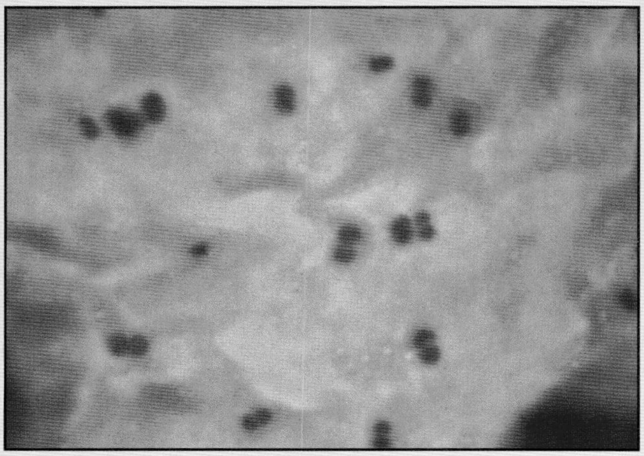

Plate 5d
Gram-negative cocci *(© Arthur M. Seigelman/ Visuals Unlimited).*

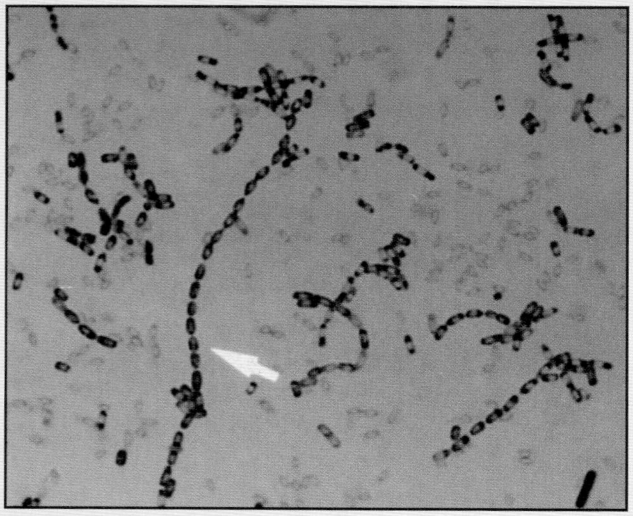

Plate 6
Endospore-forming rod *(Courtesy, Philip Stukus).*

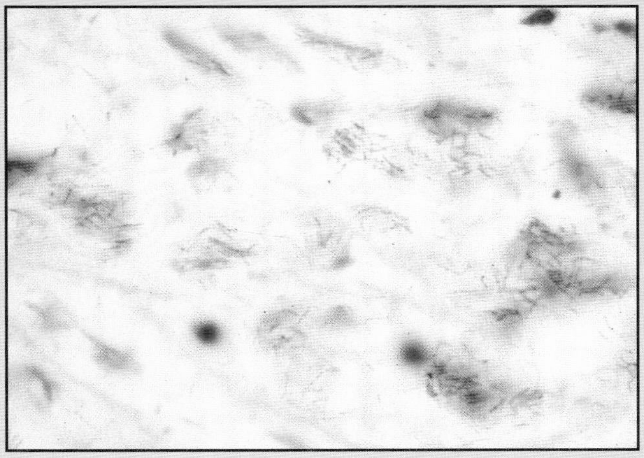

Plate 7
Acid-fast organisms *(© Biological Photo Service).*

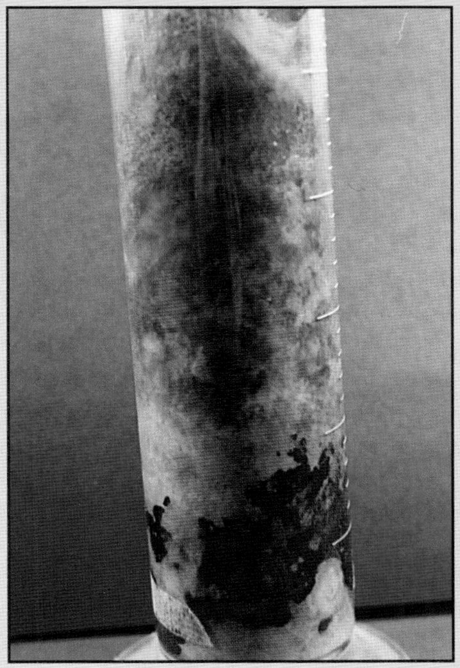

Plate 8
Winogradsky column *(Courtesy, Philip Stukus).*

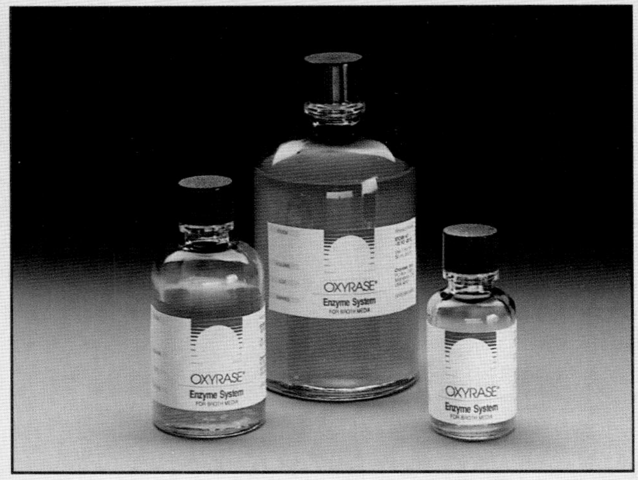

Plate 9
Oxyrase for Broth®. Used in the growth of *Bacillus fragilis (Courtesy, Oxyrase, Inc., Mansfield, OH).*

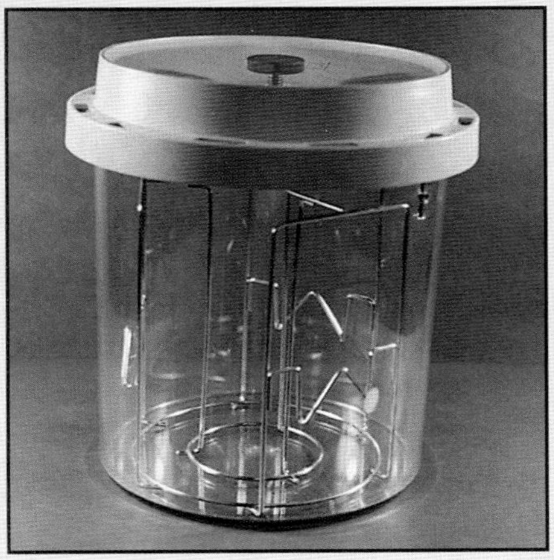

Plate 10
Brewer anaerobic jar *(Courtesy, Philip Stukus).*

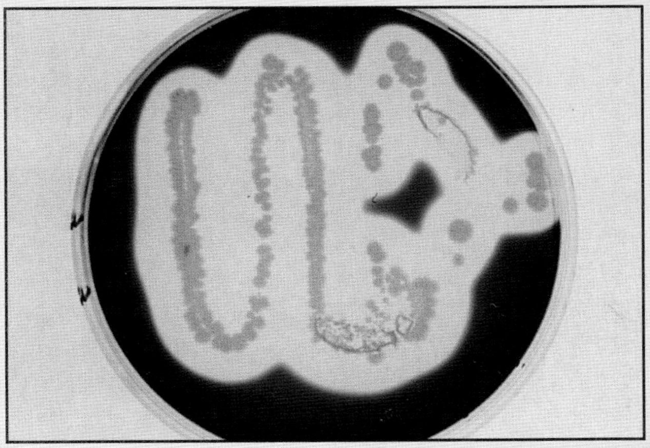

Plate 11
Positive amylase reaction *(Courtesy, Philip Stukus).*

Plate 12
Carbohydrate fermentation reactions *(Courtesy, Philip Stukus).*

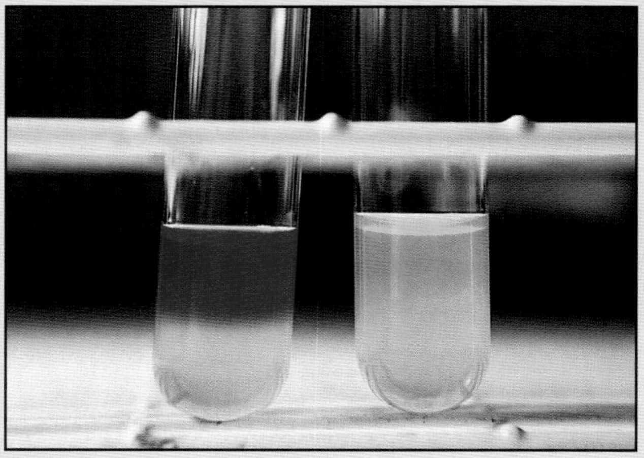

Plate 13
MR-VP reaction *(Courtesy, Philip Stukus).*

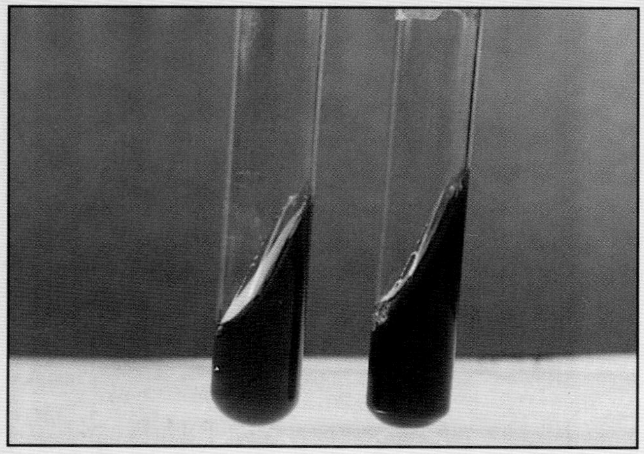

Plate 14
Citrate test *(Courtesy, Philip Stukus).*

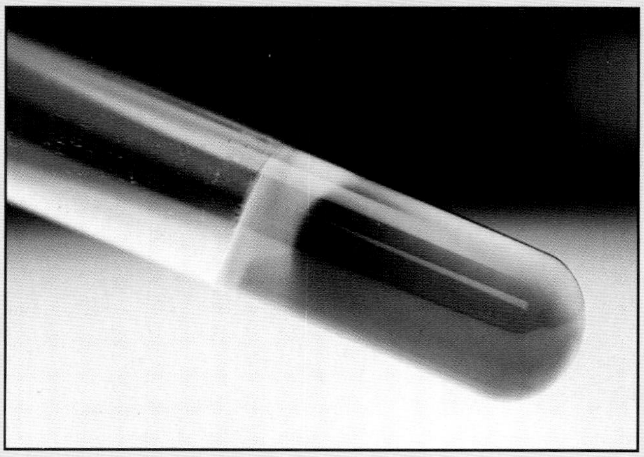

Plate 15
Production of H_2S *(Courtesy, Philip Stukus).*

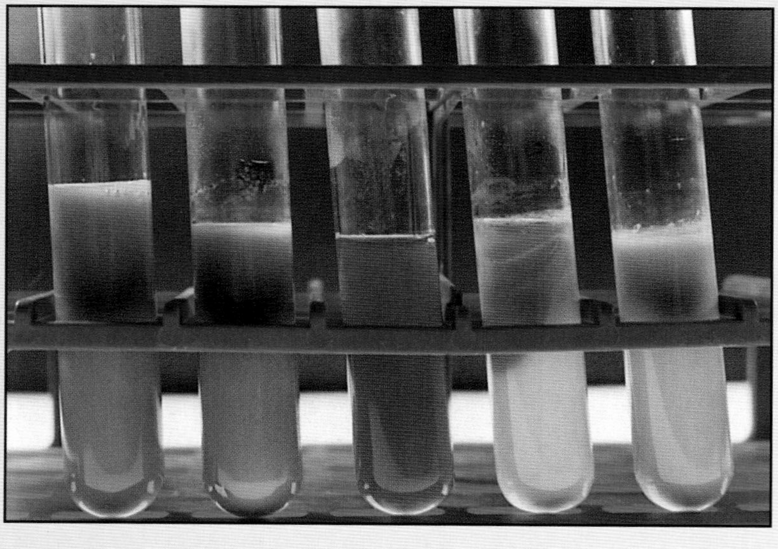

Plate 16
Litmus milk reactions *(Courtesy, Philip Stukus).*

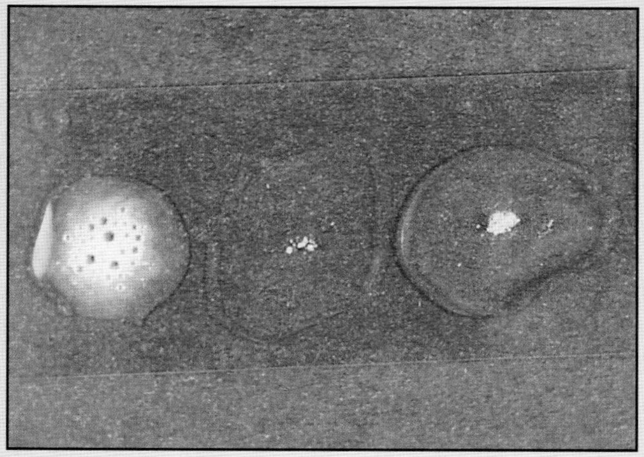

Plate 17
Catalase test *(Courtesy, Philip Stukus).*

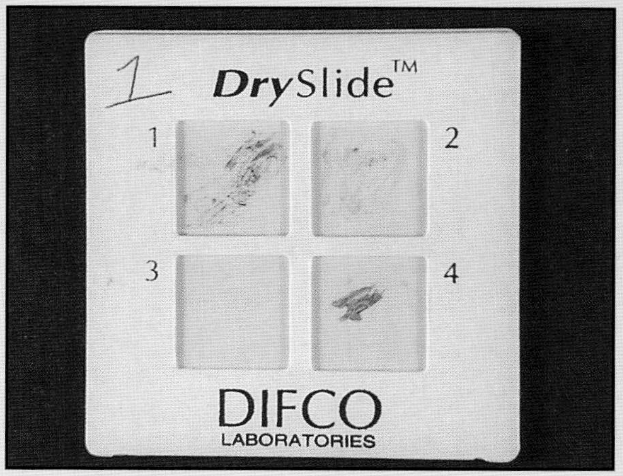

Plate 18
Difco oxidase test *(Courtesy, Philip Stukus).*

Plate 19
Coagulase test: + and - tests *(Courtesy, Philip Stukus).*

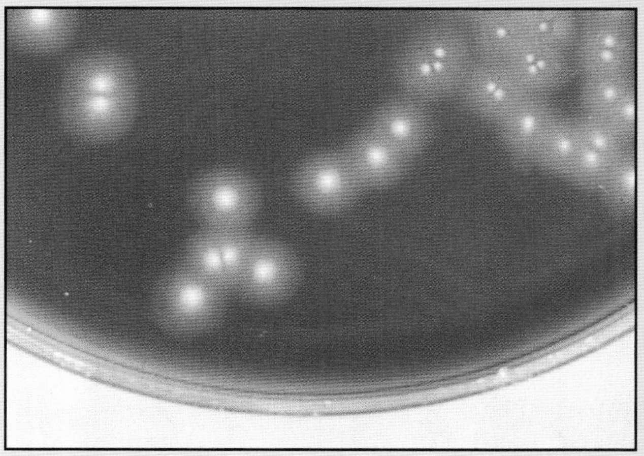

Plate 20
Beta hemolysis *(© Fred E. Hossler/Visuals Unlimited).*

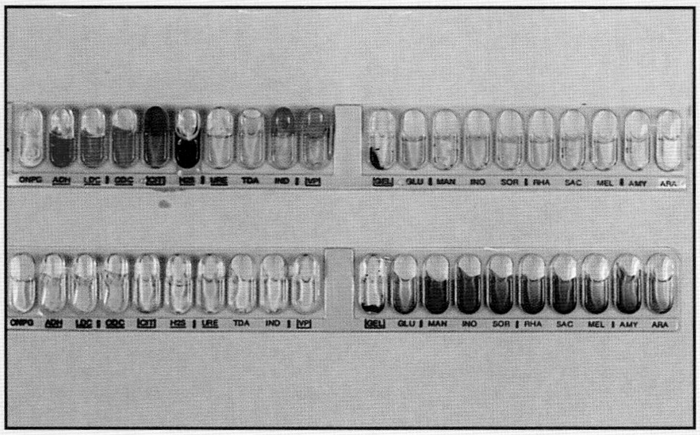

Plate 21
API-20E reactions *(Courtesy, Philip Stukus).*

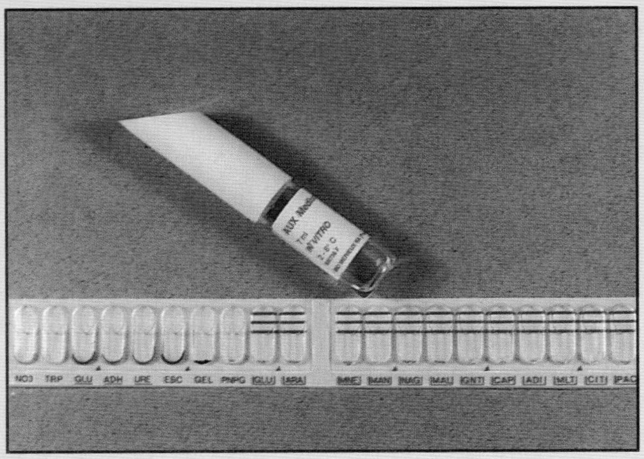

Plate 22
API-NFT setup *(Courtesy, Philip Stukus).*

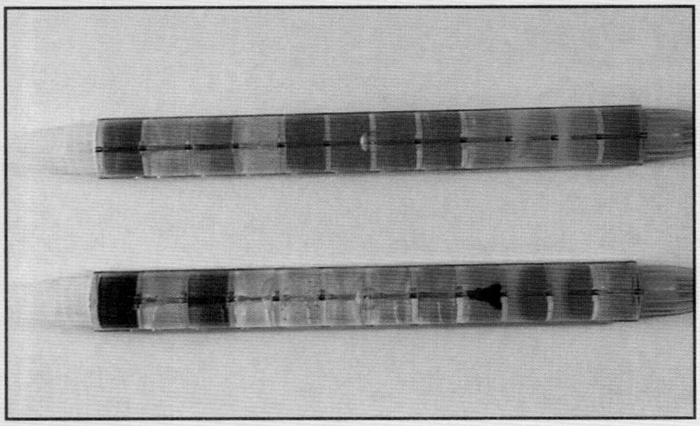

Plate 23
Enterotube II reactions *(Courtesy, Philip Stukus).*

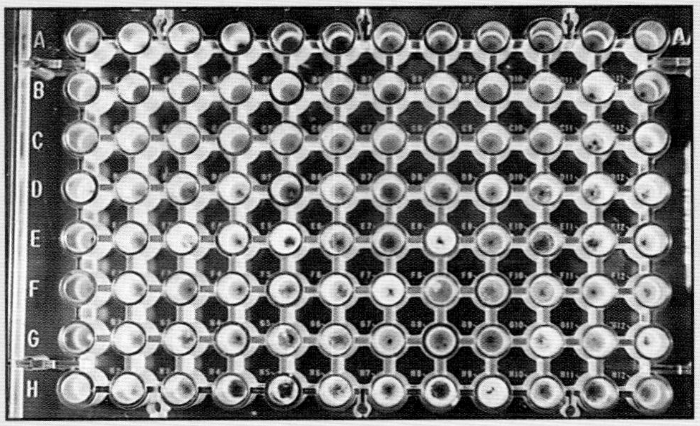

Plate 24
Biolog GN microplate *(Courtesy, Philip Stukus).*

Plate 25
Two antibiotic disk dispensers *(Courtesy, Difco Laboratories).*

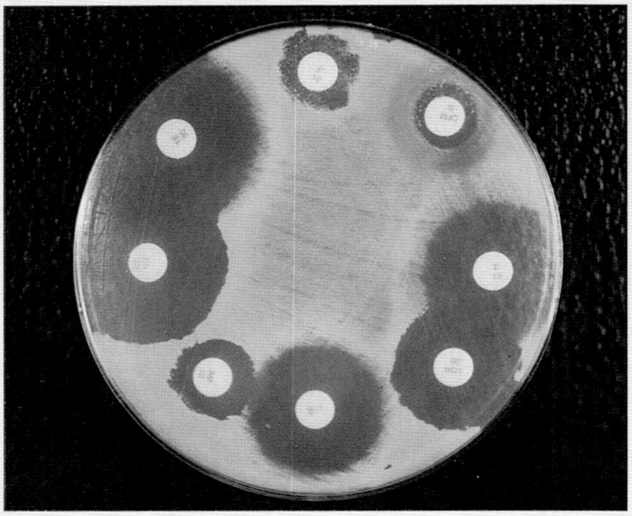

Plate 26
Zones of inhibition *(Courtesy, Philip Stukus).*

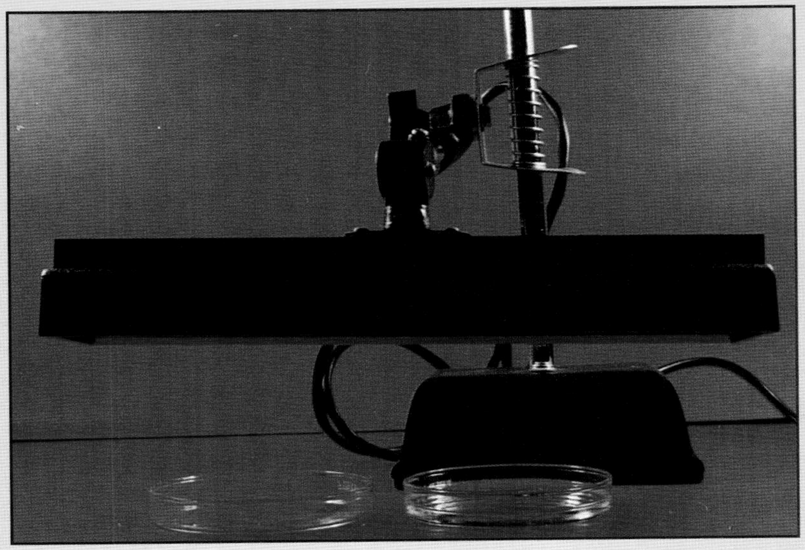

Plate 27
Ultraviolet light setup
(Courtesy, Philip Stukus).

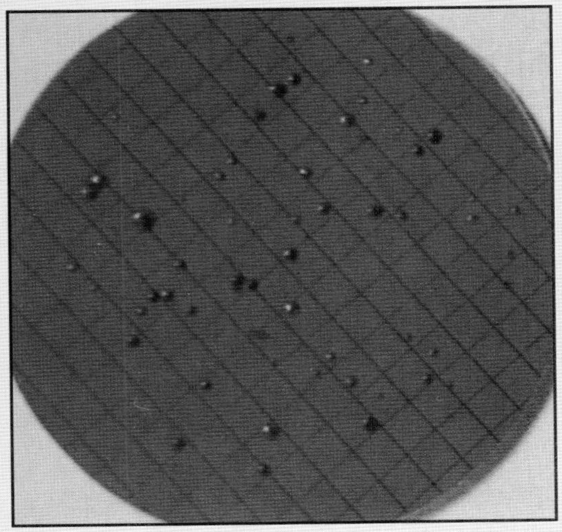

Plate 28
Coliforms *(Courtesy, Millipore Corporation).*

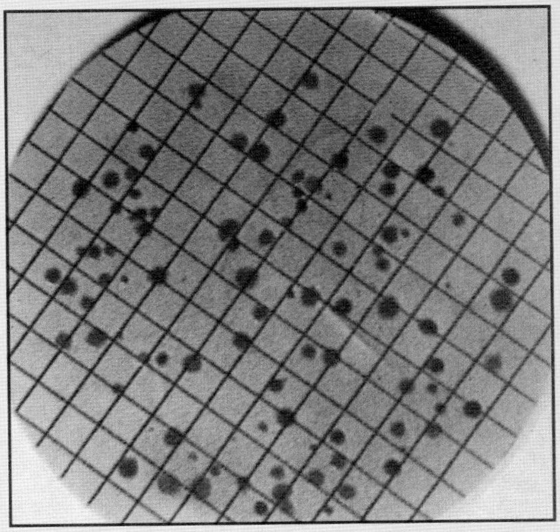

Plate 29
Fecal coliforms *(Courtesy, Millipore Corporation).*

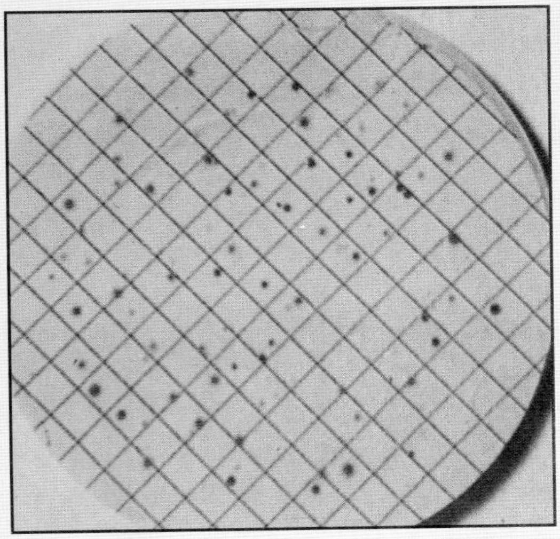

Plate 30
Fecal streptococci *(Courtesy, Millipore Corporation).*

Plate 31
SAS air sampler *(Courtesy, Philip Stukus).*

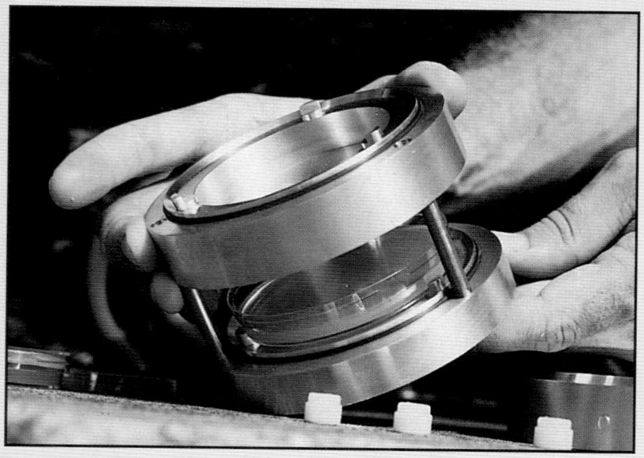

Plate 32
Anderson Two Stage Sampler *(Courtesy, Philip Stukus).*

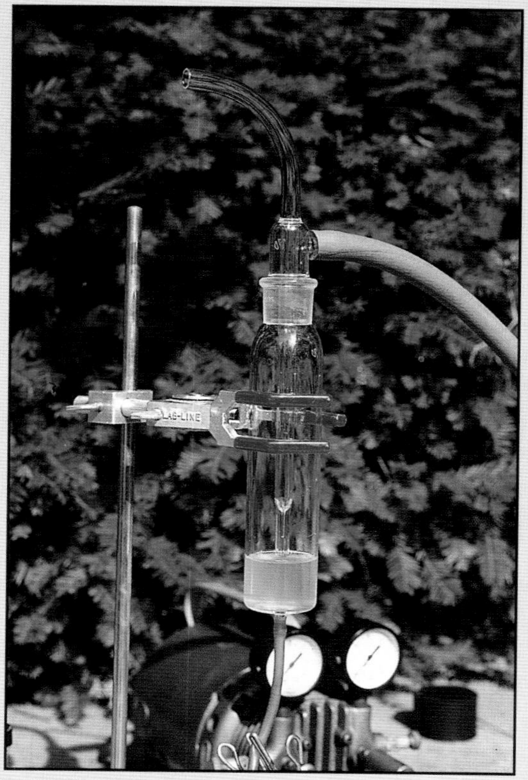

Plate 33
All Glass Impinger *(Courtesy, Millipore Corporation).*

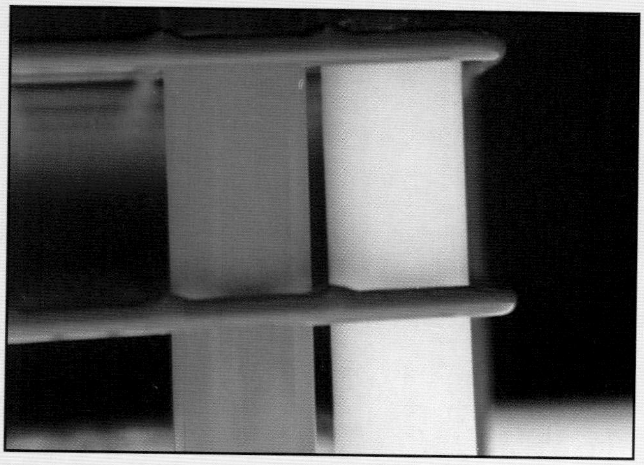

Plate 34
Methylene blue reduction test *(Courtesy, Philip Stukus).*

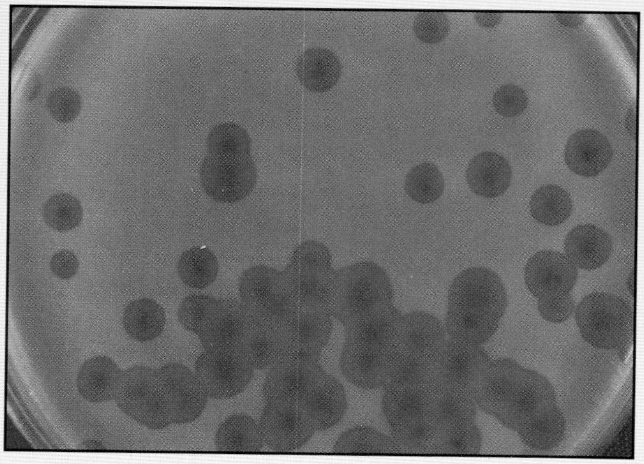

Plate 35
Bacteriophage plaques *(Courtesy, Philip Stukus).*

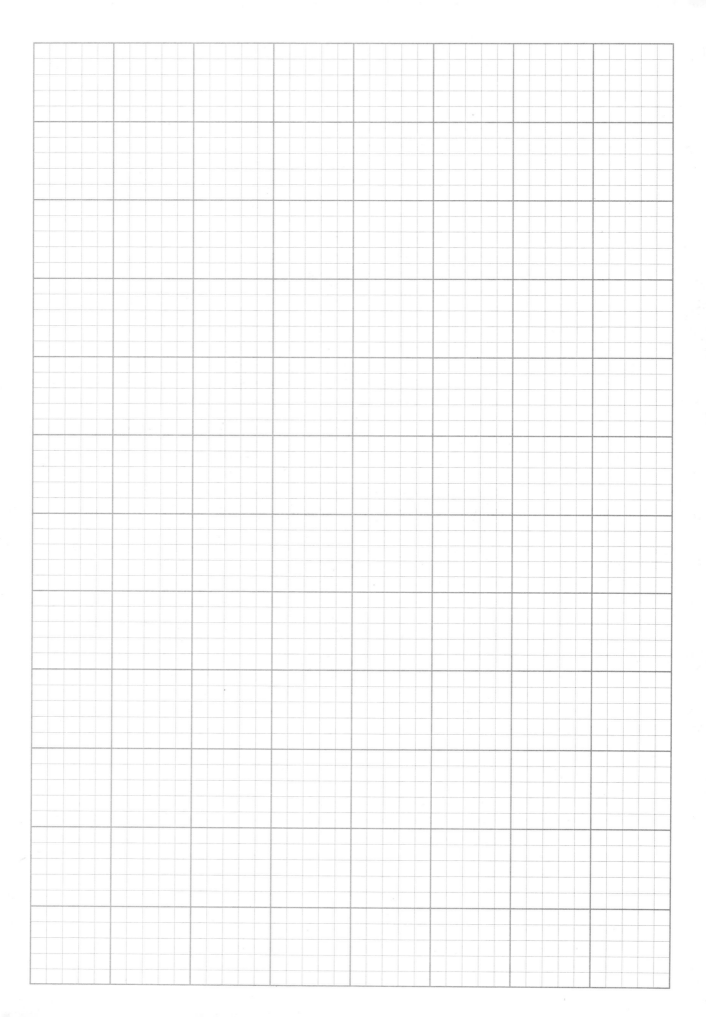

Laboratory Report 2

Name ______ Date ______ Section ______

Subculture Observations

Nutrient Broth Growth

Surface ______

Sediment ______

Overall ______

Growth on Nutrient Agar Slant

Draw pattern of organisms from Rodac or Petri plate colony.

Pure-Culture Technique

Number of mixed culture selected ______

Number of different colonies obtained from streak of mixed population ______

Colony Descriptions

	Colony 1	Colony 2	Colony 3	Colony 4
Color				
Size				
Shape				
Texture				
Margin or Edge				
Elevation				

Review Questions

1. Why is it necessary to invert Petri plates during incubation?

2. Why is it generally a good rule to label the bottom of Petri plates rather than the top?

3. Why is it necessary to heat the inoculating loop after transfer or streaking of microorganisms?

4. How could it be verified that a bacterial colony on a nutrient agar plate is a pure culture?

5. Why in most cases is an inoculating loop rather than an inoculating needle used for streak-plate isolations? Can you think of other ways to enumerate the number of different types of bacteria in a mixed culture?

6. While on a rigorous hike you have depleted your supply of drinking water. Fortunately, you come upon a mountain stream that is refreshingly cool and crystal clear. You are ready to take a satisfying drink when you remember something from your introductory microbiology laboratory. What is it that has given you pause to think?

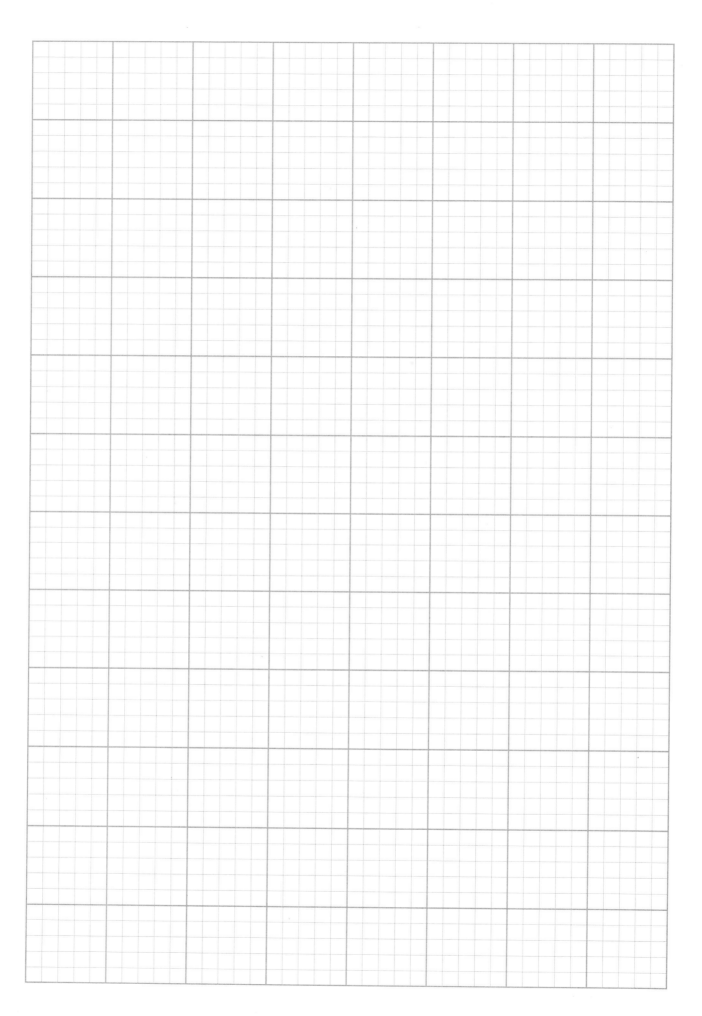

LABORATORY

3 Microscopy

Safety Issues

- Be careful not to cut yourself on any broken glass slides. Dispose of broken glass in specially marked containers.
- To prevent damage to microscope lenses make sure the immersion oil is cleaned off after use.

OBJECTIVES

1. Become proficient in the use of the microscope, particularly with the oil immersion lens.
2. Learn how to properly handle and store microscopes.
3. Appreciate the difference between resolution and magnification.

Brightfield microscopy allows magnification of images by using two different lens systems, the **ocular lens** and the **objective lens**. Refer to Figure 3.1 to locate the various parts of the microscope. The objective lenses are located on a revolving nosepiece, with usually 3 to 4 found on each microscope. Most commonly, these lenses provide magnifications of 10× (low power), 43× (high-dry), and 100× (oil immersion). The ocular lens, or eyepiece, is found at the top of the microscope and usually has a magnification of 10×.

Microscopes that have one ocular lens are referred to as monocular, and those having two ocular lenses are called binocular. A third important lens is the condenser lens found below the stage of the microscope. The condenser lens focuses the light on the specimen being studied. The substage adjustment knob controls the movement of the condenser lens. When using the 100× oil immersion objective, make sure that the condenser lens is in the uppermost position just under the stage in order to allow for maximum light capture. An iris diaphragm regulates the amount of light passing through the condenser lens from the tungsten bulb in the base of the microscope. A blue filter is usually positioned above the light source in order to improve resolution.

Magnification of the image is determined by multiplying the power of the ocular lens by the power of the objective lens. The maximum magnification using the 100× oil immersion objective is 1000× when using the normal 10× ocular lens.

Most microscopes are also equipped with a mechanical stage to allow for quick positioning of the specimen. Locate the controls for the mechanical stage, and become familiar with the directions of movement possible.

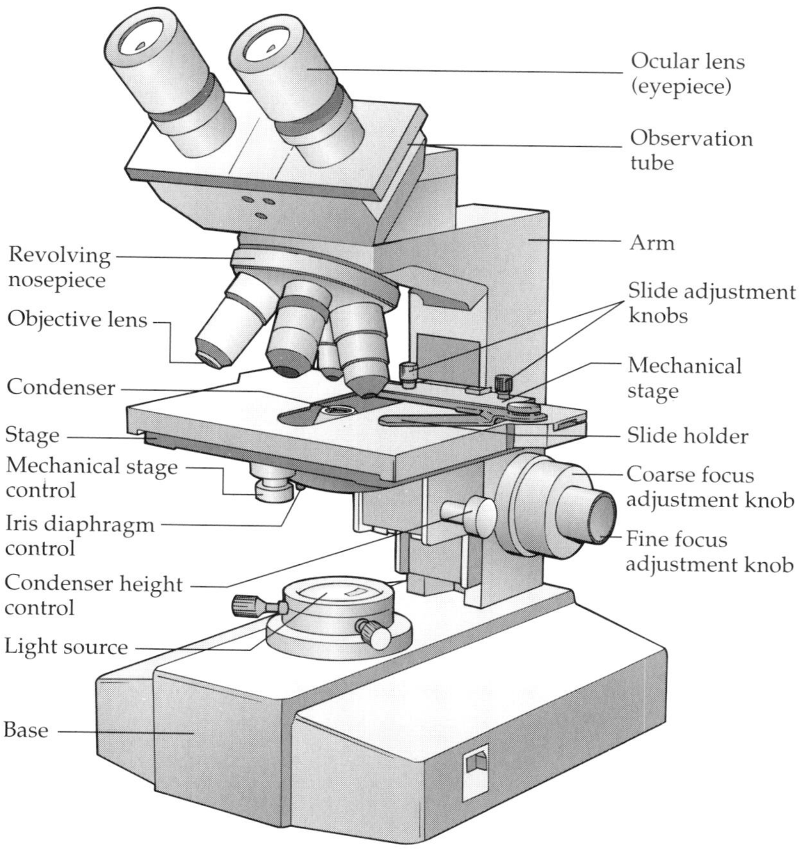

FIGURE 3.1 Diagram of microscope.

The resolving power is the ability to distinguish two images as separate. The resolving power depends on the wavelength of radiation used and the numerical aperture of the system.

$$\text{Resolving power} = \frac{\text{wavelength}}{2\ (\text{numerical aperture})}$$

The maximum resolving power of the best light microscope is 0.2 μm. Remember this when trying to distinguish the morphologies of some of the tiny "dots" of bacteria on the prepared slides. The numerical aperture describes the light-gathering capabilities of a particular lens system being used, and is influenced by the refractive index of the lens; that is, it is a measure of the deviation or bending of light rays as they pass from one medium to another. The deviation occurs at the surface of the junction of the two media, which is known as the refracting surface.

The refractive index for air is 1.00. The refractive index for glass is 1.25. In order to prevent the bending of light rays as they are passing from glass through air, immersion oil is placed directly on specimens to be examined with the 100× objective. The refractive index of the immersion oil is the same as for glass, so the path of light rays is not bent when moving from one medium to another (Figure 3.2). One unfortunate aspect of using oil is the potential damage that can be done to the 100× objective lens if the oil is

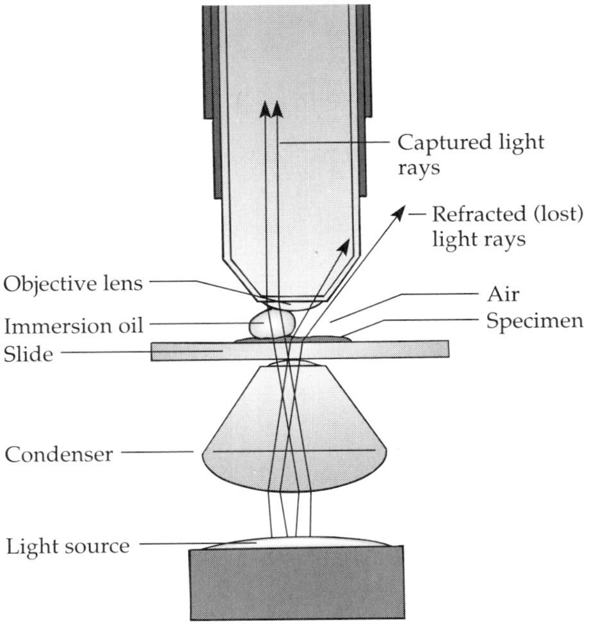

FIGURE 3.2 Use of immersion oil.

not properly removed from the lenses after use. Therefore, remember to thoroughly clean the oil objective after each use.

Materials (per 20 students)

1. Brightfield microscope (20)
2. Commercially prepared slides of protozoa, yeast, algae, and bacteria (25–40 of each)
3. Lens paper (20)
4. Immersion oil (10)

Procedure

1. Carefully remove a microscope from the storage cabinet and position it in a clear area on the lab bench directly in front of you.
2. Take a few moments to familiarize yourself with the location of the various parts of the microscope. If you have difficulty locating a part or are unclear regarding its use, ask your lab instructor for help before proceeding.
3. Be sure the microscope is plugged in and that the lamp has been turned on.
4. Position yourself at a height so that you can comfortably make observations through the ocular objectives without straining your neck or back.
5. Be sure that the microscope remains flat on the lab bench at all times. Do not tilt your microscope when you are observing specimens in this laboratory.
6. Clean all glass surfaces with lens paper.
7. Raise the condenser lens to its highest position beneath the stage of the microscope. Also check to verify that the iris diaphragm is open.
8. Turn the coarse adjustment knob until there is an ample working distance between the objective lenses and the stage of the microscope.

⚠ Dispose of any broken slides in a glass-disposal container. Severe cuts can occur from handling broken slides.

9. Obtain a slide of protozoa and secure it on the stage of the microscope either by using spring clamps, or if your microscope has a mechanical stage, by using the specimen holder clamp.
10. Swing the low-power (10×) objective into place by revolving the turret which holds the objective lenses. Be sure that the lens is locked into position.
11. While looking from the side, turn the coarse adjustment knob until the 10× objective is positioned as close as possible to the slide. Most microscopes have stops that prevent the objective lenses from touching the slides on the stage of the microscope. If the microscope does not have the safety stop feature, be sure the condenser lens does not touch the slide.
12. Look through the ocular lens and slowly increase the distance between the lens and the slide. When an image comes into focus, stop and adjust the amount of light by regulating the iris diaphragm. Use the time adjustment knob for critically focusing the specimen. When using a binocular microscope, be sure to position eyepieces so that both eyes are kept open. Using both eyes will significantly reduce eye strain.
13. Move the slide manually or by using the knobs of the mechanical stage to observe other fields. Make sketches of what you see.
14. Next, rotate the turret and swing into place a higher-powered objective (20× or 43×). Observe the specimen. It should still be in focus. Most microscopes are parfocal, which means that when a specimen is in focus using one lens, it remains in focus when switching to another objective lens. A slight adjustment of the fine adjustment knob may be necessary to bring the specimen into sharp focus.
15. Repeat the above steps to make observations and drawings of the prepared yeast and algae slides.
16. Next, obtain a prepared slide of bacteria and add a drop of immersion oil to it. Be sure to do this while the slide is on the lab bench and not on the stage of the microscope. This will prevent contamination of the microscope parts with oil.
17. Turn the coarse adjustment knob of the microscope so that there is maximum distance between the objective lenses and the microscope stage.
18. Position the slide securely in the slide holder and swing the 100× objective into place above the slide. While looking from the side, use the coarse adjustment knob to bring the lens into direct contact with the oil and the slide. Be sure that the objective has made contact with the slide.
19. While looking through the ocular lenses, **slowly** raise the 100× objective, using the coarse focus knob. At the same time, apply a slight downward pressure with your finger to a portion of the slide to prevent it from being lifted up while you are focusing. Stop when some color or an image flashes in the field of view.
20. Use the fine adjustment focus knob for critical focusing. Make sure that your condenser lens is in the uppermost position under the stage. If objects are not visible, be sure to check that the iris diaphragm is open. Reducing the amount of light illuminating the specimen will enable a better description of the cellular morphologies. Use the mechanical stage to observe different areas of the slide.
21. Record the morphology of the individual cells and note if the cells are grouped in any characteristic arrangement (clusters, filaments, groups of two, etc.). To accurately determine cell morphologies, locate areas that are not too densely populated with bacteria.

⚠ Carefully cleaning the oil from the lens will prevent permanent damage to the lens.

22. **After making all of the microscopic observations, be sure to take time to carefully clean the immersion oil off the 100× objective lens. This should be done with lens paper. The use of bibulous paper or Kimwipes will scratch the lenses.** Permanent mounts of slides may also be wiped clean.
23. Rotate the nosepiece so that the 4× or 10× objectives are in position before storing the microscope. Place the cover on the microscope and place the microscope in the appropriate position in the microscope cabinet.

This same technique will be used to observe other stains prepared in future laboratories.

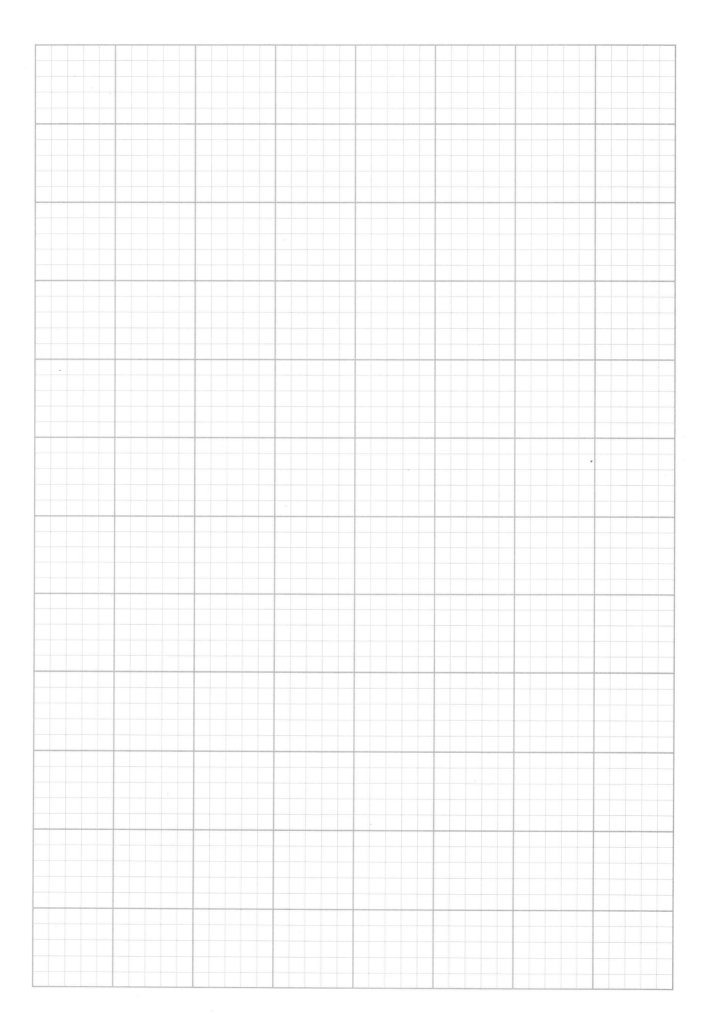

Laboratory Report 3

Name ______________________ Date ______________ Section ______________

Sketch the cellular morphologies of the organisms observed on prepared slides.

Protozoa

10× 43×

Algae

10× 43×

Yeast

10× 43×

Bacteria

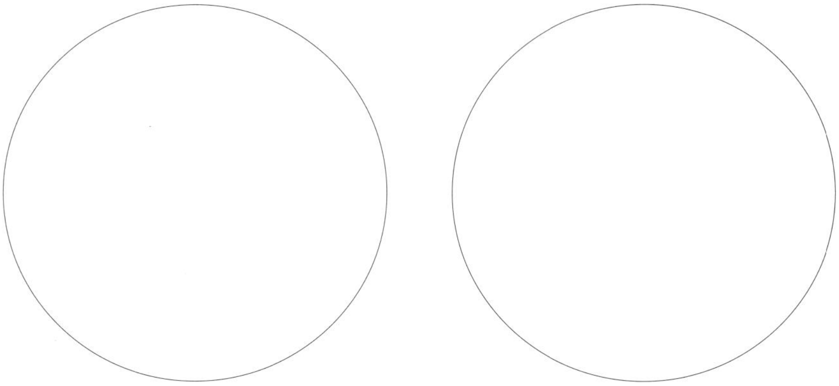

10× 43×

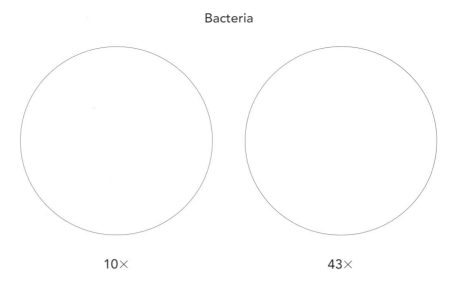

Review Questions

1. What is the maximum magnification obtainable using a light microscope? Why would increasing the magnification not necessarily lead to a more satisfying result?

2. Why must care be taken to not substitute another oil for the immersion oil when making microscopic observations?

3. How can the amount of light that passes through the slide to the ocular lens be regulated?

4. How can the resolution of the microscope be improved?

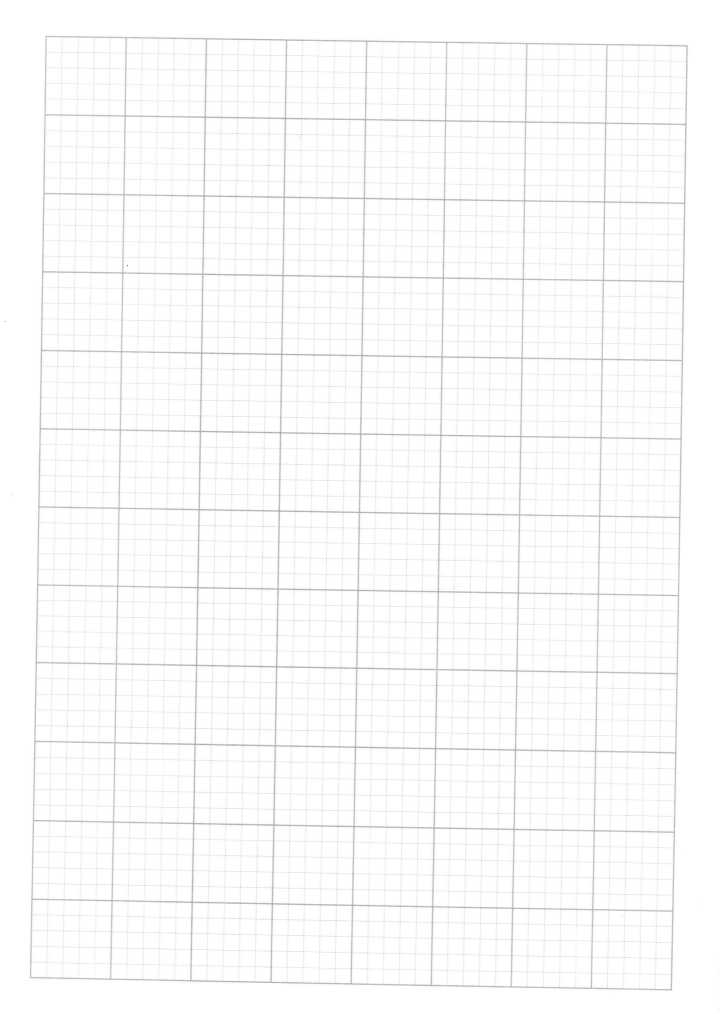

LABORATORY

4 Phase-Contrast Microscopy

Safety Issues

- Be careful not to cut yourself on any broken glass slides. Dispose of broken glass in specially marked containers.
- To prevent splattering, make sure to allow sterilized inoculating loops to cool before transferring cultures.

OBJECTIVE

Become familiar with the use of a phase-contrast microscope for observing unstained bacterial cells.

Phase-contrast microscopy allows for the visualization of bacterial cells without staining. Since unstained organisms are transparent, it is very difficult to discern structure. Fixing and staining kills cells, and is not helpful if observation of living cells is needed. Use of a phase microscope enables a very accurate description of the size and shape of cells since there is no distortion by heating or staining.

Light energy is a wave form that has a particular intensity (amplitude) and wavelength. Light energy that impacts on an object may pass through the object without changing its course—this is an example of a direct ray. Its phase has not changed. Other light rays may be bent (diffracted) as they pass through an object such as a bacterial cell—there is an alteration or phase shift of the light wave. It has a decreased amplitude and less energy compared to the light ray that passes directly through a medium, and is out of phase compared with the direct ray of light. The decrease in amplitude is proportional to the density of the observed specimen. The phase-contrast microscope is constructed to change phase shifts into differences in brightness. In bright-contrast phase microscopy, the object appears as a bright image against a dark background; and in phase-contrast, the object appears dark against a lighter background.

The phase-contrast microscope has two special features which are responsible for the images produced: the annular ring and the phase plate.

Light passage in the phase-contrast microscope is diagrammed in Figure 4.1. The annular ring is positioned below the condenser lens and regulates the amount of light entering the condenser lens. The ring restricts the light in such a way as to allow only a small hollow cone of light to pass through the specimen on the microscope slide. The phase plate is inside the objective lens and consists of a ring and an uncoated portion of the phase plate. The ring is coated with a material that advances the light rays as they pass through the ring. Transmitted (non-diffracted) light rays pass through the ring portion of the phase plate and are advanced by 1/4 wavelength, whereas

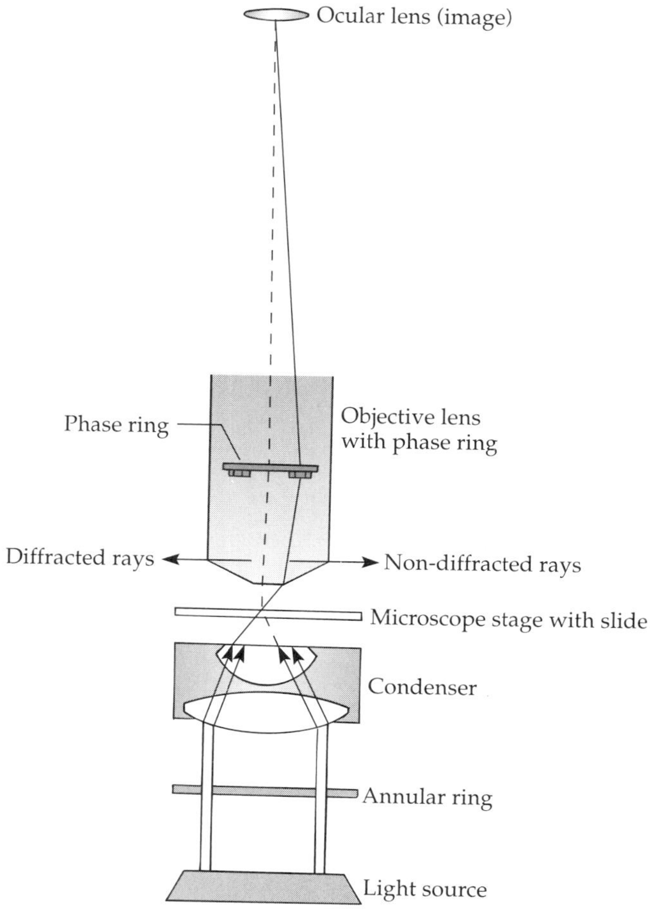

FIGURE 4.1 Phase-contrast microscope.

the light rays refracted by bacterial cells are retarded by 1/4 wavelength and pass through the uncoated portion of the phase plate. When the advanced (non-diffracted) and retarded (diffracted) rays are recombined at the image plane, differences in light intensity become apparent. These changes in light intensity vary depending upon the type of microscope used. Some microscopes will provide a brighter image (bright-phase microscopy) or a darker image (dark-phase microscopy). The differences in intensity greatly enhance contrast and allow for more critical analysis of cell structure.

It is extremely important that the annular ring and the phase plate be properly aligned (Figure 4.2). Your lab instructor will assist the adjustment of the two rings. The right of light from the annular stop must be superimposed on the phase ring image for proper operation of the phase microscope.

Materials (per 20 students)

1. Clean microscope slides
2. Cover glasses
3. Phase-contrast microscope
4. Suspension of *Paramecium*
5. Suspension of *Saccharomyces*
6. 18- to-24-hour broth culture of microorganisms (10)
7. Methyl cellulose slowing agent

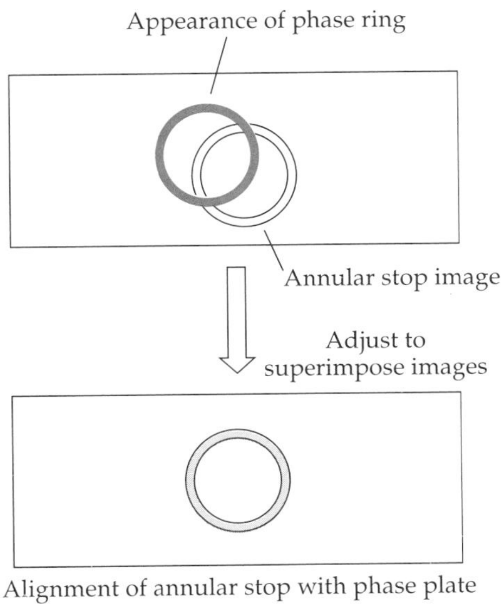

FIGURE 4.2 Alignment of annular ring.

Procedure

1. With an eye dropper, transfer a drop of the *Paramecium* suspension to a clean microscope slide. Add a drop of methyl cellulose to retard the movement of the protozoan.
2. Gently place the cover glass over the liquid.
3. Place the slide on the stage of the microscope and rotate the turret until the 10× phase-contrast objective is locked into position. There are different annular diagrams that match the phase-contrast objectives. Your instructor will assist you with the identification of the diaphragm appropriate for each of the objective lenses. Rotate the annular diaphragm to be used with the 10× objective lens into place.
4. Focus with the coarse adjustment knob and make observations. Sketch some of the organisms observed.
5. Repeat the observations, using the 40× phase objective lens and appropriate annular diaphragm.
6. Perform the same procedures, using the suspension of yeast cells.
7. Next, with a sterilized inoculating loop, transfer a drop of the bacterial culture to the center of the slide.
8. Gently place a cover glass over the drop.
9. Place the slide on the stage of the microscope and observe, using the 40× phase-contrast objective lens with the appropriate annular diaphragm in place.
10. Add a drop of immersion oil to the cover glass and rotate the turret until the 100× phase-contrast objective lens is locked into place. Rotate the 100× annular diaphragm into position.
11. While observing from the side, turn the coarse adjustment knob and position the 100× objective lens directly on the cover glass. Looking through the ocular lens, slowly increase the distance between the lens and the cover glass by turning the coarse adjustment knob. Bring image into focus by using the fine adjustment knob. Make sketches of the organisms observed.

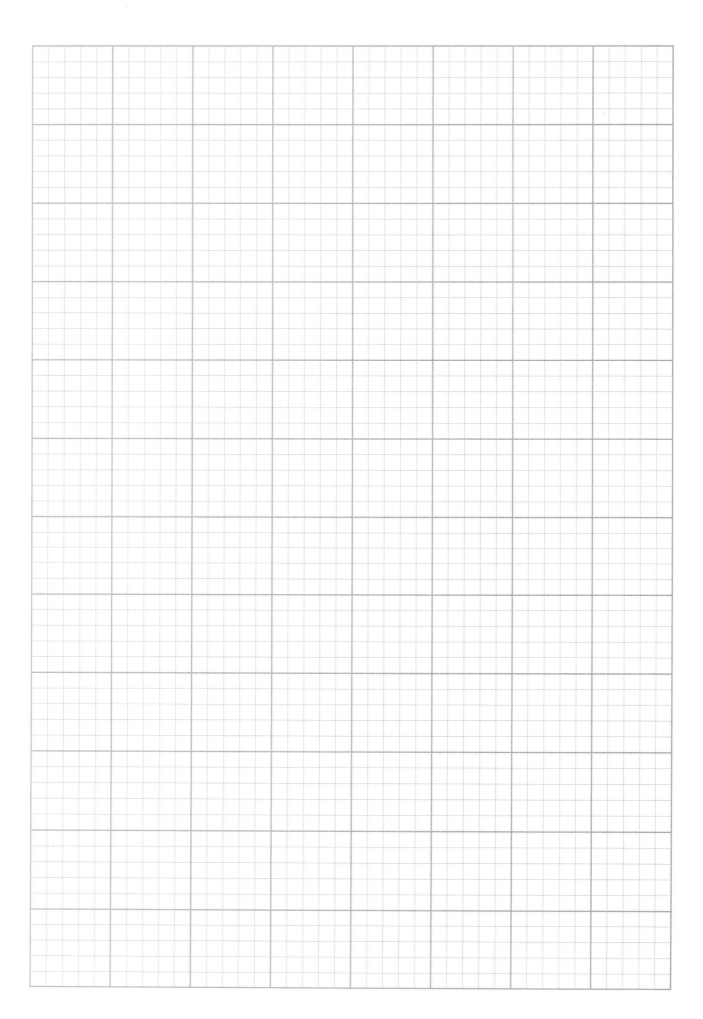

Laboratory Report 4

Name Date Section

1. What are the differences in appearance of the specimens in the light versus the phase-contrast microscope?

2. How does the diaphragm usage differ?

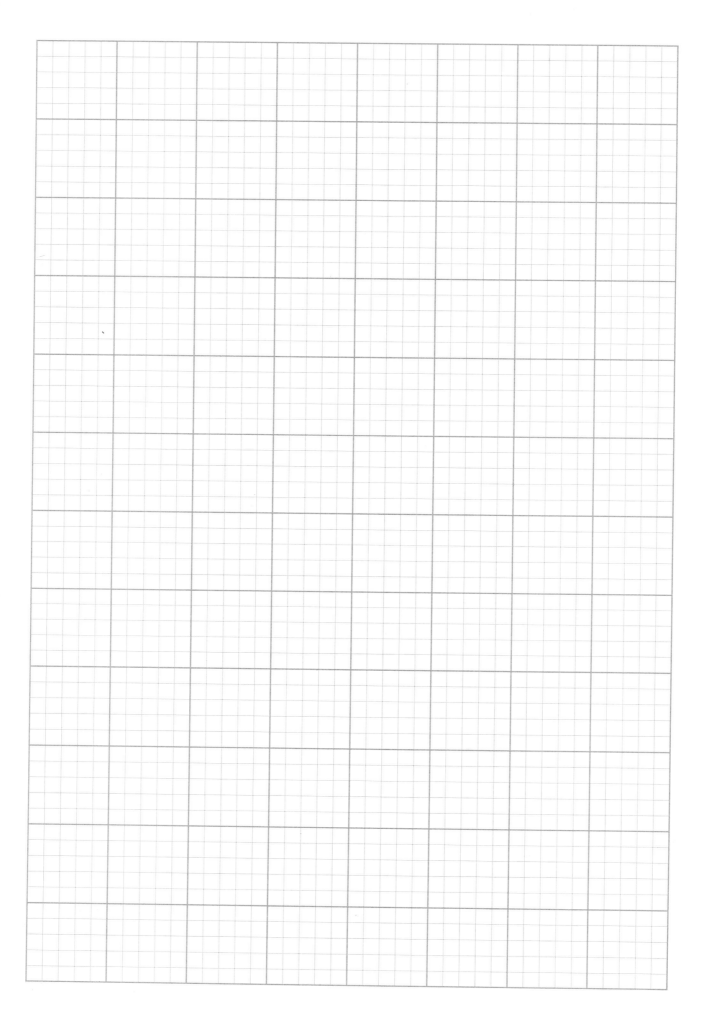

LABORATORY

5 Determination of Bacterial Motility

Safety Issues

- Be careful not to cut yourself on any broken glass slides. Dispose of broken glass in specially marked containers.
- Be sure to dispose of depression slide in the tray or container of disinfectant.

OBJECTIVES

1. Learn the application of bright-field or phase-contrast microscopy to the detection of motility in bacteria.
2. Become familiar with the techniques used to detect bacterial motility.

Bacteria can be characterized on the basis of motility. Some organisms are non-motile; that is, they show no directed, oriented movement. The erratic movement of organisms in an aqueous environment is referred to as Brownian movement. This random movement is not true motility.

True bacterial motility is attributed to the presence and functioning of flagella. There are a variety of ways in which motility can be determined. A flagellar stain could be performed to determine whether flagella are present. This technique involves layering of dye on flagella until they are able to be detected by light microscopy. This technique is quite tedious and slides must be scrupulously cleaned before use. The other drawback is that organisms may be non-motile even though they possess flagella.

Changes in cultural conditions can have a dramatic effect on the functioning of flagella. Motility determinations can also be made by stabbing a semi-solid nutrient agar medium in a straight-line inoculation technique. Any movement of the organism away from the line of inoculation can be detected by a spreading turbidity away from the line. This technique is presented in Laboratory 29: Hydrogen Sulfide Test. Motility can also be determined by direct microscopic observation of broth cultures. This can be accomplished by performing a wet-mount technique as described in the previous lab on microscopy, or by using a technique referred to as the hanging drop method. Try both the wet-mount technique and the hanging drop method.

Materials (per 20 students)

1. Depression slides and glass coverslips (20)
2. Vaseline
3. 18- to 24-hour broth cultures of:
   Proteus vulgaris (5)
   Staphylococcus epidermidis (5)
4. Broth cultures of isolates from Laboratory 2

Procedure (Figure 5.1)

1. Place a small amount of Vaseline along the edges of a clean glass coverslip using a toothpick or cotton swab.
2. With a sterile inoculating loop, transfer a loopful of the broth culture to the center of the coverglass (Figure 5.1a).
3. Take the well slide and press it onto the Vaseline-coated coverglass.
4. Carefully invert the well slide so as to have the loopful of broth hanging in the depression of the well slide (Figure 5.1b).
5. Microscopically observe the drop using a low-power objective. Most of the organisms will most likely be visible near the edge of the drop. Reduce the amount of light to better distinguish the organisms.
6. Switch to the high-dry objective and try to determine whether the organisms are moving directionally. Be careful to distinguish from Brownian

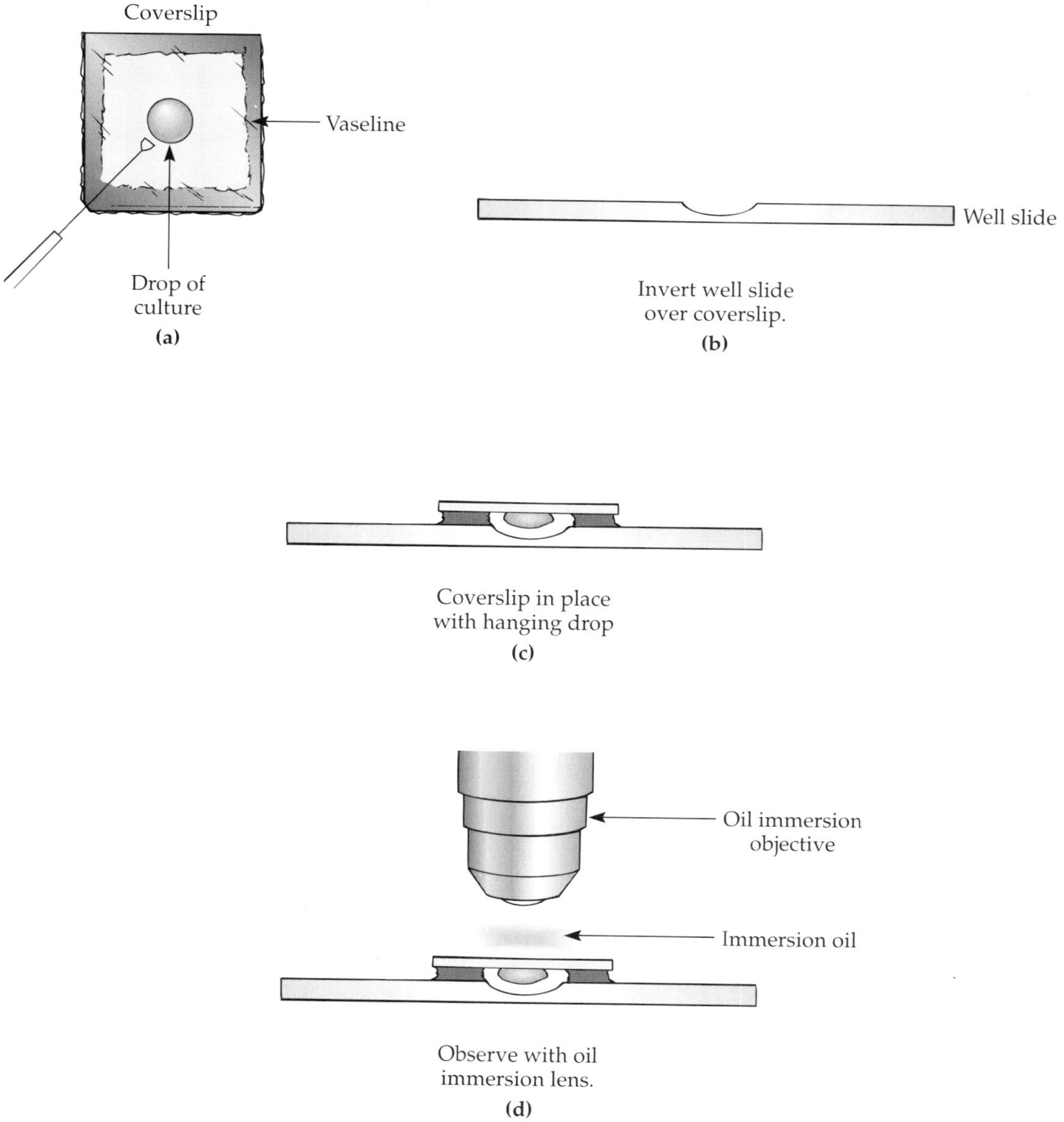

FIGURE 5.1 Hanging drop preparation.

movement, which is a vibrational movement caused by molecules interacting randomly with the bacterial cells. If only a small number of cells are moving in a directional manner, it is sufficient to score as motile.

After making observations, dispose of the depression slide in a tray containing a disinfectant solution.

7. The oil immersion objective also can be used to determine motility, but immersion oil first must be applied to the coverglass (Figure 5.1c). Carefully bring the objective into contact with the oil. Since the coverglass will have a tendency to adhere to the oil, apply a slight amount of pressure with your finger to hold the coverglass in place.

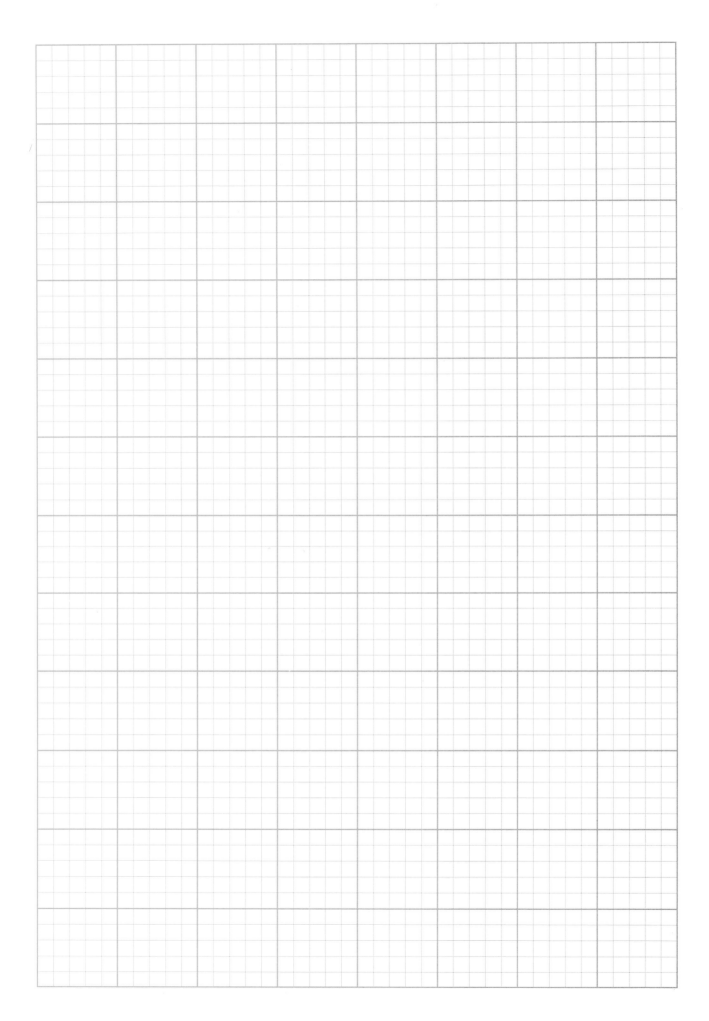

Laboratory Report 5

Name ______ Date ______ Section ______

	Morphology	Motility
Proteus vulgaris		
Staphylococcus epidermidis		
Unknown Isolate		

Review Questions

1. What are some of the environmental factors that may influence the motility of the organisms?

2. Why does the amount of light need to be decreased when performing the hanging drop technique?

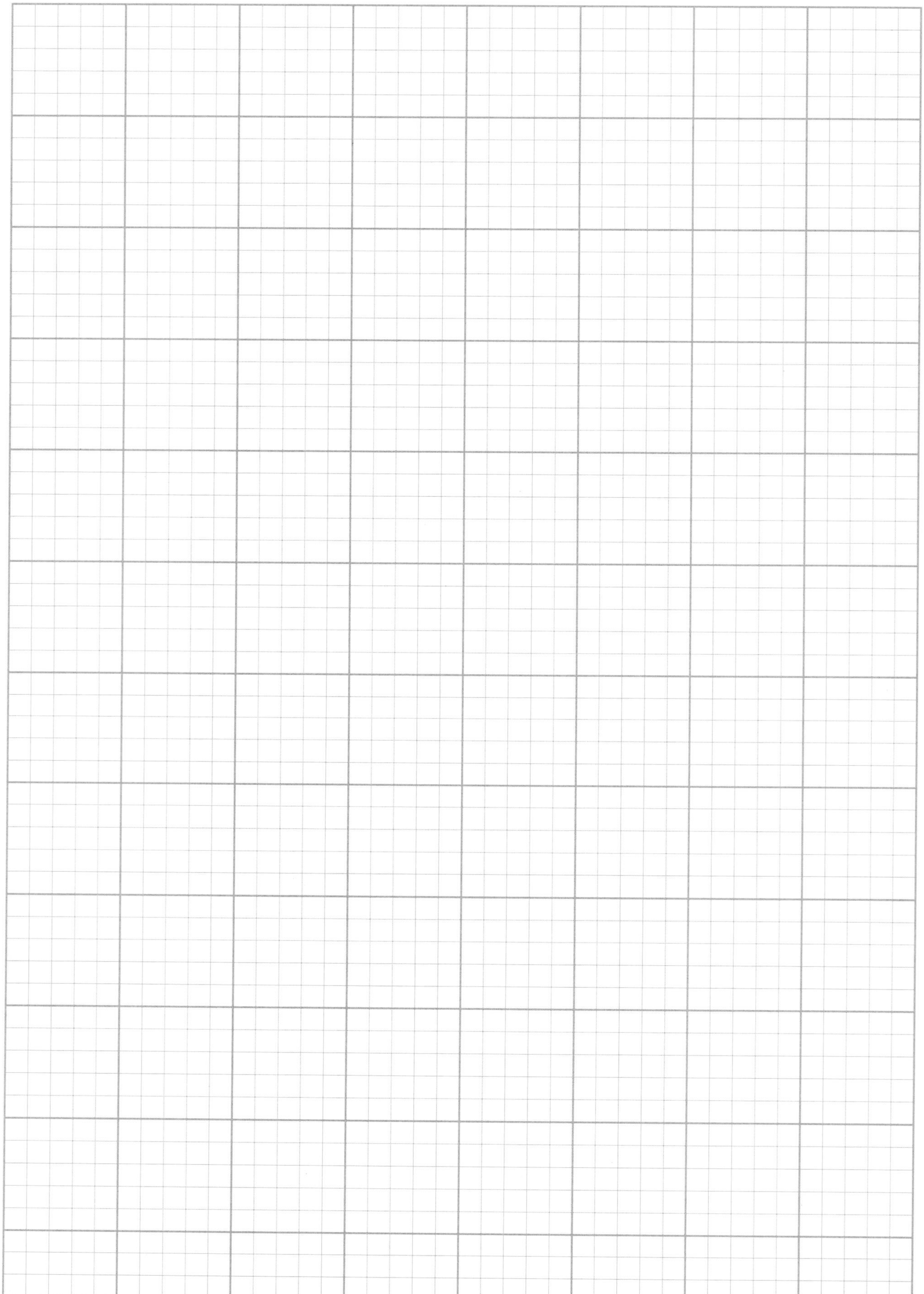

LABORATORY

6 Preparation of Smears and Simple Stain

Safety Issues

- Be careful not to touch any of the colonies on the plates with your fingers.
- Do not place slides containing wet smears on any notebook or book. Slides should be positioned on the lab bench while they are drying.
- To prevent burns to fingers use a clothespin or forceps to heat-fix slides.
- Exercise caution when using dyes as they can permanently stain fabrics. Stained fingers can be cleaned using stain-removal cream found near the laboratory sinks.

OBJECTIVES

1. Learn the techniques of heat-fixed smear and simple stain.
2. Discern shapes of unknown organisms from agar plates in Laboratory 1.

Materials (per 20 students)

1. Methylene blue or crystal violet stain (5 bottles)
2. 20 microscopes with oil immersion lenses
3. Immersion oil (10 bottles)
4. Wash bottles (5)
5. Clean glass slides
6. Staining racks (5)
7. Blotting or bibulous paper
8. Rodac or Petri plate from Laboratory 1.

Procedure

A. Preparation of Smears

Following the outline in Figure 6.1, prepare smears from at least three colonies from the Rodac or Petri plate in Laboratory 1.

Be sure to cool inoculating loop before making transfers in order to prevent splattering of organisms.

1. With a sterilized loop, place a **small** drop of tap water on a **clean** glass slide. If the drop of water beads, the slide probably is dirty and should be discarded. Using a small amount of water initially will facilitate rapid drying of the smear.
2. Perform three separate stains on three slides by using a sterilized inoculating loop to individually remove a **scant** amount of each of the three

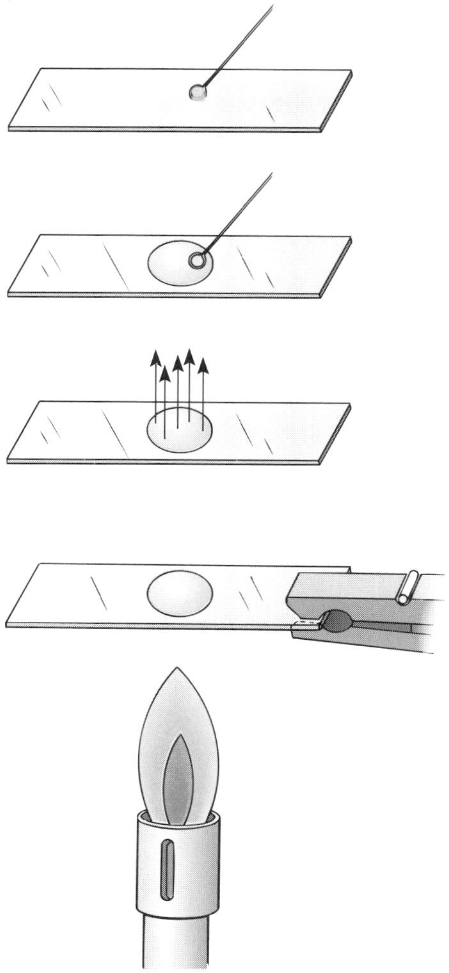

FIGURE 6.1 Smear preparation.

colonies, and emulsify in separate drops of water. If the smear is too dense (i.e., contains a large number of organisms), staining and differentiating the organisms will be difficult. Make sure to spread out the liquid on the surface of the slide to facilitate drying and distribution of the organisms.

3. Allow the smears to dry at room temperature. Do not blow on the slide. Remember, we are laden with "germs" and are not interested at this point in determining the types of microbes in human breath!
4. Gently heat-fix by passing the dried smears through the flame of the Bunsen burner. A clothespin or forceps should be used to prevent burning of fingers. Caution must be exercised not to overheat the slide. Remember to gently heat-fix and not cook the bacteria on the slide.
5. If a smear is prepared from a broth culture rather than from an agar surface, a slightly different procedure needs to be used. When using a broth culture the organisms are more easily washed off the slide, so it is necessary to apply more culture fluid at the start. This is accomplished by allowing the fluid to dry and applying additional fluid until a number of dry layers have been formed.

B. Simple Staining

The simple staining procedure uses a single stain to help visualize the individual cells of microorganisms as deposited on a slide using the smear-preparation technique. Most dyes used in this procedure (crystal violet, methylene blue, or basic fuchsin) contain chromophoric (colored) cations (positively charged ions). Since the surfaces of bacterial cells have a slightly negative charge, the positively charged dyes have an affinity for the cells. The dyes that interact in this fashion with the bacterial cells are referred to as **basic dyes**. Other dyes that are anionic, that is, those in which the chromatophore carries a negative charge, are useful in other staining techniques.

Stain all of the smears using the following procedure: (Figure 6.2)

1. Place a slide on a staining rack either on a staining dish or over the sink.
2. Flood the smear with the crystal violet dye.
3. Allow the dye to remain in contact with the slide for 2 to 3 min.
4. Using a wash bottle, rinse the crystal violet from the slide. Use a clothespin to prevent staining fingers.
5. Blot the slide between the pages of a tablet of blotting paper referred to as bibulous paper. Wipe the bottom portion of the slide to remove excess dye. Do not wipe the top part of the slide as that would remove the smear!
6. Observe each of the smears using oil immersion microscopy as described in Laboratory 3.

Examples of morphological types are pictured in Color Plates 2 and 3.

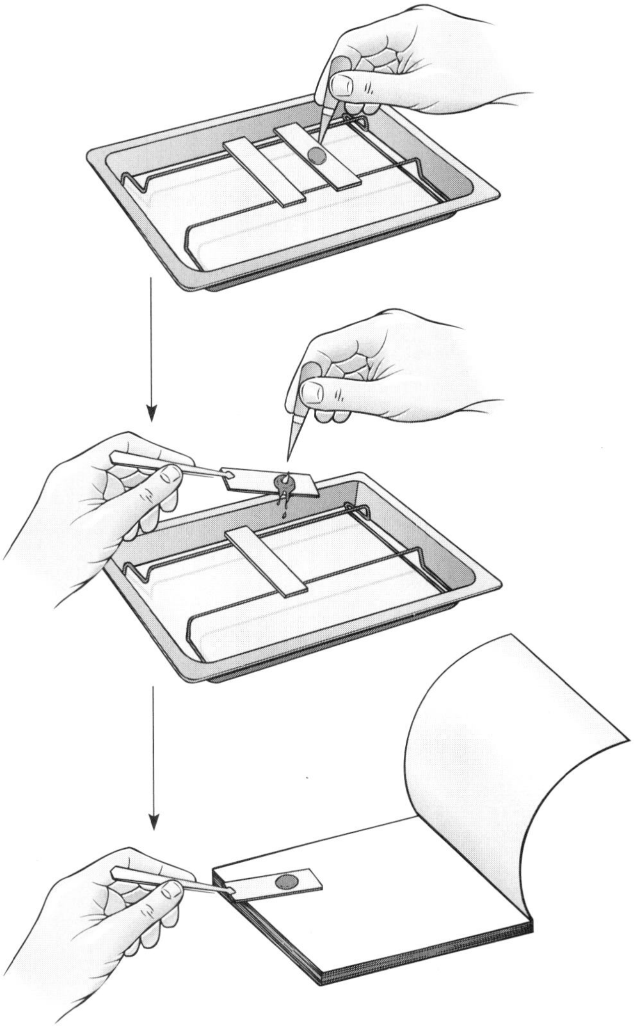

FIGURE 6.2 Simple staining procedure.

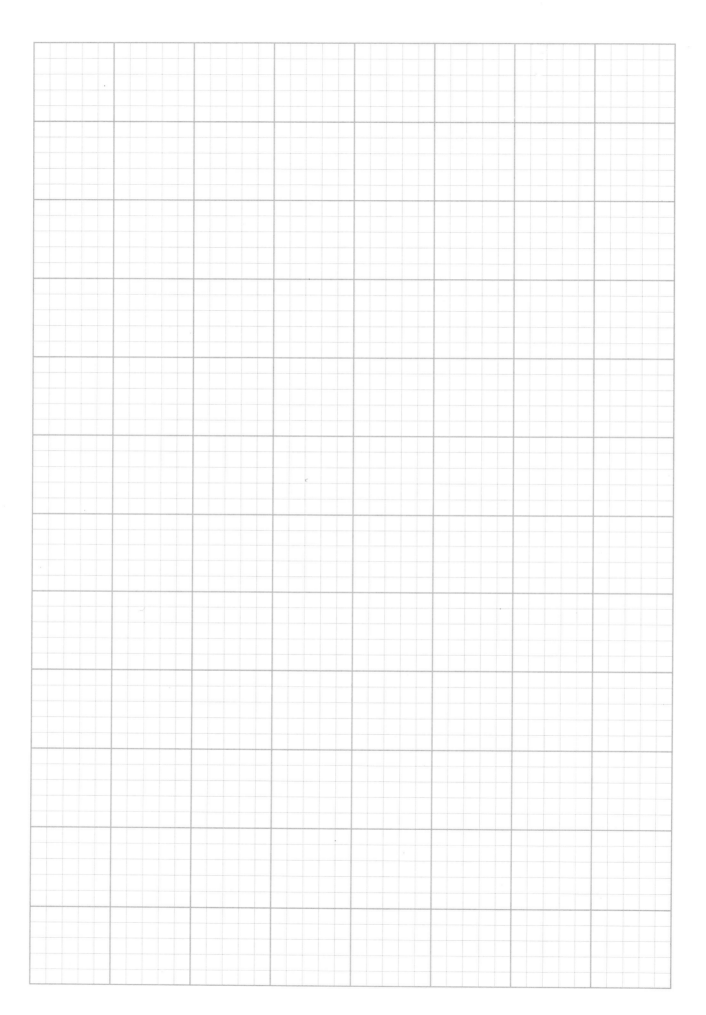

Laboratory Report 6

Name ____________ Date ____________ Section ____________

Simple Stain

Microscopic description of cellular morphologies of cells simple stained from Rodac plate colonies

Draw Shape of Cells and Arrangement

Colony 1

Magnification __________

Draw Shape of Cells and Arrangement

Colony 2

Magnification __________

Draw Shape of Cells and Arrangement

Colony 3

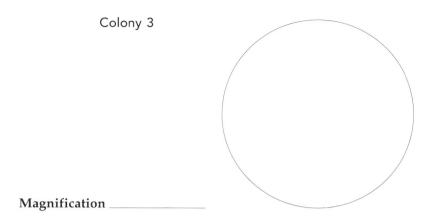

Magnification ________________

Review Questions

1. Why is the number of organisms present in a smear preparation of such critical importance?

2. In the preparation of a smear for staining, why is heat-fixation important?

3. What would be the result of overheating the bacterial smears?

4. Does it make a difference whether tap water or distilled water is used to prepare smears? Explain.

LABORATORY

7 The Negative Stain

Safety Issues

- To prevent splattering of organisms, cool the inoculating loop after sterilization before attempting to transfer any organisms.
- Place slide used to spread out nigrosin-bacteria mixture in a beaker of disinfectant when smear preparation has been completed.
- Discard the toothpick used to obtain dental plaque in a beaker of disinfectant.

OBJECTIVES

1. Learn the negative staining technique to better visualize small cells or those resistant to staining.
2. Understand how negative stain can be used for the observation of capsules.

When performing the simple stain, use a basic dye (crystal violet or methylene blue). The chromophores in basic dyes are positively charged. Since the net charge on the surface of most bacterial cells is negative, the dyes are attracted to the cells. The negative stain uses an acid dye, nigrosin. It has negative charges on its chromophore and therefore is repelled by negatively charged bacteria. As a result, cells remain unstained with the background being colored. The particles of the stain are also too large to penetrate the capsular material of cells and, as a result, the capsules appear as clear zones surrounding the cells.

There are a number of advantages in using a negative stain. (1) Some bacteria such as *Mycobacterium* have a high concentration of lipid in their cell walls and are resistant to staining. When using a negative stain these bacteria become more readily visible, standing out as unstained against a dark background. (2) The simple staining procedure uses heat-fixation which can distort the cells, either causing a shrinking or a disruption of the arrangement of cells. Since negative staining does not involve heating, little distortion of cells occurs. (3) Bacterial capsules and slime layers can be visualized by using the negative staining technique. They are resistant to staining and can be seen as a halo surrounding the nigrosin-staining bacteria. Gin's method of staining capsules is similar to the procedure of negative staining and uses Loeffler's methylene blue to stain the bacteria. With this method, the cells will be stained blue, the capsular material will remain unstained, and the background will be dark.

Materials (per 20 students)

1. Cultures of *Enterobacter aerogenes* (5)
2. Loeffler's methylene blue stain (5 bottles)
3. Nigrosin stain (5 bottles)
4. Clean glass slides
5. Sterile distilled water (20 tubes)
6. Sterile toothpicks (5 foil packages)
7. Brightfield microscopes (20)
8. Experimental isolates from Laboratory 2

Procedure

Cool inoculating loop before mixing to prevent splattering of organisms.

1. Using an inoculating loop, add one loopful of nigrosin stain to a clean glass slide. Then add a loopful of distilled water to the drop.
2. With a sterilized inoculating loop, transfer a portion of one of the colonies from an agar slant (Laboratory 2) to the drop of distilled water and mix to prepare an emulsion. Use the loop and a rotary motion to mix the bacterial suspension with the nigrosin.
3. Position the edge of a second glass slide over the drop (Figure 7.1). Then push the slide over the bottom slide to spread out the nigrosin-bacteria mixture.
4. A fine film should result. Let it air dry, and **do not fix with heat**.
5. Place a drop of immersion oil on the slide and observe microscopically using the oil immersion objective.
6. Prepare another slide, this time using the *Enterobacter* culture. Once again add a loopful of nigrosin stain to a clean glass slide. Place a loopful of distilled water next to the drop of stain. With an inoculating loop, transfer some of the *Enterobacter* culture to the drop of water and mix

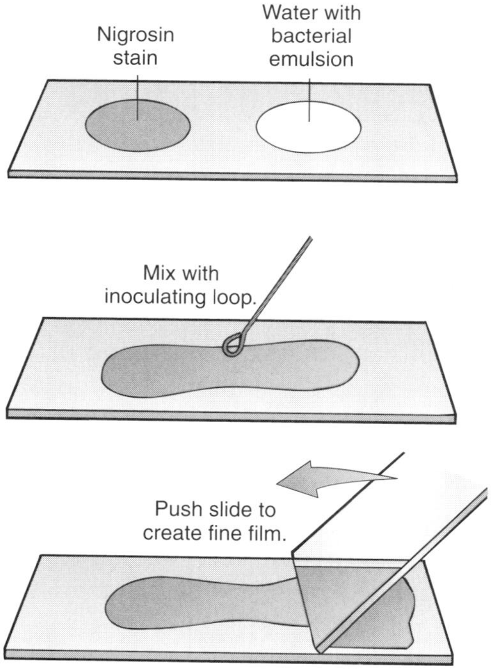

FIGURE 7.1 Preparation of negative stain.

with the drop of stain. Use the technique mentioned previously for preparing the thin film. Allow to air dry and observe.

7. Stain the dried smear for 1 min, using Loeffler's methylene blue stain.
8. Rinse the slide with water.
9. Allow to air dry. Do not blot the slide.
10. Place a drop of immersion oil on the slide and observe under the 100× objective.
11. The bacterial cells should be stained blue. Capsules appear as halos or transparent areas around the cells.
12. Using an inoculating loop, place a drop of nigrosin and a drop of water at the end of a third glass slide. Use the sterilized toothpick to remove some plaque from your teeth. Mix the plaque with the nigrosin and water to form an emulsion.
13. Repeat steps 3 through 5 and record your observations.

Discard toothpick in container of disinfectant to prevent spread of organisms.

Procedure

The following procedure, developed by Woeste and Demchick,* is an alternative technique for performing the negative stain.

1. Prepare a bacterial smear in a normal fashion, allowing it to air dry and heat-fix.
2. Use a blunt tip of a water-resistant black ink marker to apply a coat of ink in a straight line over the smear.
3. Allow to dry and, using the oil immersion objective of the microscope, observe the cells (see Color Plate 4a, b).

*Woeste, S. and P. Demchick (1991). New Version of the Negative Stain. *Applied and Environmental Microbiology 57:* 1858–1859.

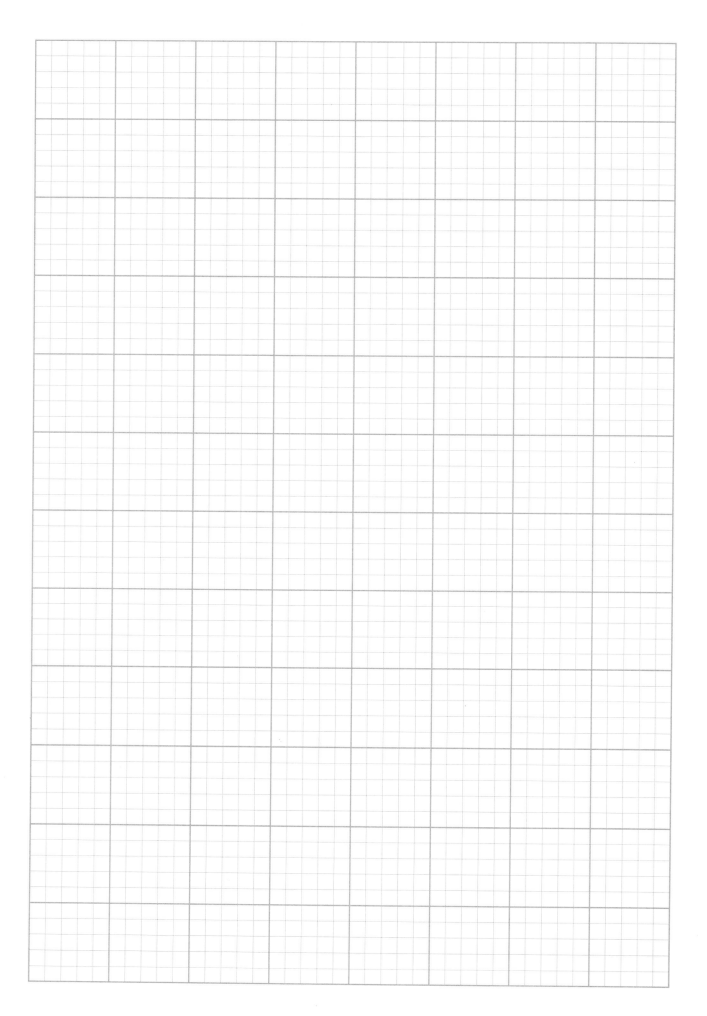

Laboratory Report 7

Name ______________________ Date ____________ Section ____________

Negative Stain

Sketch appearances of cells, noting the shape and arrangement of cells of *Enterobacter* and the morphology of organisms from plaque.

Cells from Agar Colonies

Cells of *Enterobacter*

Shape ____________________

Arrangement ______________

Plaque Sample

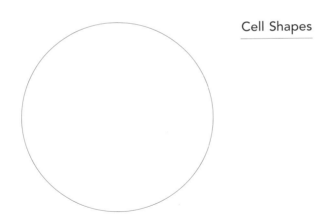

Cell Shapes

Arrangements

Review Questions

1. How many different morphological types of organisms are evident in the plaque sample?

2. Why is the negative stain not heat-fixed?

3. If nigrosin stain is not available, can another stain be used?

LABORATORY

8 The Gram Stain

Safety Issues

- Hold all slides with a clothespin or forceps when heat-fixing.
- The decolorizer used in the Gram stain procedure contains alcohol and thus is flammable. Keep the bottle away from any open flame.
- The liquid coverslip is highly flammable and should be kept a distance from any flame.
- Be careful not to cut yourself on any broken glass slides. Dispose of broken glass in specially marked containers.
- Exercise caution when using dyes as they can permanently stain fabrics. Stained fingers can be cleaned using stain-removal cream found near the laboratory sinks.

OBJECTIVES

1. Master the Gram stain procedure.
2. Discern morphology and Gram reaction of "unknown" organisms and "known" cultures.

The Gram stain was developed in 1884 by Hans Christian Gram, a Dutch bacteriologist, to help study a group of spherically shaped bacteria isolated from human lung tissue. The stain differs from the simple stain performed previously in that the Gram stain is used to differentiate types of bacteria depending on their abilities to retain a particular stain. It is therefore referred to as a **differential staining** technique.

The Gram stain is one of the most frequently used techniques in microbiology, and its mastery can be difficult. In order to become proficient, pay careful attention to the procedure and **practice it often**. The technique separates bacteria into two groups: gram-positive and gram-negative.

The first step of the Gram stain involves staining the fixed smear of organisms with a **primary stain** of crystal violet. This is followed by the application of Gram's iodine stain, also known as the **mordant** (a substance capable of intensifying or deepening the reaction of the specimen to a stain). The next step in the procedure is the most critical and involves washing the stained smear with a **decolorizing agent**, usually 95% ethanol or isopropyl alcohol. The last step employs a **counterstain** known as safranin. Figure 8.1 depicts the reactions that are taking place at each step depending on whether the organisms are gram-positive or gram-negative. Gram-positive organisms will not be easily decolorized and thus retain the purple stain of crystal violet. On the other hand, the gram-negative organisms will be decolorized by the alcohol so that they then can take up the counterstain. They will appear

Step	Reagent	Time	Response: Gram-positive	Response: Gram-negative
Primary stain rinse	Crystal violet	1 minute	Purple	Purple
Mordant rinse	Gram's iodine	1 minute	Purple	Purple
Decolorizer rinse	95% Ethanol	Brief	Purple	Colorless
Counterstain rinse	Safranin	1 minute	Purple	Pink

FIGURE 8.1 Steps of the Gram stain.

pinkish or red. If enough alcohol is applied, it is possible to decolorize almost all cells. Careful adherence to the procedure will help to ensure successful results.

The age of the culture is an important factor in the outcome of the Gram stain. Most gram-positive organisms will lose their gram-positivity with age. The Gram stain should always be performed on vigorous, actively growing cultures (18- to 24-hour cultures usually will give excellent results). To further complicate interpretation of results, some organisms turn out to be **gram-variable** following Gram staining. This means that some of the cells in the population will stain purple (gram-positive) and other cells will stain red (gram-negative). Only repeated staining of the same culture at different times will verify the existence of a truly gram-variable culture. Gram-variability is relatively rare; mixed results in the Gram stain most probably occur from working with an impure culture, or either underdecolorizing or overdecolorizing with ethanol.

The basis of the Gram stain resides in the differences in cell wall composition of gram-positive and gram-negative bacteria. Be sure to read your textbook for a complete discussion of the topic. Briefly, gram-positive organisms have a thick cell wall composed of a sugar-protein substance referred to as peptidoglycan, and a molecule called teichoic acid. Very little lipid is associated with the gram-positive cell wall. On the other hand, the gram-negative cell wall has an abundance of lipopolysaccharide and lipoprotein. Two mechanisms have been proposed to explain the basis of the Gram reaction. One mechanism advanced is that the alcohol decolorizer extracts lipid from the cell wall of gram-negative bacteria, thus facilitating the loss of the crystal-violet complex. A second mechanism proposed is the shrinkage of pores in the thick peptidoglycan layer of cell walls of gram-positive organisms due to exposure to the alcohol decolorizer. The shrinkage of the pores leads to retention of the dye complex in gram-positive cells; whereas, the higher porosity of the thin peptidoglycan layer of gram-negative organisms leads to more rapid loss of the dye complex. Differences in cell wall composition of gram-positive and gram-negative organisms explains the differences in susceptibility to various antibiotics.

Critical precautions to take when performing Gram stains:

1. Do not overdecolorize slides with alcohol.
2. Make sure all stains are fresh.

3. Use actively growing cultures.
4. Prepare thin smears containing a sparse number of organisms.
5. Do not overheat the slide during the fixation step as structural morphologies of the cells may be altered. Overheating during the fixation step may influence the structure of the cell wall and, as a result, alter the outcome of the stain.
6. Do not allow the stains to react with the smear for longer than the specified time.
7. Use as little water as possible when rinsing your slides.

Materials (per 20 students)

1. Streak-plate isolates of mixed culture from Laboratory 2.
2. Cultures:
   Escherichia coli (5)
   Staphylococcus sp. (5)
   Branhamella sp. (5)
   Bacillus sp. (5)
3. Gram stain reagents (5 sets)
   (Crystal violet, Gram's iodine, alcohol decolorizer, and safranin)
4. Clean glass slides
5. Microscope with oil immersion lens (10)
6. Wash bottles (5)
7. Commercially available Gram✔™ slides (Fisher Scientific) (20)
8. Liquid coverslip solution (1 bottle)

Procedure

☣ To prevent burns to fingers or hands, hold slides with a clothespin, forceps, or slide holder while heat-fixing.

1. Use the smear preparation technique from Laboratory 6. Remember to prepare a thin smear, allow enough time to air dry completely, and do not excessively heat-fix the slide. Prepare smears from the streak-plate isolates reserved on plates from Laboratory 2. Also prepare smears of four freshly grown cultures of *Bacillus sp.*, *Staphylococcus sp.*, *Branhamella sp.*, and *Escherichia coli.* Once becoming proficient and confident in performing Gram stains, a number of smears can be prepared on the same slide, as shown in Figure 8.2. Caution should be exercised to not mix cultures on the slide during preparation. Sectored slides or slides marked with a wax marking pencil are sometimes used to prevent such mixing. Alternatively, a commercially prepared gram check slide can be used.

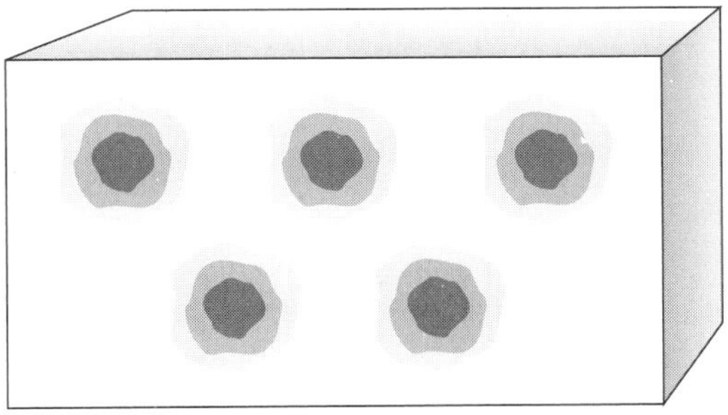

FIGURE 8.2 Placement of multiple smears on slide.

These slides have known gram-positive and gram-negative cells fixed to them. Sectors are available for positioning of smears of other organisms. Heat-fix all smears on the slide.

⚠ Dyes can cause permanent stains to fabrics. Be careful not to splash or spill.

2. Place slides on a staining tray rack or over a sink rack and cover the smear with the crystal-violet stain. Allow the stain to react for **1 min**.
3. Using the wash bottle provided, rinse the slide with a small amount of water. Make sure to remove excess water by shaking the slide.
4. Cover the smear with the Gram's iodine stain and allow to react for **1 min**. Wash off the stain with water.
5. Holding the slide at a 45° angle over the sink or staining tray, apply the decolorizer dropwise at the top of the slide and allow the alcohol to run off the slide. Add a maximum of 4 to 5 drops of alcohol to the slide to prevent overdecolorization. Immediately rinse the slide with water. Remove the excess water by shaking.
6. Counterstain by flooding the smear with safranin for **1 min**.
7. Rinse the water and blot dry by placing between pages of the bibulous pad. Wipe the **bottom** of the slide to remove excess stain that may decrease visibility.

⚠ The material is highly flammable. Keep away from any open flame.

8. To preserve any slide, apply liquid coverslip or equivalent to uniformly coat your slide. In order to do this, place a few drops of the liquid coverslip on the stained slide. Spread uniformly with another clean glass slide. Allow to dry for at least 15 min before proceeding.
9. If not making a permanent slide, apply a drop of immersion oil directly to the smear(s) and place the slide on the stage of the microscope. Following the procedure used when performing the simple stain, observe the stain organisms.
10. Record the Gram stain reaction of each of the organisms, the morphology of the individual cells (i.e., rod, spherical, or spiral shaped), and the arrangement of the cells (clusters, chains, or pairs). (Known gram-positive and gram-negative microorganisms are pictured in Color Plate 5 a–d.)

Laboratory Report 8

Name Date Section

1. Gram stains of streak-plate isolates from Laboratory 2.

Streak-Plate Isolate	Colonial Morphology	Gram Stain	Cellular Morphology	Sketch
1				
2				
3				
4				
5				

2. Gram stains of "known" cultures

BACILLUS

Stain Morphology Arrangement Sketch

Magnification ______

STAPHYLOCOCCUS

Stain Morphology Arrangement Sketch

Magnification ______

Brahamella

Stain	Morphology	Arrangement	Sketch

Magnification ________

Escherichia

Stain	Morphology	Arrangement	Sketch

Magnification ________

Review Questions

1. A student completed the Gram stain technique, but neglected to use the Gram's iodine. Describe the appearance of the cells.

2. What would a smear of *E. coli* look like if you were to observe it after the decolorization step in the Gram stain procedure? What about the *Bacillus*? Explain.

3. List some of the significant differences of gram-positive and gram-negative organisms other than their retention of dye complexes.

LABORATORY

9 Endospore Stain

Safety Issues

- Be careful not to touch the hot plate or slides with your unprotected hands. Wear gloves and use forceps when manipulating hot objects.
- If using the ring stand method, wear hot gloves to grasp the Bunsen burner. Make sure the flame is directed carefully so that only the underside of the slide is heated.
- Wear eye protection while heating the stains to protect your eyes.
- To minimize splattering of stains, do not allow stains to boil.
- Exercise caution when using dyes as they can permanently stain fabrics. Stained fingers can be cleaned using stain-removal cream found near the laboratory sinks.

OBJECTIVES

1. Become proficient in endospore staining to determine shape and position of spores.
2. Contrast vegetative cells and endospores.
3. Be able to draw representative pictures of the various stages in the "spore cycle" and place them in an appropriate order.

Endospore formation is an example of differentiation in prokaryotic cells. Members of the genera *Bacillus* and *Clostridium* are the most commonly encountered spore-forming bacteria.

Endospores (see Color Plate 6) allow bacteria to be resistant to heat, drying, ultraviolet radiation, and disinfectants. The size and location of endospores within cells are important determinants for taxonomic differentiation. Endospores may be central, terminal, or subterminal. Some species form endospores that are larger than the vegetative cell and are referred to as swollen. The technique used in this laboratory for observing spores is the Schaeffer–Fulton endospore stain.

Various sporulation media can be used to induce spore formation. In this experiment, a medium is used which has an elevated concentration of Mn^{2+}. A low-nutrient medium such as R2A (Difco Laboratories) may also be used.

Materials (per 20 students)

1. Tryptic soy agar slant cultures of *Bacillus subtilis* (5)
2. Tryptic soy agar + Mn^{2+} slant culture of *Bacillus subtilis* (5)
3. 5% aqueous solution of malachite green (5 bottles)

4. 0.5% aqueous solution of safranin (5 bottles)
5. Paper towels
6. Ring stand or hot plate with heavy aluminum foil (5)

Procedure

Prepare smears of the two *Bacillus* cultures, using either a hot plate or ring stand to stain.

⚠ Exercise caution so as not to burn your hands or fingers on the hot plate.

A. Hot Plate Method

1. Prepare a boat of heavy aluminum foil and place on a hot plate (Figure 9.1).
2. Place your slides with the smears on the aluminum foil and cover them with a small piece of paper towel. Flood the slides with malachite green stain.
3. Turn the hot plate to a low heat setting and heat the slides until they steam. **Do not boil**. Add more stain as evaporation occurs. Continue for about 3 to 5 min.
4. Allow the slides to cool and then rinse with water until no green color emerges in drops falling off the slide.
5. Place the slides on a staining rack and flood with safranin stain. Allow to react for 1 min.
6. Wash with distilled water and blot dry.

⚠ Handle burner with gloved hands and apply flame carefully to the underside of the slides. Do not expose flame to paper or any other combustible or flammable material.

B. Ring Stand Method

1. Position slides on a ring stand in a tray to catch drops of dye (Figure 9.2).
2. Place a small piece of paper towel on the slides and flood with malachite green stain.
3. Carefully grasp a Bunsen burner (Figure 9.3) and heat the underside of the slides until they begin to steam. **Do not allow the dye to boil on the slides**. Replenish the dye and continue to steam for 3 to 5 min, being careful to not allow the slide to dry out.
4. Allow the slides to cool.

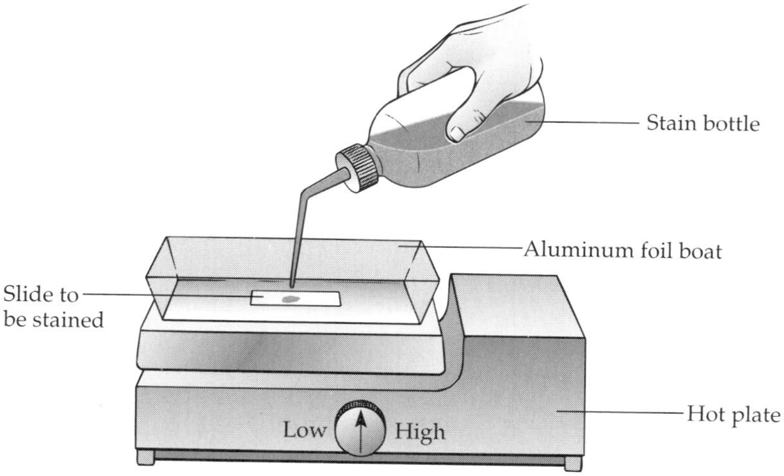

FIGURE 9.1 Aluminum-foil boat.

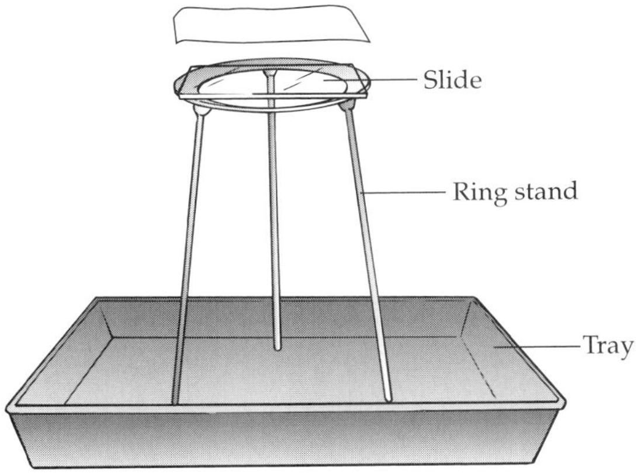

FIGURE 9.2 Ring stand method for spore stain.

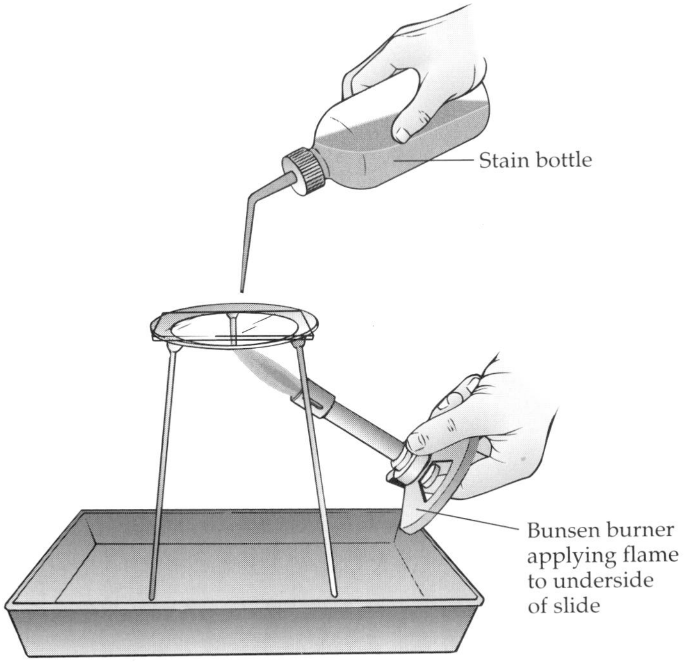

FIGURE 9.3 Holding of Bunsen burner.

5. Rinse with distilled water until the drops coming off the slides show no green color.
6. Place the slides on a staining rack and flood with the safranin stain. Allow to react for 1 min.
7. Rinse with distilled water and blot dry.
8. Observe slides microscopically using the oil immersion objective.

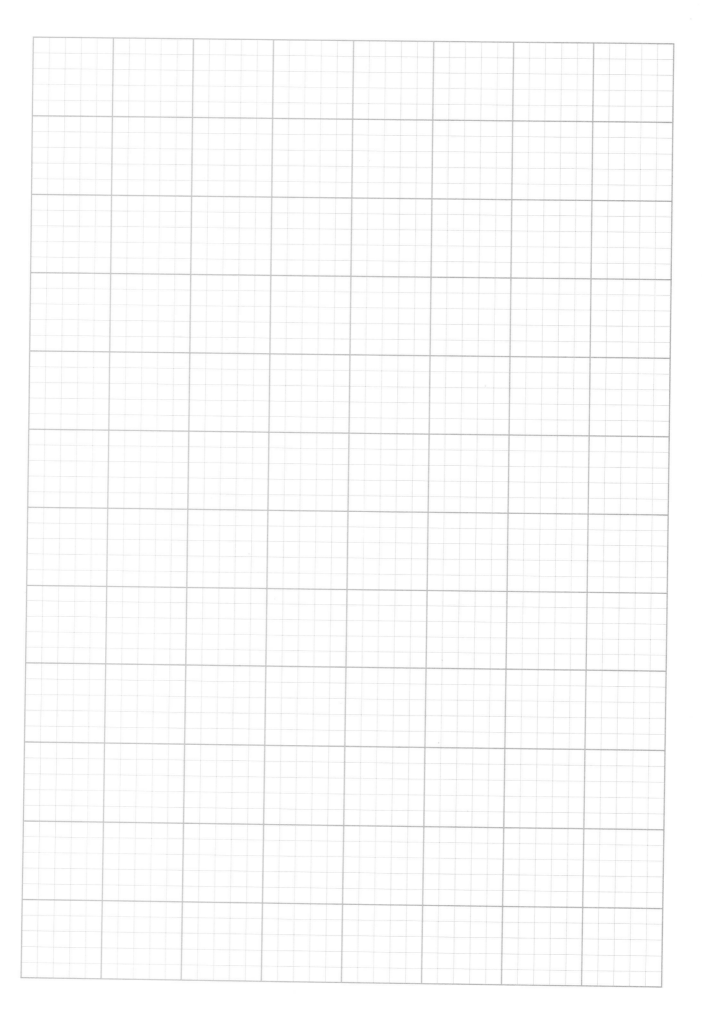

Laboratory Report 9

Name Date Section

Endospore Stain

Sketch observations of the cells from the two cultures of *Bacillus*. Indicate the color of the vegetative cells and the endospores. Make sure to note the location of the endospores and whether they are swollen.

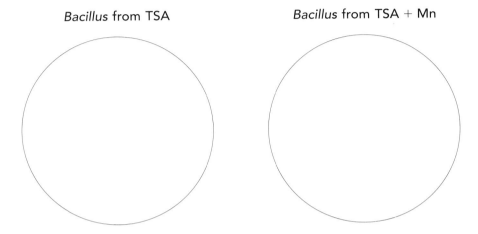

Bacillus from TSA

Bacillus from TSA + Mn

Which of the cultures has the greatest number of sporulating cells?

Review Questions

1. What functions do bacterial endospores perform in the cells that produce them?

2. How does the composition of the growth medium influence the production of spores?

3. Does the spore location within the cell help differentiate species? Explain.

4. How would cells possessing endospores appear if the Gram stain was performed?

LABORATORY

10 Acid-Fast Stain

Safety Issues

- Extreme caution should be exercised to prevent spilling boiling water from the beaker. Severe burns could result to your skin.
- Use eye protection and hot gloves while steaming the slides.
- Since the acid-alcohol decolorizer is flammable, make sure to keep it at a distance from any open flame.
- Exercise caution when using dyes as they can permanently stain fabrics. Stained fingers can be cleaned using stain-removal cream found near the laboratory sinks.
- Be careful not to cut yourself on any broken glass slides. Dispose of broken glass in specially marked containers.

OBJECTIVES

1. Learn the procedure for acid-fast staining and how it is used to detect members of the genus *Mycobacterium*.
2. Distinguish between acid-fast bacteria and non–acid-fast bacteria.

Similar to the Gram stain, the acid-fast stain is a differential stain. In this case, acid-fast organisms, those retaining carbolfuchsin, stain red and non–acid-fast organisms stain blue. Microorganisms that are acid-fast have a very high percentage of lipid in their cell walls. The high lipid content renders the cells resistant to staining, and, if stained, the cells are resistant to decolorization. Heat must be applied to the carbolfuchsin staining step to force stain into the cells. To decolorize the organisms, a solution of acid alcohol is used. All non–acid-fast organisms are decolorized. A counterstain is then applied (usually methylene blue) to stain all decolorized non–acid-fast organisms.

Materials (per 20 students)

1. Agar slants of *Mycobacterium smegmatis* (5)
2. Agar slants of *Mycobacterium phlei* (5)
3. Agar slants of *Bacillus subtilis* (5)
4. Agar slants of *Escherichia coli* (5)
5. Fisher Acid Fast (AFB) slides (20)
6. Carbolfuchsin stain (5)
7. Acid alcohol (2.5% sulfuric acid/95% ethanol) (5 bottles)
8. Methylene blue (5 bottles)
9. Ring stands with boiling water bath (3)

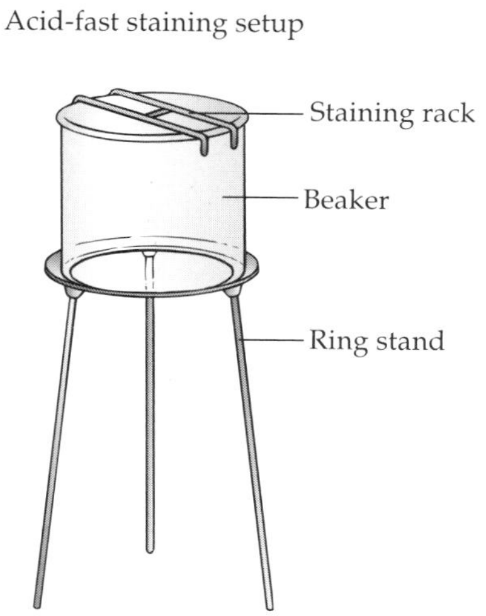

FIGURE 10.1 Acid-fast staining setup.

Procedure

1. Place a beaker of water with a staining rack on a ring stand, as shown in Figure 10.1. Place a Bunsen burner under the beaker.
2. Prepare smears of the four known cultures on the Fisher AFB slides.
3. Air dry and heat-fix the slide.
4. Cover the slides with a piece of paper towel cut to fit.
5. Flood the entire smear with carbolfuchsin stain.
6. Light the Bunsen burner and steam the slides gently for at least 10 min (Figure 10.2). Do not allow the paper towel to dry during staining. Apply additional stain as evaporation occurs.

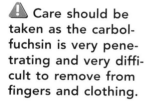

⚠ Care should be taken as the carbolfuchsin is very penetrating and very difficult to remove from fingers and clothing.

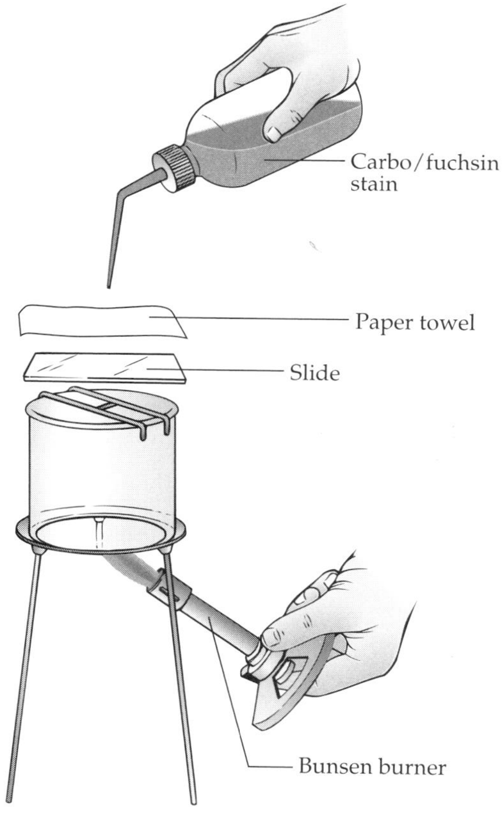

FIGURE 10.2 Steaming slides for acid-fast staining.

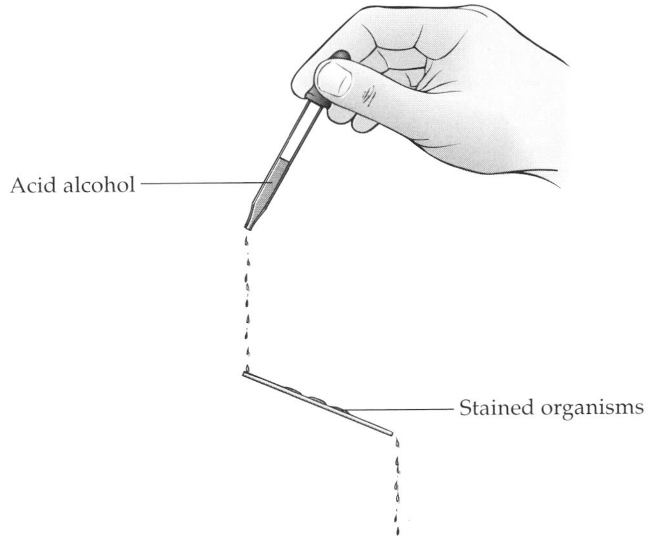

FIGURE 10.3 Decolorization step in acid-fast stain.

7. Carefully remove the paper towel with forceps and place it in a discard beaker containing disinfectant.
8. Gently wash the slide with distilled water to remove excess stain.
9. Decolorize by holding the slide at a 45° angle (Figure 10.3) and adding the acid alcohol to the top edge of the slide. Decolorize for 20 to 30 sec or until the drops falling from the slide are no longer tinged with color.
10. Wash the slide with distilled water.
11. Flood the slides with the methylene blue counterstain and allow to react for at least 1 min.
12. Wash the slide with distilled water and blot dry.
13. Microscopically examine the slides, using the oil immersion objective, and record observations. Acid-fast organisms will stain red and non–acid-fast organisms will stain blue. The positive control area on the AFB slide should contain red-stained rod-shaped bacteria. The negative control portion should contain blue-stained coccoid forms.
14. Compare stains to the acid-fast organism pictured in Color Plate 7.

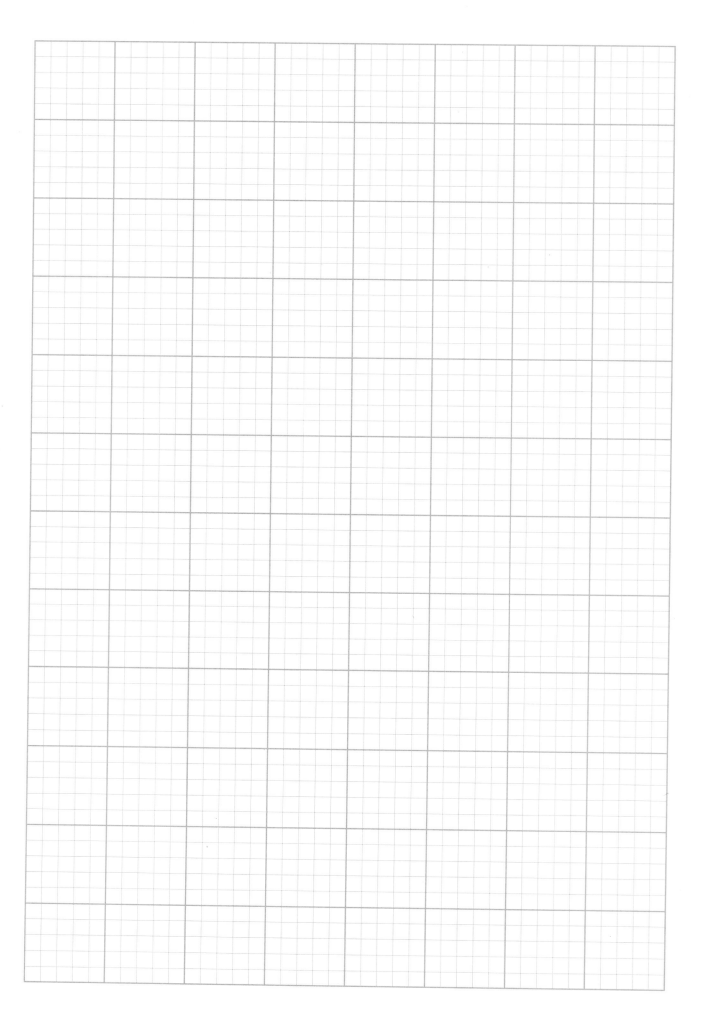

Laboratory Report 10

Name Date Section

Sketch the morphological appearance of each of the stained cultures.

M. smegmatis *M. phlei* *E. coli*

Bacillus Positive Control Negative Control

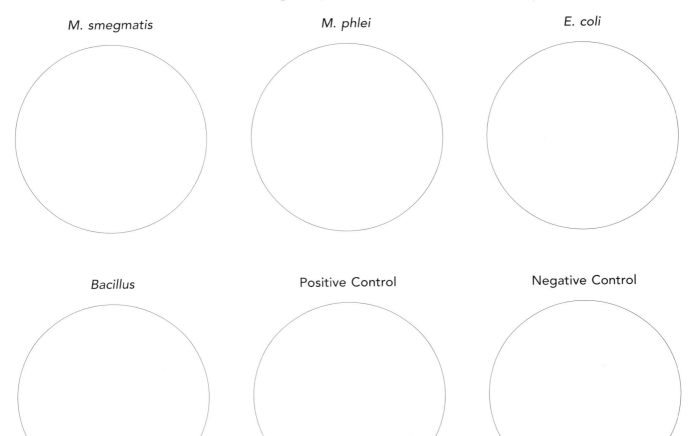

Describe the stain of each culture. Which are acid-fast?

Review Questions

1. How would a culture of *Mycobacterium* stain if the Gram stain were used? Would the organisms be gram-positive or gram-negative?

2. How would the cultures stain if steaming did not occur for a long enough period of time?

3. Is there any association between the degree of acid-fastness and the potential pathogenicity of an organism?

UNIT II

Maintenance of Stock Cultures, and Media Preparation

In the first unit, we observed the microscopic growth of organisms and characterizations of cellular morphologies using a variety of staining techniques. In this second unit methods of preservation of microorganisms and preparation of various types of media are described. Culture preservation can be accomplished by regular transfer of microorganisms onto fresh media and maintenance at lower temperatures. Long-term preservation requires freezing in stabilizing solutions at lower temperatures. Preparation of stock cultures for short-term or for long-term storage is crucial to any systematic microbiological investigation.

Every student of microbiology should feel very confident about the preparation of culture media. Future experiments call for the preparation of various types of media.

LABORATORY

11 Maintenance of Stock Cultures

Safety Issues

- To prevent splattering of organisms, cool inoculating loop before making any transfers.

OBJECTIVE

Learn procedures for maintaining stock cultures and techniques for long-term storage.

When performing experiments in microbiology, it is always important to have actively growing cultures available. Cultures could be purchased each time when needed, but this would be quite expensive. An alternative is to maintain a supply of **stock cultures**. Most laboratories will maintain two types of stock cultures: one is referred to as a **working stock**, used for daily inoculations or slide preparations; the second is called a **reserve stock** and is usually stored in the refrigerator, to be used to prepare future working stocks.

It is critical to learn the techniques involved in maintaining viable cultures. Some isolates may be irretrievable, and, as a result, reliable stocks are needed for future experimentation. There is no one best strategy for maintaining such a stock. The choice of technique depends upon a number of factors: (1) frequency of use of the stock, (2) the type of organism (some bacteria will accumulate toxic end products even at refrigeration temperatures and as a result will die more quickly than others), and (3) availability of appropriate freezers. The most often used technique is to inoculate a nutrient agar slant and allow the microorganism to grow at the optimum growth temperature for 18 to 24 hours. Usually the culture is then placed in a refrigerator (4°C) for long-term storage. The type of medium used for stock culture maintenance can also influence survival. A rich medium that contains abundant protein and carbohydrate sources will support abundant growth. This growth can lead to the accumulation of higher concentrations of waste products, which can cause death of microorganisms. To prevent this from happening, some laboratories store cultures on ½- to ¼-strength media or grow organisms on a medium containing a single carbohydrate source with an inorganic source of nitrogen.

Many organisms are quite fastidious and die after being refrigerated for a few days. It is wise to maintain slant stock cultures at room temperature and in the refrigerator until the survival of the organism can be determined. Most cultures should be maintained for no longer than 3 to 4 weeks before they are subcultured to fresh agar slants.

For longer-term storage, cultures can be freeze dried or frozen. Freeze drying (lyophilization) is a popular method of preservation of foods (e.g., coffee); however, the technique involves the use of a heavy duty vacuum pump and accessory equipment that may not be available in every laboratory. As an alternative, cultures may be preserved by freezing at −70°C in a cryoprotective environment containing glycerol or skim milk. When the glycerol procedure is used, the organisms are grown in nutrient broth for 16 to 24 hours and then transferred to a sterile cryovial. Sterilized glycerol (30%) is added and mixed with the broth culture. The cryovial is labeled and stored at −70°C. If skim milk is used, the organisms are grown on the appropriate agar medium, and then a loop or swab is used to transfer some of the organisms to cryovials containing 4.0 ml of sterilized skim milk. The contents of the cryovial are mixed and the vial stored at −70°C. Another alternative method of preservation involves the use of Microbank (Pro-Lab Diagnostics, Austin TX). This system consists of a vial of porous beads in a cryovial which can be stored at −70°C. When organisms need to be recovered, a bead is removed from the vial and either streaked onto the appropriate agar plates or dropped into a liquid medium. Many researchers make duplicate freezer cultures and ask a friend at another institution to store the cultures in the event of a local power outage, or have a backup generator available.

In order to become familiar with the preparation and maintenance of stock cultures, examine the streak plates prepared from the mixed bacterial cell culture in Laboratory 2. You should have been able to determine the number of different types of bacteria present if your streaking technique was successful. Select one of the isolated colonies for preservation.

Materials (per 20 students)

1. Streak-plate isolates from Laboratory 2 or another culture
2. Sterile cryovials (20)
3. Sterile 30% glycerol (five 50-ml bottles)
4. Nutrient agar slants (20)
5. Nutrient broth tubes (20)
6. Microbank cryopreservation vials (20)
7. McFarland Standards (#3 and #4)

Procedure

1. Using a sterilized loop, transfer some organisms from the slant culture reserved from Laboratory 2 to the surface of a nutrient agar slant. Incubate at 30°C for 24 hr and then refrigerate at 4°C. Every three weeks, streak some of the stock culture on a fresh nutrient agar plate to check for viability of the organism.
2. Using an inoculating loop, inoculate a nutrient broth tube with organisms from the original slant culture and incubate at 30°C for 16 to 24 hr.
3. Transfer the 2.0-ml volume of the broth tube to a 4.0-ml cryovial (containing 2.0 ml of 30% glycerol); mix, label, and store at −70°C. Alternatively, a 2.0-ml cryovial may be used. In that case, mix 1.0 ml of the broth culture and 1.0 ml of glycerol (Figure 11.1).
4. Every three weeks, remove some of the sample by scraping the surface of the frozen culture with a loop or needle. Then streak for isolation on a nutrient agar plate.

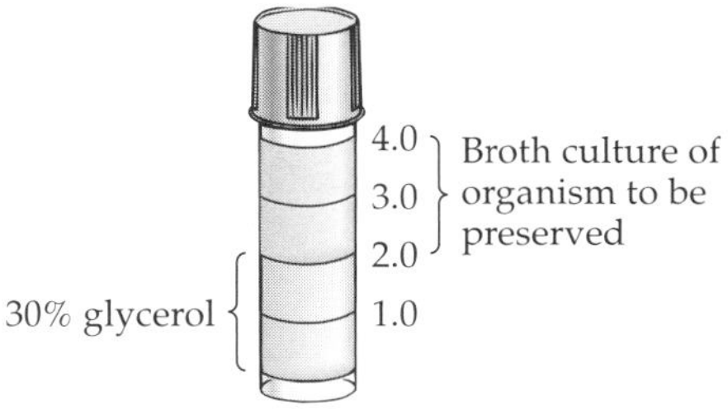

FIGURE 11.1 Cryovial preservation system.

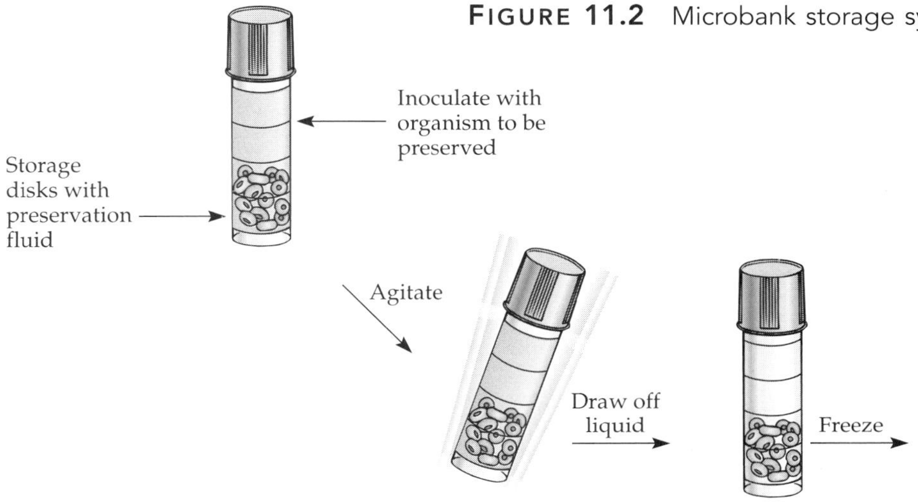

FIGURE 11.2 Microbank storage system.

If the Microbank (Figure 11.2) system is employed, use the following procedure:

1. Inoculate the cryopreservation fluid with loopfuls of a young culture (18–24 hr) to a 3 to 4 McFarland turbidity level.
2. Close the vial tightly and invert 4 to 5 times to emulsify the organisms.
3. The excess cryopreservation fluid should be removed using a sterile pipette, leaving the inoculated beads as free of liquid as possible.
4. Reclose the lid and store the cryovial at −70°C.
5. Check for viability by aseptically removing a bead from the vial with a sterile needle or forceps and streaking onto a nutrient agar plate, or by placing the bead in a nutrient broth tube.
6. Incubate for 24 hr and check for growth.

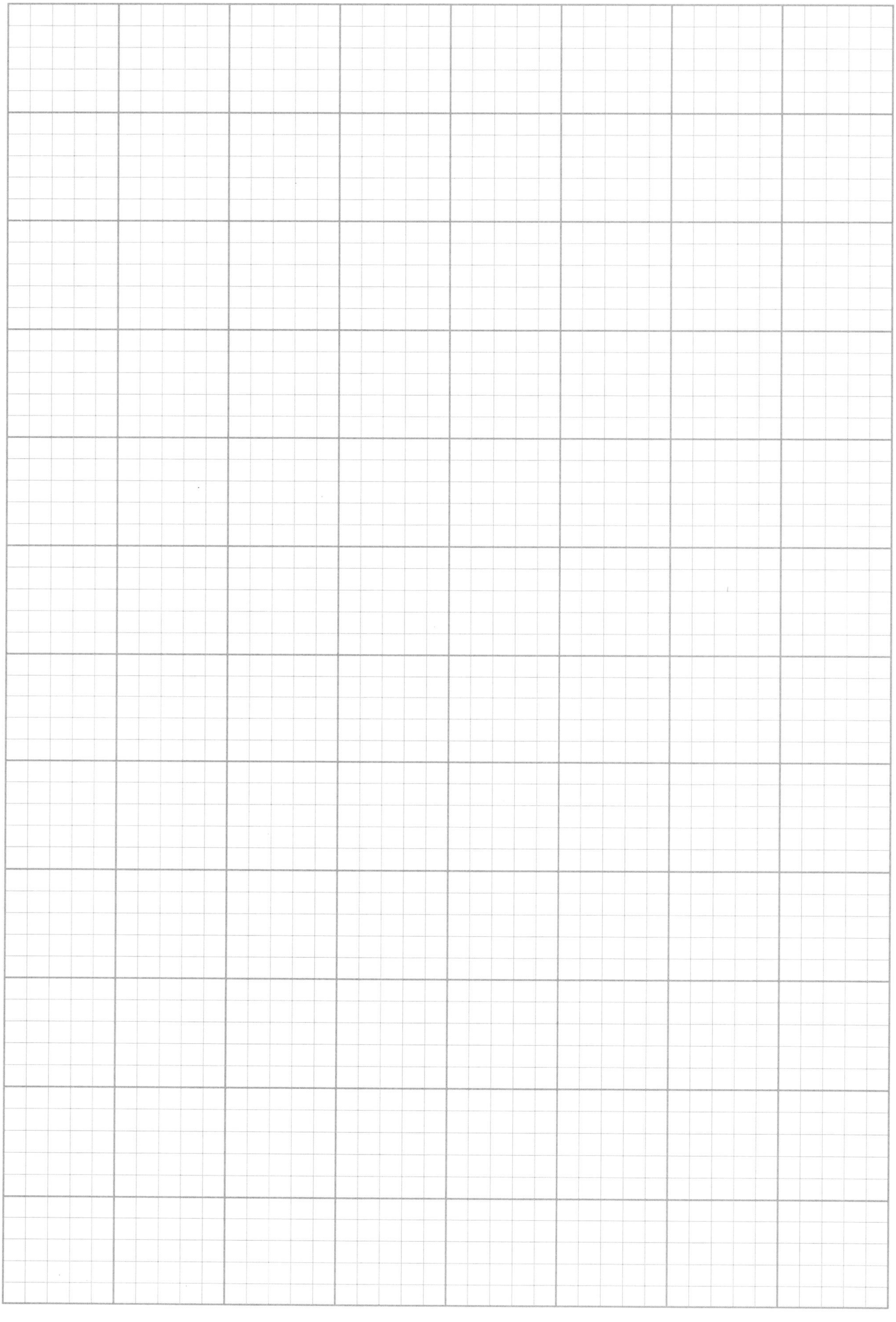

Laboratory Report 11

Name ______ Date ______ Section ______

Maintenance of Stock Cultures

Viability of refrigerated and frozen cultures.

A. Description of Cellular and Colonial Morphology of the Organism

Cell Morphology ______

Colonial Morphology ______

B. Viability and Morphology After Storage

Time	Refrigerated Culture			Frozen Culture		
Weeks	**Viability**	**Cellular Morphology**	**Colonial Morphology**	**Viability**	**Cellular Morphology**	**Colonial Morphology**
3						
6						
9						
12						

Review Questions

1. What is the role of glycerol in this experiment?

2. What other types of solidifying agents could be used in place of agar? What limitations do these other agents have when compared to agar?

3. What controls should be implemented to determine whether repeated freezing and thawing of the frozen culture influences the viability of the culture?

LABORATORY

12 Media Preparation

Safety Issues

- An autoclave or microwave oven should be used only when supervised by the lab instructor. Serious burns can result if proper lab techniques are not followed.
- The autoclave should be used only after proper instruction in its use. Never work in the laboratory alone and never attempt to use the autoclave when supervisory personnel are absent.
- Use hot gloves to protect hands when removing items from the autoclave or microwave oven.

OBJECTIVES

1. Become familiar with various types of media.
2. Learn how to prepare and sterilize media for agar plates and slants.

There are many different types of media on which bacteria can be grown. These media may be categorized as either **chemically defined** or **complex**. In using a chemically defined medium, the exact amount of pure chemicals being used is known. A complex medium is composed of a mixture of proteins and other extracts where the exact amount of a particular amino acid or sugar is not known. Examples of the formulation of both of these types of media are shown in Table 12.1. In addition to being defined or complex, media may also be classified in several other ways. These include:

1. **Enriched**—Enriched medium usually contains some important growth factor such as a vitamin or amino acid or blood component and is used for the cultivation of fastidious organisms. Tryptic soy agar with 5 to 10% sheep red blood cells for growing *Streptococus* is an example.
2. **Selective**—Using this type of medium allows for the selection of a particular microorganism that may be present in a mixed population of microorganisms. Examples of such media are eosin methylene blue (EMB) agar and *Salmonella–Shigella* (SS) agar which select for gram-negative enteric organisms. EMB is used for the isolation and detection of members of the *Enterobacteriaceae* or other coliforms from specimens containing mixed populations of bacteria. SS agar selects for the growth of *Salmonella* and *Shigella* from environmental and clinical specimens by inhibiting the growth of other coliforms.
3. **Differential**—As the name implies, this medium allows for the separation of organisms depending upon coloration changes in bacterial colonies or other changes in the appearance of the medium. Mannitol salt agar supports the growth of the genus *Staphylococcus* and allows

TABLE 12.1 EXAMPLES OF DEFINED AND COMPLEX MEDIA

Defined Medium (Medium for nitrite-oxidizing bacteria) (g/L)
0.1 $NaNO_2$
0.5 K_2HPO_4
5.0 $CaCO_3$
0.2 $MgSO_4 \cdot 7\ H_2O$
0.005 $FeSO_4 \cdot 7\ H_2O$
0.5 NaCl
Complex Medium (Plate Count Agar) (g/L)
5.0 tryptone
2.5 yeast extract
1.0 glucose
15.0 agar

for separation of various species depending on their ability to metabolize mannitol. Those species utilizing mannitol produce acids which change the indicator dye phenol red from red to yellow. Appendix I lists the indicator dyes most often used in microbiology. Blood agar is also considered to be a differential medium, in that organisms can be differentiated on the basis of changes in the color of the agar due to alteration of the blood cells.

4. **Minimal Salts Medium**—This is a chemically defined medium (a medium in which the exact amount and nature of the chemical is known), which contains all of the inorganic salts necessary for the growth of most organisms. The appropriate carbon source can be added depending on the type of organism you wish to isolate and grow.

Any single medium may be a combination of the categories listed above. Mannitol salt agar medium is differential, selective, and complex. Explain.

The preparation of media is fairly straightforward and can be successfully accomplished if care is taken to follow appropriate rules.

1. Always use clean glassware that has been rinsed with distilled water to prevent contamination with soap or other chemicals.
2. Make sure to clean spatulas before weighing out any medium or chemical.
3. The labels on most bottles of dehydrated media contain directions for preparation as well as information about the specific composition of the medium. In most cases, the directions indicate the amount of material to be dissolved in a liter of distilled water.
4. Weigh out the material in a weigh boat or weighing dish. **Never, never** return any unused chemical to the bottle from which you removed it. Discard it to the waste container. Do not be wasteful by removing large quantities of chemical that will not be used.
5. It is usually preferable to add distilled water to a flask and then dissolve chemicals in the water by either swirling or using a magnetic

stirrer. This is done to prevent excessive clumping of the powder or chemicals on the bottom of the container. If the medium has many components, dissolve each one separately. Add any agar component last, as any medium containing agar will have to be heated to completely melt the agar.

6. **Do not prepare any medium unless you are prepared to sterilize it immediately**. Why is this important?
7. Never autoclave a container that is more than half filled. Make sure any caps on containers are kept loose during autoclaving.

To avoid contamination of media by organisms in the environment, it must be sterilized in some fashion. **Sterilization** is the complete destruction or elimination of all living microorganisms. There are a number of techniques that are currently used for sterilization purposes. For sterilizing liquid media, **moist-heat sterilization** is the most widely employed method. This procedure utilizes an instrument referred to as an **autoclave**. It is essentially a large pressure cooker. In order to be sterilized, the medium must be held at 121°C at 15 lb of pressure for 15 min. Care must be exercised when using an autoclave as severe burns are possible. Before using the autoclave on your own you must receive certification by your instructor or laboratory assistant. Only those media which can withstand high temperatures and pressures can be sterilized in this fashion. Sterilization directions can be found on the labels of most dehydrated media. If you are uncertain, please consult such references as *The Handbook of Chemistry and Physics* or *The Merck Manual* for stability.

If media have components that cannot withstand heating at 121°C, an alternative method of sterilization must be used. A commonly used procedure is the **membrane filter technique** which employs bacteriological filters that physically remove all microorganisms. These filters are composed of a variety of materials, but the ones most commonly used in a microbiology lab are made of cellulose acetate or polycarbonate. These filters come in a number of different pore sizes. To assure sterility of a medium, a filter should be used that has a pore size of 0.22 μm. Other commonly used bacteriological filters have a pore size of 0.45 μm; however, some bacteria known as ultramicrobacteria are small enough to pass this size of filter. If a medium has any visible turbidity (cloudiness), it more than likely will clog a 0.22-μm filter, and a prefilter may be needed. Membrane filter sterilization can be accomplished by using either a filter flask setup with a vacuum pump or water aspirator (Figure 12.1), or a syringe device containing the filter holder (Figure 12.2).

Groups of 3 to 4 students are responsible for preparing at least three types of media as directed. The types of media to be prepared are:

a. Tryptic soy agar (TSA) plates (10)
b. Nutrient agar (NA) plates (10)
c. Tryptic soy agar slants (10)
d. Eosin methylene blue (EMB) plates (10)
e. Mannitol salt agar (MSA) plates (10)
f. *Salmonella–Shigella* (SS) agar plates (10) (*Note: Salmonella–Shigella* agar should NOT be sterilized in the autoclave as is the case with the other media. Rehydrate the powder in cold distilled water and heat to boiling to dissolve the medium completely.)

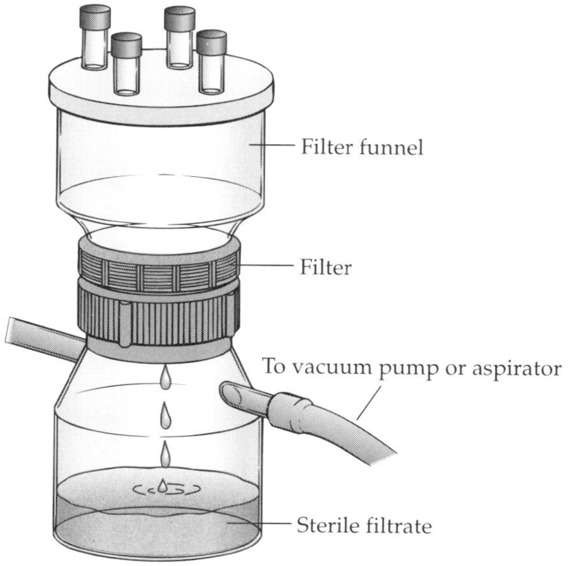

FIGURE 12.1 Membrane filter sterilization system.

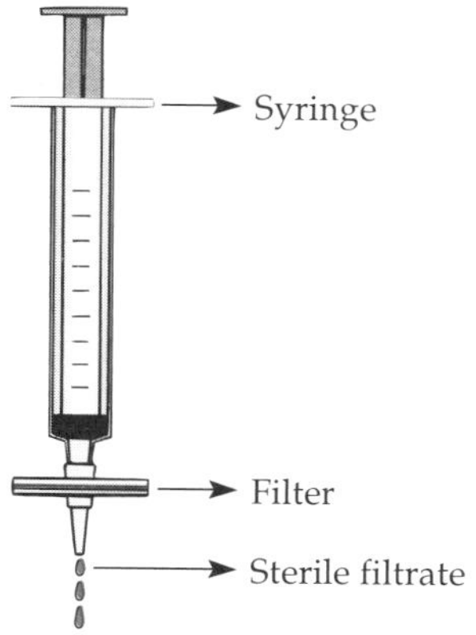

FIGURE 12.2 Syringe membrane filter system.

Materials (per group)

1. Clean 1-L flask
2. Dehydrated tryptic soy agar or nutrient agar
3. Distilled water
4. Spatula
5. Foam stoppers or plugs
6. Weigh boats
7. Balance

Procedure

A. Plate Preparation

1. If using standard size Petri plates (15 × 100 mm), allow 20 to 25 ml of agar per plate.
2. Calculate the total volume of agar you will need, and weigh out the appropriate amount. Mix with distilled water that has been placed in a flask.

⚠ When autoclaving any liquid material in a flask, make sure that the flask or bottle is never more than half filled and that the stopper has been loosened.

⚠ Melting agar in a microwave oven can be very dangerous. Superheated solutions are formed and if agitated rapidly, they will "pop" from the vessel and potentially cause serious burns. Placing the flask containing the agar in a beaker with water slows down the heat transfer to the agar and makes "popping" less likely to occur. Use a microwave oven only under the careful supervision of the laboratory instructor.

3. Mix to dissolve the dry components. The agar will not dissolve unless it is boiled or autoclaved. Agar melts at 100°C and resolidifies at 43 to 45°C. You have a choice to make here, either: a) boil the agar in a boiling water bath to completely dissolve the agar, b) heat it in a microwave oven to melt the agar, or c) autoclave immediately. Remember that the **SS agar is not to be autoclaved**. If using method c, the agar will melt when autoclaved since the standard autoclaving temperature is 121.5°C.
4. Autoclave agar at 121.5°C for 15 min. When preparing large volumes of agar (greater than 500 ml), it is necessary to increase autoclaving time to 25 to 30 min. **Caution**: You will receive specific instructions on the use of the autoclave. Do not attempt to use without supervision. If the agar was not previously melted before autoclaving, be sure you have a homogeneous solution by gently mixing the liquid before pouring the plates.
5. To prevent significant contamination, plates should be poured in a hood. If none is available, the plates can be poured in a quiet portion of the laboratory.
6. Allow the agar to cool to the point where the flask is hot but can be handled without burning your hands. (Cooling the agar in a water bath before pouring decreases the amount of condensation in the plates.)
7. Flame the lip of the flask and proceed to pour appropriate volumes into the sterile Petri plates. Remember to quickly open and close the plates during the pouring process. Some laboratories flame the surface of the plates prior to hardening to remove bubbles that form during pouring.
8. Allow 25 to 30 min for the agar to harden before attempting to move and store the plates.
9. Upon solidification, place the plates in a basket. Reserve for use in a subsequent laboratory session.

B. Preparation of Slants

1. Allow 8–10 ml (16 × 125-mm tube) of agar per slant.
2. Weigh out the appropriate amount of agar into the desired volume of distilled water in a clean flask.
3. Mix to dissolve the nutrient.
4. Heat in a boiling water bath or microwave oven to dissolve the agar.
5. Mix the flask to make sure a homogeneous solution has been achieved. Dispense the agar to screw-capped tubes and **cap loosely**.
6. Autoclave in a test tube rack for 15 min at 121.5°C.
7. After autoclaving, slant the rack and allow the slants to form in the tubes as the agar cools.

Review Questions

1. Why should you not weigh out and mix media unless you are prepared to autoclave it immediately?

2. It has been over 1 hr since a student has poured agar plates. They are cool, but they have not solidified. Can you give this student some advice as to what may have gone wrong?

3. You have dispensed agar into tubes for the preparation of slants. Upon cooling, it is discovered that half of the slants look good, but the other half are semi-solid. What might have gone wrong?

4. Why isn't the *Salmonella–Shigella* agar autoclaved prior to pouring? What are some other types of media that cannot be sterilized by autoclaving?

UNIT III

Investigative Projects

The laboratories in this unit present opportunities to design experiments, conduct literature surveys, execute a research plan, and formulate a manuscript. Information concerning the scheduling and integration of these laboratories can be found in the instructor's manual accompanying this laboratory manual.

In Laboratory 13 microorganisms are isolated from an environment, are maintained in culture, and are further studied in labs of Units V and VI. A collaborative research effort is necessary for execution of Laboratory 14. This group-designed experiment fosters cooperation and interaction among lab groups. Laboratory 15 provides information on a semester-length laboratory project which may be assigned in the course.

LABORATORY

13 Microbiological Characterization of an Environment

Safety Issues

- Maintain all culture tubes in an upright position to prevent spillage.
- Always use a pipetting device. Never attempt to mouth pipet any culture material.
- Do not overheat the lip of a dilution tube when opening and closing it. Severe burns to fingers and hands can result from grasping an overheated tube.
- Keep the beaker of alcohol that is used for the sterilization of glass spreaders away from the burner flame. If the alcohol is ignited, smother the flame by placing a book or other solid object over the flame. If flaming alcohol should spill on the lab bench, **do not** attempt to extinguish the flames with water, as that would likely cause the flames to spread. Smother the flame with a cloth, or, if the volume of the spill is small, just allow the alcohol to burn off.

OBJECTIVE

Learn experimental design strategies and apply them to the microbiological characterization of an environment.

Any research in microbiology starts with a good experimental design. The following experiment provides insight into the process and gives the opportunity to design and execute an experiment.

The basic techniques acquired in the previous laboratories should provide sufficient background and direction.

Steps in the process include the following:

1. Select an environment to be sampled.
2. Develop a hypothesis concerning the microbiological quality of the environment sampled.
3. Develop a sampling strategy.
4. Determine the appropriate growth medium.
5. Sample the site.
6. Enumerate as many of the organisms present as you can, and isolate them in pure culture.
7. To determine the Gram reaction and morphology of the isolated organisms, perform Gram stains on all of the pure cultures obtained.
8. Preserve all cultures as stock cultures for identification in subsequent laboratories.

You will need to refer to previous experimental techniques in your manual for assistance in developing the techniques for your characterization.

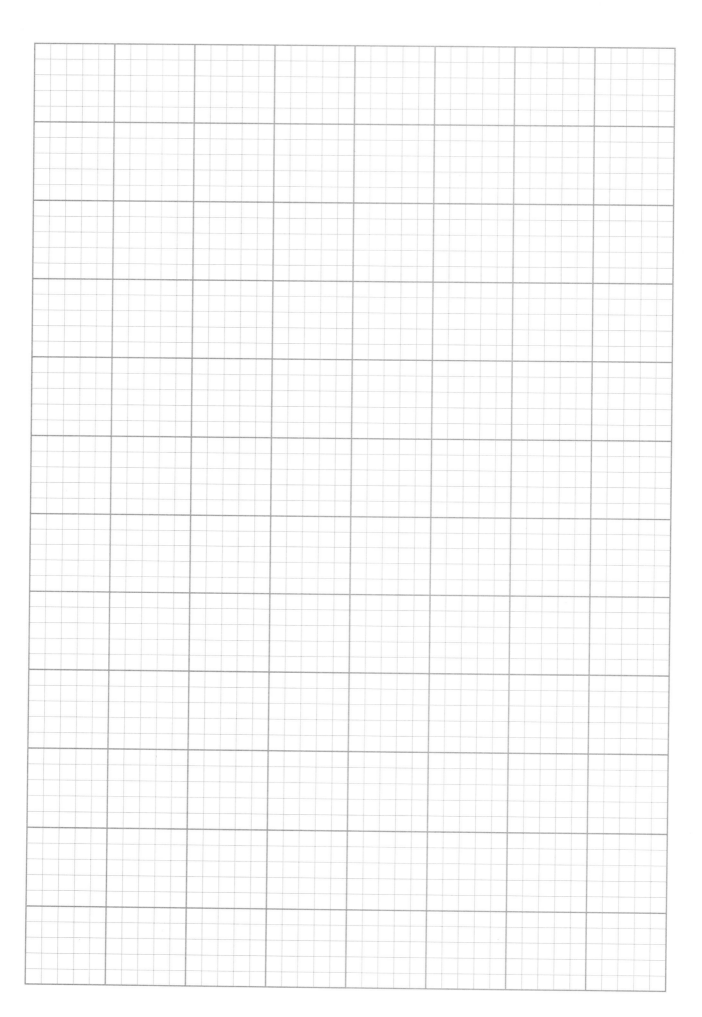

Laboratory Report 13

Name Date Section

Write a preliminary draft of a final report using the following format. A final report is to be prepared after identifications have been made using information gained from laboratories in Unit V.

1. Title of project
2. Investigator
3. Introduction. Include information on the environmental site characterized and hypothesis formed concerning the microbiological quality of the site.
4. Materials and methods. List all media and stains and the method of sampling used.
5. Results. Include macroscopic description of the isolated organisms, and the Gram stains and cellular morphologies of all isolated microorganisms.
6. Discussion. Do your experimental results affirm or refute your original hypothesis concerning the microbiological quality of the site? How was the choice of sampling methodology and the choice of isolation media made?

Review Questions

1. How do the colony morphologies of these new isolates compare to those previously studied?

2. What types of microbial organisms, aside from bacteria, would you expect to find in the environment sampled?

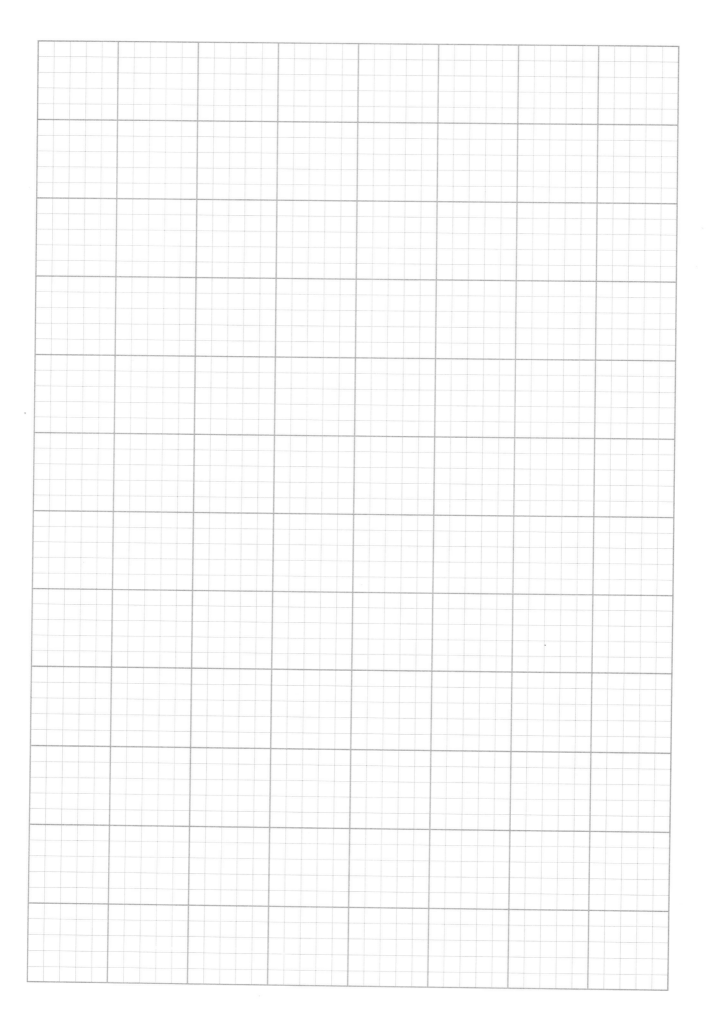

LABORATORY

14 Group-Designed Experiment

OBJECTIVE

Learn experimental design strategies and apply them to a group-designed project.

This experiment will be done as a **group project**. Each student has to select one of the following projects on which to work:

A. Efficacy of contact lens cleaners and disinfectants*
B. Effectiveness of hand soaps
C. Survival of organisms on wooden vs. synthetic cutting boards
D. Isolation of *Escherichia coli* from food
E. Effectiveness of surface disinfectants
F. Types and distribution of microorganisms in salad-bar foods
G. Survival of organisms on wash cloths or towels

After each student has made a selection, research teams will be formed. Each group will design an experimental protocol and also will have the opportunity of critiquing each other's designs. Execution of the designed experiment will take place in a subsequent laboratory period. Before the completion of lab, each group should submit a list of the supplies and media needed to complete the experiment.

*Idea for this project was suggested by Judith Kandel, California State University, Fullerton.

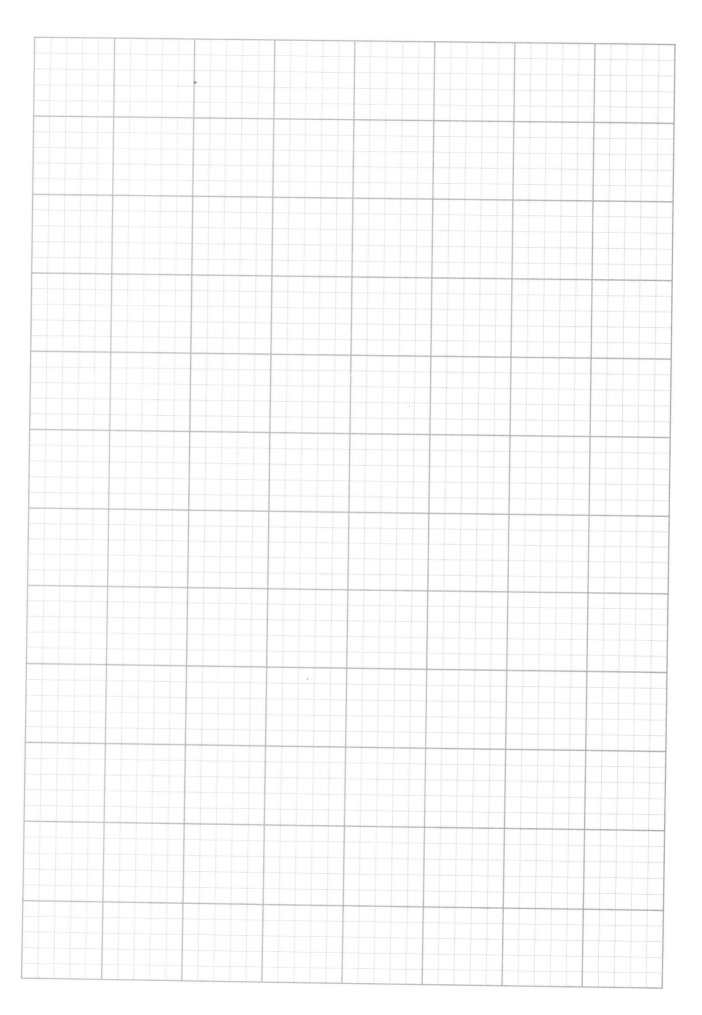

LABORATORY

15 Semester-Length Project*

One of the requirements of this course is the completion of a project in which each group or individual will attempt the isolation of a specific microorganism. This project has been staged so that you do not procrastinate in completing your assignments, and so that you are provided with constructive feedback throughout the term. (Projects are listed in **Appendix M.**)

The first step in the project is to choose an "Organism to be Isolated" card. This card contains an initial reference to get you started. It is not necessarily the definitive reference on the subject, but should assist you in the initial search of the literature. The first important decision you should make before you draw your card is whether you wish to work alone or with a partner. The choice is yours. Selection of a lab partner is critical in that the project will receive a single grade, and you would not want to be penalized for the inadequate effort of a partner. Draw a card and good luck.

In two weeks you will be asked to submit three additional references on the subject. Try to keep in mind that these references should be selected judiciously. You will use these references to help formulate your isolation technique.

You will be provided with assistance from your instructor and from the reference librarians. In subsequent weeks you will be asked to submit additional references and an isolation strategy. The protocol should be very specific and should include all of the equipment and materials you will need. A final paper will be submitted at the end of the semester, and should include the following sections:

Title
Abstract
Introduction
Materials and methods
Results
Discussion
Literature cited

The format that will be used is that specified by the "Instructions to Authors" section from the journal *Applied and Environmental Microbiology.* A copy of these instructions is included for your benefit (approval for use has been granted by The American Society for Microbiology). It also would be valuable for you to carefully examine a number of papers from this journal so that you can come to appreciate the appropriate writing style and format. A slant and a streak plate of the isolated culture along with a Gram stain slide will be required with your final paper. You will be evaluated on the quality of your paper as well as on the effort expended during the semester. The project will constitute a significant portion of your final evaluation for the course. Please refer to your lecture syllabus for further evaluation information.

*Stukus, P. and J. Lennox. 1995. Use of an Investigative Semester-Length Laboratory Project in an Introductory Microbiology Course. *Journal of College Science Teaching* 25: 135–139.

UNIT IV

Nutrition and Growth

Bacterial nutrition can be defined as the determination of the types and amounts of chemicals required for optimal growth of microorganisms. All microorganisms must have sources of carbon and energy. If the carbon source utilized by the organism is organic, the organism is known as a **heterotroph**. If carbon dioxide is the carbon source, the organism is an **autotroph**. Under many conditions the carbon source turns out to be the energy source for most heterotrophs. If the energy source is chemical, the organism is called a **chemotroph**. A **phototroph** utilizes light as its source of energy. All of these terms can be used to construct a nutritional classification of microorganisms.

Photoautotrophs—Use light energy and carbon dioxide as carbon source [examples are photosynthetic green and purple sulfur bacteria].

Photoheterotrophs—Use light energy and organic compounds as carbon sources [examples are purple and green non-sulfur bacteria].

Chemoautotrophs—Use inorganic compounds as energy sources and carbon dioxide as carbon source [examples are sulfur-oxidizing, nitrifying, and hydrogen-oxidizing bacteria].

Chemoheterotrophs—Use organic compounds as carbon and energy sources. The chemotrophs comprise the majority of bacteria.

When defining the growth environment of a microorganism or examining a culture medium, it is important to determine what are serving as carbon and energy sources.

All organisms require a source of organic or inorganic nitrogen in addition to carbon. Other essential elements include phosphorus (P), sulfur (S), magnesium (Mg), Copper (Cu), and iron (Fe). Some microorganisms require smaller amounts of other elements. These are called trace elements and include nickel (Ni), cobalt (Co), Zinc (Zn), and manganese (Mn).

Microorganisms are quite versatile. Some can use either organic carbon or carbon dioxide as carbon sources. These organisms are called **facultative chemoautotrophs**. In fact, some of them can simultaneously use organic and inorganic carbon. This type of metabolism is referred to as **mixotrophic**.

Measuring the growth of bacteria is quite different than measuring the growth of plants or animals. Microbial growth is not measured by examining the height, girth, or weight of a single bacterium. Rather, microbial growth is measured by examining the increase in the population of bacteria. This can be accomplished either directly or indirectly. In direct measurements the number of cells in a population can be determined by actually counting under a microscope. Alternatively, only living cells are enumerated when performing an agar plate count. A viable cell will divide and form a population of cells in or on an agar medium. This can be accomplished by using the spread-plate or pour-plate method. An indirect way of measuring the population of cells is to measure the mass of cells. A simple way of doing this is by the use of turbidity measurements. Turbidity can be measured with the use of a colorimeter or spectrophotometer.

In the experiments that follow, you will learn the role that various nutrients play in supporting the growth of microorganisms. The growth will be monitored spectrophotometrically. The influence that temperature has on growth will also be examined. Some media can be selective for certain types of organisms; while others can allow for the differentiation of different types of organisms. One of the experiments utilizes eosin methylene blue as a selective medium and mannitol salt agar as a selective and differential medium.

Another experiment deals with the cultivation of anaerobic microorganisms. Anaerobic culture methods are used to isolate the obligate anaerobes belonging to the group of sulfate-reducing bacteria. The Winogradsky column experiment is a long-term experiment that selects for a variety of microorganisms including photosynthetic, anaerobic, and autotrophic microorganisms. An experiment is included that involves the cultivation of the autotrophic bacteria *Thiobacillus thiooxidans.*

LABORATORY

16 Nutrition and Growth of Bacteria

Safety Issues

- Maintain all culture tubes in an upright position to prevent spillage.
- Always use a pipetting device. Never attempt to mouth pipette any culture material.
- Do not overheat the lips of dilution tubes when opening and closing. Severe burns to fingers can result if such an overheated tube is grasped.
- Keep the beaker of alcohol used for sterilizing spreaders away from the flame of the burner. If the alcohol in the beaker is ignited, smother the flame by placing a book or other solid object over the flame. If flaming alcohol spills on the lab bench, do not attempt to extinguish with water as this will likely further spread the flame. Smother the flame with a cloth or just allow the alcohol to burn off if the volume of the spill is small.

OBJECTIVES

1. Learn the role of various media components in the growth of bacteria.
2. Understand the use of selective and differential media.
3. Become proficient in pipetting and in performing reliable plate counts.
4. Be able to plot and interpret bacterial-growth curves.

Growth of microorganisms can be measured in a number of ways. In some cases growth is associated with an increase in the size of an organism. This is the case for many animal and plant species. However, bacterial growth is normally measured as an increase in population size. Direct methods involve (a) total count, which is the microscopic enumeration of cell numbers using a counting grid (similar to doing a human blood cell count with a hemocytometer), or (b) viable counts, where bacterial colonies are counted on agar plates. Indirect methods do not enumerate the number of organisms but rather involve measurements of cell mass (dry or wet weight) or turbidity measurements of a liquid suspension.

This lab demonstrates the role that various nutrients play in the metabolism and growth of bacteria. A viable count spread-plate technique (direct method) and a spectrophotometric technique (indirect method) are used to monitor bacterial growth. The effect of temperature on growth will also be observed. Lastly, there will be a familiarization with the use of selective and differential media.

An inorganic salts medium provides the essential inorganic ions and buffering capacity. This medium is referred to as a **minimal medium** or **basal salts medium**.

A. MEASUREMENT OF GROWTH ON VARIOUS MEDIA

Materials (per group)

1. Broth Media (10.0 ml in 18 × 150 mm tubes)
   (A) Minimal medium (MM) (25 tubes)
   (B) MM + 1% glucose (25 tubes)
   (C) MM + 1% glucose + 1% nutrient broth (25 tubes)
   (D) MM + 1% glucose + 1% nutrient broth + 1% yeast extract (25 tubes)
   (E) MM + 1% nutrient broth (25 tubes)
   (F) MM + 1% yeast extract (25 tubes)
   (G) MM + 1% nutrient broth + 1% yeast extract (25 tubes)
   (H) MM {minus $(NH_4)_2SO_4$} + 1% glucose (25 tubes)

Minimal Medium (MM)	*Grams/Liter of Distilled Water*
NaCl	5.0
$MgSO_4$	0.2
KH_2PO_4	1.0
K_2HPO_4	1.0
$(NH_4)_2SO_4$	1.0
$FeSO_4$	0.05

2. Cultures (18- to 24-hr broth cultures)
   (a) *Escherichia coli* (100 ml)
   (b) *Streptococcus faecalis* (100 ml)
   (c) An unknown microorganism (100 ml)
3. Spectrophotometers (3)
4. Incubators set at 37 and 55°C
5. Nutrient, plate count, or tryptic soy agar plates (30)
6. Sterile 1.0-ml pipettes (5 canisters)
7. 4.5-ml sterile saline dilution tubes (25)
8. Turntables, glass spreaders, and alcohol (5)

Since the class will work in groups for this experiment, data should be pooled for analysis and discussion.

Procedure

1. Select 8 tubes of each individual medium from those in group A–D or those in group E–H and, using a sterile 1.0-ml pipette, inoculate each set with 0.1 to 0.2 ml of a different microorganism. Incubate tubes of each organism at one of the following temperatures:
   (a) Room temperature
   (b) 37°C
   (c) 55°C
2. Monitor growth in all tubes by measuring the apparent absorbency at 540 nm on a spectrophotometer. See Appendix D for instructions if you are unfamiliar with the use of a spectrophotometer. Frequency of readings

will be determined by the growth rates of the organisms. If no growth is apparent, readings should be taken less frequently. When there is vigorous growth, readings should be taken every 30 to 45 min. Will each of the media support growth? Accumulate the data and plot growth curves for all organisms that showed a growth response on the graph paper in the Report section. Remember to plot absorbency on the y-axis and time on the x-axis.

3. Calculate the generation times under conditions where growth occurred. Check your textbook if you have forgotten how to do this. For this laboratory, use a doubling of the turbidity as a measure of the generation time.
4. After the cultures have incubated for at least 90 min, select one tube for enumeration of bacterial population size by performing a viable plate count using serial dilutions and the **spread-plate technique**.
5. Using aseptic technique, prepare serial ten-fold dilutions from the culture by pipetting 0.5 ml of the culture into 4.5 ml of sterile saline. (See Appendices E and F for use of pipettes and dilution techniques.) This gives a 10^{-1} dilution. Mix this tube well and, with a fresh pipette, deliver 0.5 ml to a second sterile saline tube. This will give a 10^{-2} dilution of the original culture. Continue the same procedure until you have made dilutions through 10^{-5} (Figure 16.1).
6. Plate duplicate samples of the 10^{-3}, 10^{-4}, and 10^{-5} dilutions by pipetting 0.1 ml onto the surfaces of the agar plates.
7. With an alcohol-sterilized glass spreader and the aid of a turntable, spread the liquid on the surface of the plate until it is absorbed into the agar. Make sure to allow time for the spreader to cool before spreading to avoid cooking the bacteria while spreading them!

⚠ Keep the beaker of alcohol used for sterilizing spreaders away from the flame of the burner.

8. Invert the plates and incubate at the appropriate temperature for 48 hr. If plates are incubated at 55°C, place the plates in a plastic bag containing wet paper towels, or seal the plates with a large rubber band to prevent dehydration.
9. Count the colonies and report the total population as the number of colony-forming units per milliliter (cfu/ml) in the original culture.

$$\frac{\text{Number of colony-forming units}}{\text{ml}} = \text{(Number of colonies)} \times \left(\frac{1}{\text{dilution factor}}\right) \times \text{(plating factor)}$$

For example, if a 10^{-5} dilution plate has 20 colonies, then number of cfu/0.1 ml $= 20 \times 1/10^{-5} = 20 \times 10^5 = 2.0 \times 10^6$. Since only 0.1 ml was spread plated, the number has to be multiplied by 10 to provide

$$\frac{\text{Number of cfu}}{\text{ml}} = 2.0 \times 10^6 \times 10 = 2.0 \times 10^7$$

B. USE OF SELECTIVE AND DIFFERENTIAL MEDIA

Materials (per 20 students)

1. Broth Cultures (18- to 24-hr)
   (a) *Staphylococcus aureus* (ATCC #25923) (4)
   (b) *Staphylococcus epidermidis* (ATCC #14990) (4)

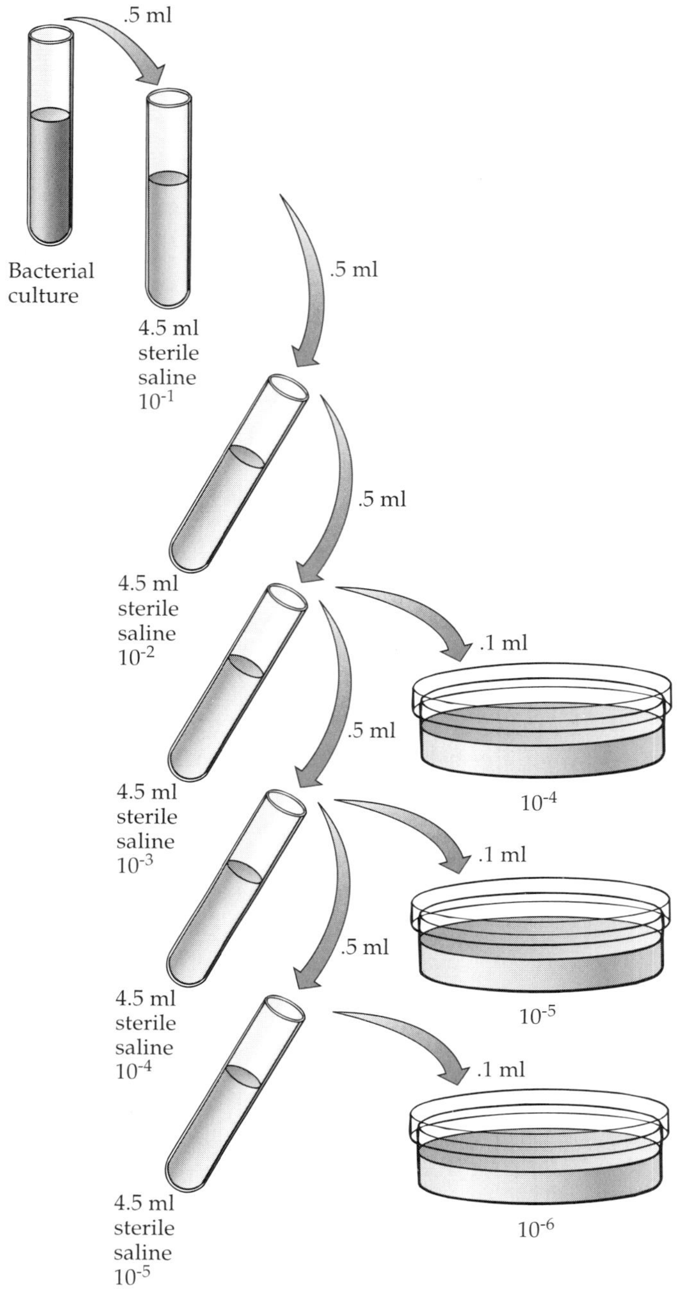

FIGURE 16.1 Preparation of serial solutions.

   (c) *Escherichia coli* (ATCC #11229) (4)
   (d) *Enterobacter aerogenes* (ATCC #13048) (4)
   (e) *Salmonella typhimurium* (ATCC #29631) (4)
   (f) *Shigella dysenteriae* (ATCC #11456a) (4)
2. Media
   (a) Eosin methylene blue agar (EMB) (40 plates)
   (b) Mannitol salt agar (MSA) (40 plates)
   (c) *Salmonella-Shigella* (SS) agar (40 plates)

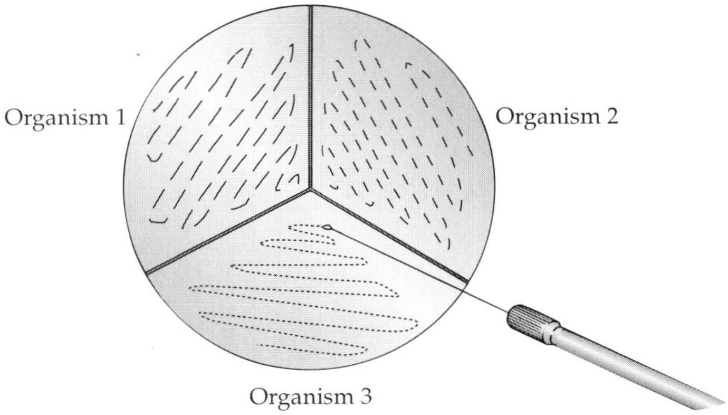

FIGURE 16.2 Streaking differential and selective media.

Procedure

1. Obtain two EMB, two MSA, and two *SS* plates and streak the six cultures onto their surfaces (three cultures/plate) (Figure 16.2).
2. Incubate the plates in an inverted position at 37°C for 48 hr.
3. Record whether growth occurs and make note of the color of the colonies and the media surrounding the colonies.

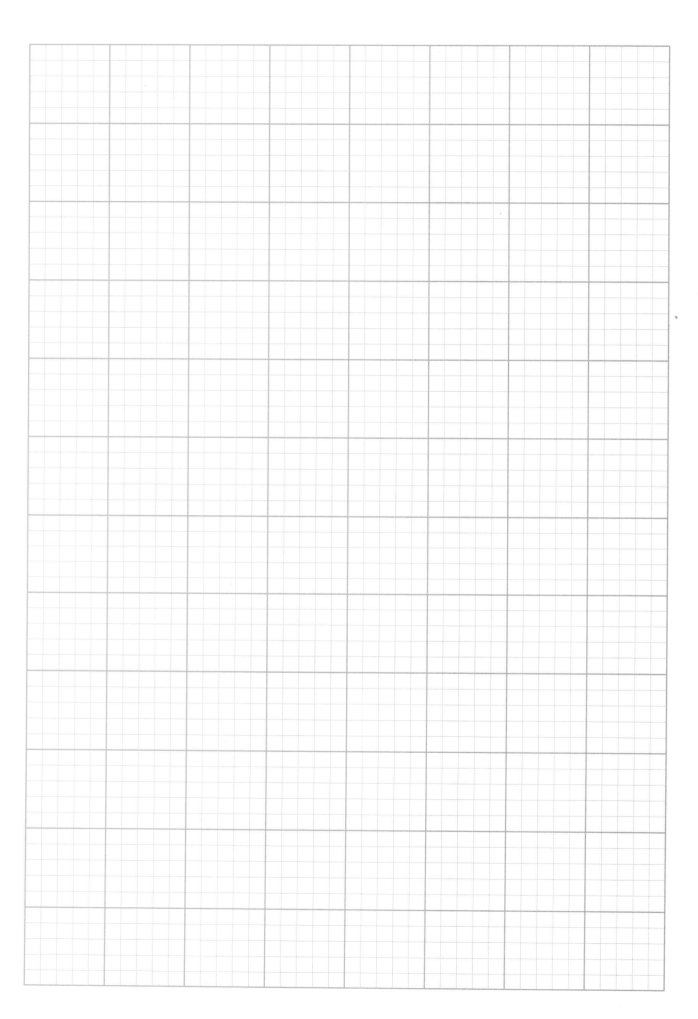

Laboratory Report 16

Name Date Section

Culture: *Escherichia coli*

GROWTH AT ROOM TEMPERATURE

Incubation Time (hr)	Absorbencies on Growth Media							
	A	B	C	D	E	F	G	H

GROWTH AT 37°C

Incubation Time (hr)	Absorbencies on Growth Media							
	A	B	C	D	E	F	G	H

Growth at 55°C

Incubation Time (hr)	Absorbencies on Growth Media							
	A	B	C	D	E	F	G	H

Culture: *Streptococcus faecalis*

Growth at Room Temperature

Incubation Time (hr)	Absorbencies on Growth Media							
	A	B	C	D	E	F	G	H

Growth at 37°C

Incubation Time (hr)	Absorbencies on Growth Media							
	A	B	C	D	E	F	G	H

Growth at 55°C

Incubation Time (hr)	Absorbencies on Growth Media							
	A	B	C	D	E	F	G	H

Culture: Unknown

Growth at Room Temperature

Incubation Time (hr)	Absorbencies on Growth Media							
	A	B	C	D	E	F	G	H

Growth at 37°C

Incubation Time (hr)	Absorbencies on Growth Media							
	A	B	C	D	E	F	G	H

Growth at 55°C

Incubation Time (hr)	Absorbencies on Growth Media							
	A	B	C	D	E	F	G	H

Generation Times (hr):

Culture: *Escherichia coli*

Medium	Generation Time
A	
B	
C	
D	
E	
F	
G	
H	

CULTURE: *STREPTOCOCCUS FAECALIS*

Medium	Generation Time
A	
B	
C	
D	
E	
F	
G	
H	

CULTURE: UNKNOWN

Medium	Generation Time
A	
B	
C	
D	
E	
F	
G	
H	

Viable Plate Count

Culture: ______________________ Time: ______________________

Growth Condition: ______________________

A_{540} at Sampling Time: ______________________

Dilution	# of Colonies	Microbial Count

Calculation of Microbial Titer in Culture Tube Sampled:

Selective and Differential Media

Growth Response

	Eosin Methylene Blue Agar Appearance		Mannitol Salt Agar Appearance		Salmonella-Shigella Agar Appearance	
	Colony Color	**Medium Color**	**Colony Color**	**Medium Color**	**Colony Color**	**Medium Color**
Culture						
Staphylococcus aureus						
Staphylococcus epidermidis						
Escherichia coli						
Enterobacter aerogenes						
Salmonella typhimurium						
Shigella dysenteriae						

Review Questions

1. What are the components of nutrient broth and yeast extract? What purpose do they serve in the growth media?

2. Identify the roles of the inorganic components of the mineral salts medium.

3. Explain the differences in the growth rates for each of the organisms.

4. What is the correlation of absorbencies with the actual viable count? Can you assign a population number for a particular absorbency? What are some of the factors that influence the correlation of population and absorbency?

5. Explain why mannitol salt and eosin methylene blue agars are considered to be both selective and differential. Do the growth responses of the cultures used verify that fact?

6. How does *SS* serve as a selective medium?

7. What are the advantages and disadvantages of using spectroscopy for monitoring the growth of microorganisms?

8. If you had all of the following items, how would you prepare 10^{-4}, 10^{-5}, and 10^{-6} dilutions of a bacterial culture?
   (a) two 9.0-ml saline tubes
   (b) two 9.9-ml saline tubes
   (c) four 1.0-ml pipettes
   (d) a bacterial culture

9. A student has inoculated an organism into the following medium and has not observed any growth after incubating for five days. What is the problem? What would you suggest that the student do?

Medium g/L	
K_2HPO_4	1.2
KH_2PO_4	1.2
glucose	2.0
pyruvic acid	0.5
$MgSO_4$	0.8
citrate	0.5
sucrose	1.0

The culture is incubated at 30°C with shaking, bubbling an atmosphere of 70% N_2, 20% O_2, and 10% CO_2 into the medium.

LABORATORY

17 The Winogradsky Column

Safety Issues

- When removing the top of the plastic pop bottles, exercise caution so as not to cut your hand with the razor blade or scissors.

OBJECTIVE

To build a soil column and observe the emergence of anaerobic and photosynthetic microorganisms.

The Winogradsky column is named after the Russian microbiologist, Sergei Winogradsky, who first used the column to describe the growth of various soil microorganisms, particularly photosynthetic anaerobic bacteria. Aerobes and microaerophiles will grow near the top of the column, and anaerobes will grow in the remainder of the column. By adding the appropriate organic and inorganic components, and by adjustment of the pH, you can select for many different types of organisms. By placing the column in sunlight (subdued, not direct) you can also select for phototrophs in the column. Check your textbook for a further description of the organisms and biochemistry occurring in the column. Organisms that may be present at each layer are shown in Figure 17.1. The classic Winogradsky column selects for organisms involved in the sulfur cycle. Cellulose is added to the column as the electron donor, and sulfate is used as the acceptor.

Materials (per group)

1. 100–500-ml graduated cylinders (5) or 1-L clear plastic pop bottles (If bottle is used, cut 2–3 in. off top of bottle.)
2. Mud from bottom of lake or pond (bucketful)
3. $CaSO_4$
4. $CaCO_3$
5. K_2HPO_4
6. Cellulose slurry or strips of filter paper
7. Aluminum foil or plastic film
8. Pond or lake water (10 L)

Procedure

1. Add 1.0 g of $CaSO_4$, 0.5 g of $CaSO_3$, 1.0 g of K_2HPO_4, and 0.1 g yeast extract to every 50 g of mud.

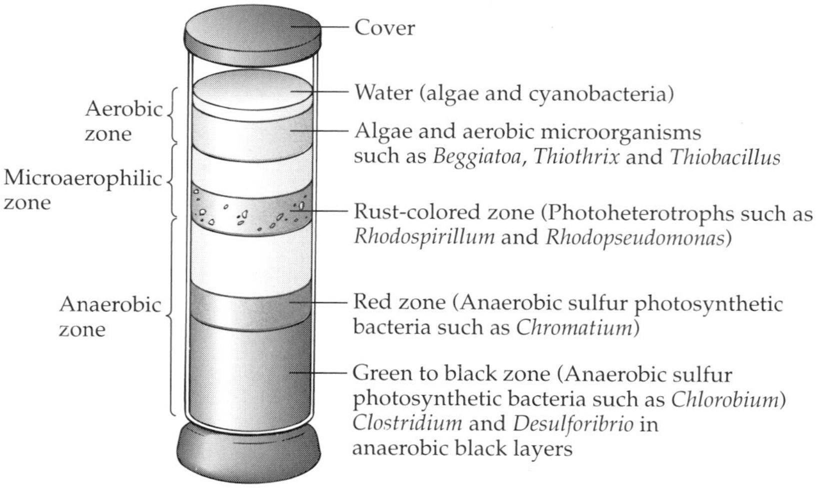

FIGURE 17.1 Distribution of organisms in Winogradsky column.

2. Pack the graduated cylinder or bottle with the mud/salts mix. Add strips of shredded filter paper or some of the cellulose to pack the column. The secret here is to obtain a nice, tightly packed column with no air bubbles. If bubbles are present, add some water and attempt to work out the bubbles with a glass rod or pipette.
3. After the column is tightly packed, add about 2 to 3 cm of pond or lake water to the top of the column.
4. Cover the column with foil or plastic. Do not allow the water at the top of the column to evaporate during the experiment. Pack the mud down if any bubbles start to form.
5. Incubate at room temperature in a window that has subdued, but not direct, sunlight. Be careful not to overheat the column. An alternative method is to use a tungsten lamp that provides light energy in the appropriate infrared range.
6. Make careful observations noting color changes that appear in the column.

At the top of the column, cyanobacteria, algae, and aerobic bacteria develop. Aerobic microbes remove traces of oxygen and the water will rapidly become anaerobic. Anaerobic microorganisms attack the cellulose with the evolution of carbon dioxide, hydrogen, and other fermentative end products. Other anaerobic bacteria reduce sulfate to sulfide which is manifest in the column as a black iron precipitate. The sulfide has a tendency to diffuse upward and a gradient is formed. The photosynthetic bacteria assimilate carbon dioxide while oxidizing sulfide to sulfur or sulfate. Regions of purple sulfur bacteria usually appear on the wall of the cylinder in the upper anaerobic layer, whereas the green sulfur bacteria appear sporadically or in the liquid. In the aerobic and microaerophilic zones, sulfur oxidizing bacteria such as *Thiobacillus* may be observed (see Color Plate 8).

Refer to your textbook and other references for a complete description of the types of organisms found in the column and the metabolism that is occurring. After four to five weeks, sample various areas of the column using a pipette and report the Gram stain and morphology of the organisms from the various parts of the column.

Laboratory Report 17

Name Date Section

Indicate color changes in the column over time.

Time	Appearance of Column	Type of Organisms
1 week		
2 weeks		
3 weeks		
4 weeks		
5 weeks		

Review Questions

1. What types of microorganisms are apparent at the top of the column?

2. Where does hydrogen sulfide appear in the column? How would you enrich for those organisms involved in its production?

3. What are some of the genera of organisms present in the column?

4. Using *Bergey's Manual* or other references, try to list at least five organisms you might expect to isolate from the column, and describe media used to isolate and maintain these organisms.

LABORATORY

18 Anaerobic Culture Techniques

Safety Issues

- All tubes of broth should be kept in an upright position to prevent spillage.
- Allow any heated inoculating loop time to cool before removing growth from a colony or liquid from a broth tube. Contact with a hot loop may cause splattering of material and the aerosol produced may be harmful.

OBJECTIVE

Learn and compare methods for the cultivation of anaerobic bacteria.

Those microorganisms that are capable of growing in the absence of oxygen are called anaerobes. **Facultative anaerobes** can grow in the presence or absence of oxygen. **Strict (obligate) anaerobes** only grow in the absence of oxygen, with the presence of oxygen actually being toxic for the organisms. As might be expected, the strict anaerobes are found in areas devoid of oxygen, such as in the mud in the bottom of ponds or bogs, or in human intestines or deep-wound infections. The ability of anaerobes to grow in human tissue is dependent on the availability of oxygen. If decreased blood flow starves an area of tissue damage from oxygen, conditions become favorable for the proliferation of anaerobes.

A number of methods have been devised to exclude oxygen from an environment and facilitate the growth of anaerobic microorganisms. One method incorporates reducing agents into liquid media. Agents such as cysteine and sodium thioglycollate contain sulfhydryl groups (—SH) and can donate hydrogen to other organic molecules. Thioglycollate has been used for a number of years to cultivate anaerobes. A newer method, which also incorporates a chemical into either agar or liquid medium, is called **Oxyrase** (see Color Plate 9). The Oxyrase system is an enzymatic system which in the presence of suitable hydrogen donors reduces dissolved oxygen directly to water. The reaction does not produce hydrogen peroxide. Hydrogen donors are succinic acid, formic acid, lactic acid, and alpha-glycerol phosphate, which at millimolar concentrations are able to remove all dissolved oxygen in most applications. The source of enzymes is *Escherichia coli*. Oxyrase has a pH range of 6.8 to 9.0. A second Oxyrase system is under development and has a pH range of 3.5 to 7.0. Sterile Oxyrase is added to the appropriate broth medium or to agar that has been allowed to cool to 55°C before pouring into special plastic anaerobic cultivation plates. Oxyrase For Broth maintains an anaerobic environment in bacteriological media for at least 16 days at 35 to 37°C in a culture vessel incubated without shaking or mixing. The oxygen concentration is reduced to less than ten parts per billion.

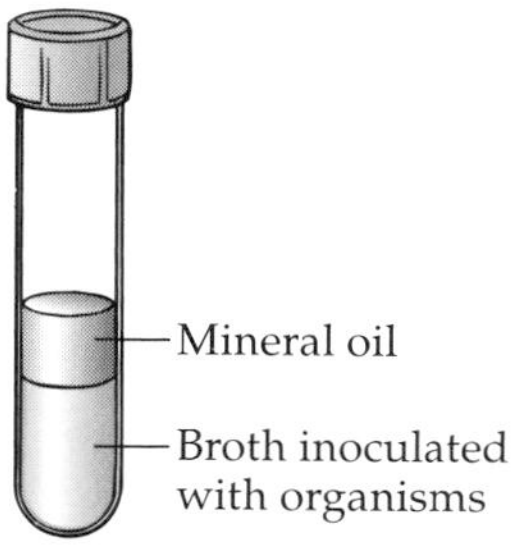

FIGURE 18.1 Application of mineral-oil layer.

Another simple method that can be used to grow anaerobes involves layering sterile mineral oil over an inoculated tube of broth. The layer of oil keeps the liquid from absorbing oxygen from the atmosphere (Figure 18.1).

Another popular method used is the Brewer anaerobic jar, which contains a gas generation package and produces hydrogen and carbon dioxide when water is added (see Color Plate 10). In the presence of a catalyst (palladium pellets), the hydrogen combines with any oxygen present to produce water. The carbon dioxide produced stimulates the growth of some anaerobes. A methylene blue indicator strip is placed in the chamber to monitor the level of oxygen present. The dye will remain blue in the presence of oxygen and white when oxygen has been removed from the system.

To ensure the growth of strict anaerobes, an anaerobic incubator may be necessary. These chambers are expensive and are used in laboratories where a significant amount of research is conducted on anaerobes.

Materials (per 20 students)

1. Slant cultures of *Escherichia coli* (5)
2. Slant cultures of *Staphylococcus epidermidis* (5)
3. Slant cultures of *Pseudomonas aeruginosa* (20)
4. Broth cultures of *Clostridium butyricum* (ATCC #8260) (5)
5. Broth cultures of *Bacteroides fragilis* (ATCC #23745) (5)
6. Broth cultures *Peptostreptococcus magnus* (ATCC #29328)
7. Tubes of thioglycollate broth (120)
8. Tubes of tryptic soy broth (120)
9. Oxyrase for Broth (2 bottles)
10. Brewer anaerobic jars with GasPak envelopes (3)
11. Plate-count agar plates (60)
12. Oxyrase for Agar (2 bottles)
13. OxyDish

Procedure

1. Inoculate six tubes of thioglycollate broth with each of the organisms. Use loops to inoculate *E. coli*, *Staphylococcus*, and *Pseudomonas*, and use sterile 1.0-ml pipettes for inoculating the *Clostridium*, *Bacteroides*, and *Peptostreptococcus*.
2. Do not shake the tubes. Incubate at 37°C and observe growth.
3. Add 0.1 ml of the Oxyrase for Broth to 5.0-ml tubes of tryptic soy broth. Inoculate six tubes with the test organisms and incubate the tubes without shaking at 37°C. Make observations when growth becomes apparent in tubes.

4. Inoculate six tubes of tryptic soy broth with each of the organisms. These will serve as aerobic controls.
5. Obtain three plate-count agar plates and streak two of the test organisms on each plate.
6. Place the plates in an anaerobic jar. Obtain a gas generator pack, cut open, and fill the pouch with 10 ml of distilled water. Place upright in the chamber. Tear open a methylene blue indicator strip and position in the chamber. Replace the lid and screw tightly in place. Incubate at 37°C and examine at regular intervals for growth.

⚠ Caution should be exercised in the use of GasPak envelopes, for the hydrogen generated is a flammable gas. A mixture of hydrogen gas with oxygen or air in a confined area will explode if ignited by a spark, flame, or other source of ignition.

7. Obtain an Oxyrase for Agar dish and open it by breaking the seal. This is accomplished by gripping the dish at the edge opposite the cutouts and gently flexing the plastic.
8. Streak a loopful of either the *Bacteroides* or *Peptostreptococcus* culture onto the surface of an Oxyrase for Agar dish. Be careful not to streak any organisms on the outside sealing ring.
9. Place the dish top cover on the base and insert the complete dish so that the dish base rests on the top.
10. Incubate the plate at 37°C for 48 hr.
11. Observe the growth and record results.

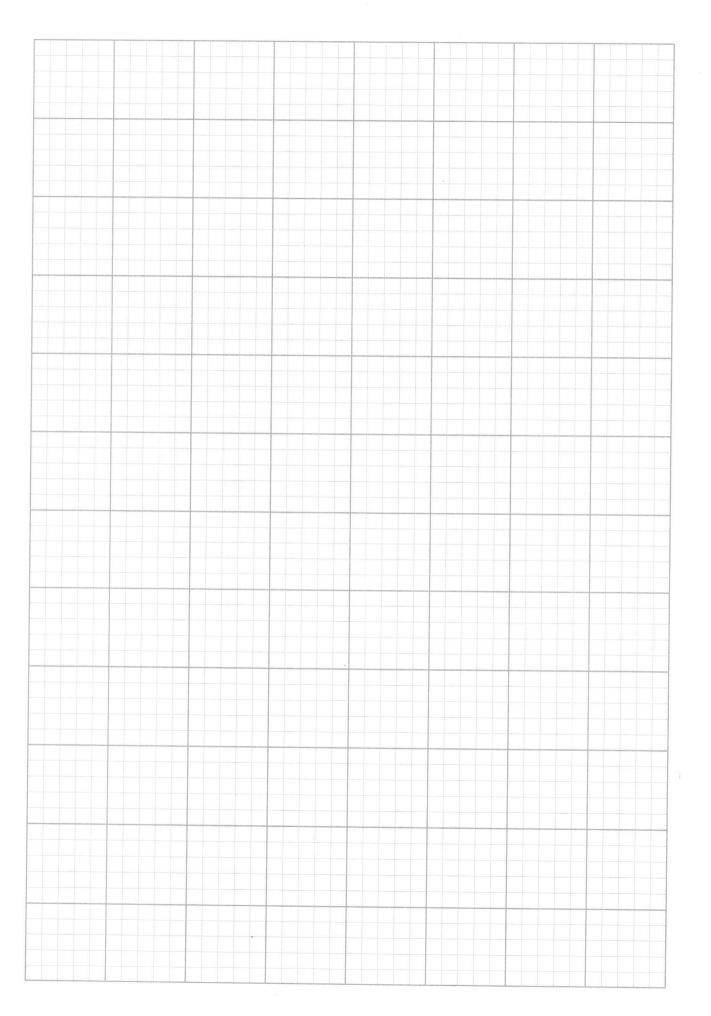

Laboratory Report 18

Name Date Section

1. Examine the thioglycollate tubes and the tryptic soy broth tubes. Make note of the growth and sketch its location in each of the tubes.

Description of Growth and Location

Organism	Tryptic Soy Broth	Thioglycollate Broth	Tryptic Soy Broth + Oxyrase
E. coli			
Staphylococcus			
Bacteroides			
Clostridium			
Pseudomonas			
Peptostreptococcus			

2. Sketch growth on plates in anaerobic jar.

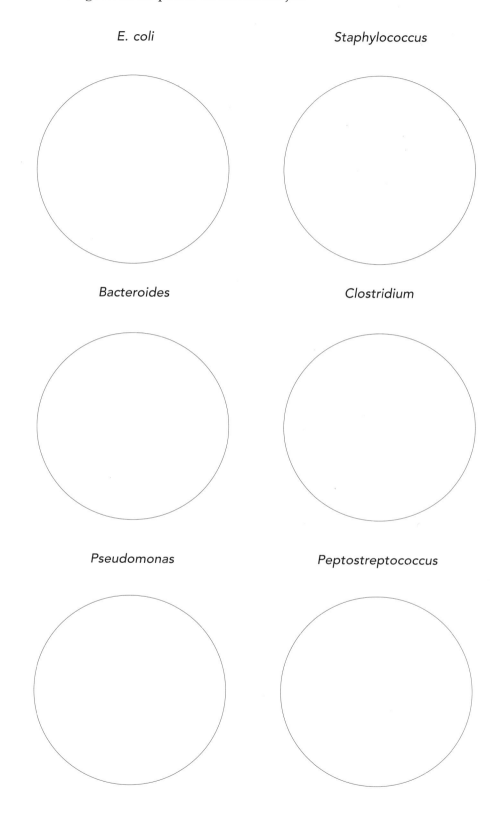

E. coli

Staphylococcus

Bacteroides

Clostridium

Pseudomonas

Peptostreptococcus

3. Sketch growth of organisms on Oxyrase dishes.

Bacteroides

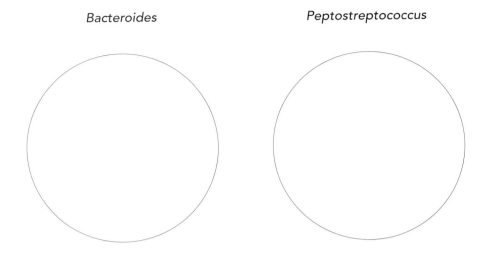

Peptostreptococcus

Review Questions

1. Which of the organisms is classified as anaerobic, based upon the growth in broth and on plates?

2. Was there any discrepancy in results of the three tests? How can differences be explained?

3. Why won't a strict anaerobe grow near the top of thioglycollate tubes?

4. How can you account for growth near the top of thioglycollate tubes?

LABORATORY

19 Isolation of Obligate Anaerobic Sulfate-Reducing Bacteria*

Safety Issues

- Use a pipetting device for all transfers.

OBJECTIVE

To apply anaerobic culture methods to the isolation of sulfate-reducing bacteria from pond sediments.

Sulfate-reducing bacteria (SRB) are gram-negative anaerobic microorganisms, which are widespread in nature. They are found both in aquatic and terrestrial environments. They reduce SO_4^{2-} to S^{2-}. They are biologically diverse, varying in morphology from cocci to regular curved, or sigmoid-shaped, bacilli. Some are flagellated and others non-flagellated. SRB can often be distinguished by the appearance of black colonies on an agar medium containing $FeSO_4$, which upon reduction forms a blackened precipitate. A characteristic hydrogen sulfide odor is also present. The genera *Desulfovibrio* and *Desulfotomaculum* are the best known SRB.

SRB are known to corrode iron pipes and can decay the inner surfaces of transportation canals in sewage systems. SRB are a major concern of the offshore oil industry because their growth and respiration can cause souring of oil and gas.

Materials (per group)

1. Soil sediment from a pond (bucketful)
2. (18 × 150 mm) tubes containing 10.0 ml of SRB medium (35)
3. SRB agar plates
4. 250-ml flasks containing 100 ml of SRB broth (35)
5. Oxyrase reagent (1 bottle)
6. Brewer anaerobic jar (2)
7. Micropipettor
8. Sterile micropipette tips or sterile 1.0-ml pipettes

*This laboratory is patterned after an experiment of two students, Leanne Watson and Nicole Rashid, who performed the experiment as their semester-length project.

Procedure

1. Set up the SRB broth tubes using the protocol in Table 19.1.

TABLE 19.1

Tube	Soil Sediment	Oxyrase/Reagent
1	0.5 g	200 μl
2	0.5 g	200 μl
3	0.5 g	—
4	1.0 g	200 μl
5	1.0 g	200 μl
6	1.0 g	—
7	—	200 μl

2. After setup, incubate tubes at 30°C and observe daily for color changes in media.
3. Withdraw samples from any tube showing a blackening and transfer to an SRB agar plate. Streak two loopfuls of the blackened broth onto the surface of the plate and stab the plate five or six times. Place the plate in a Brewer anaerobic jar and incubate at 30°C.
4. Remove 1.0 ml of blackened broth from any of the tubes and transfer to flasks of SRB broth according to the protocol in Table 19.2.

TABLE 19.2

Flask	Oxyrase/Reagent	Blackened Broth from Tubes
1	2.0 ml	1.0 ml
2	—	1.0 ml
3	—	1.0 ml
4	2.0 ml	—
5	2.0 ml	—
6	—	Uninoculated control
7	—	Uninoculated control

5. Incubate flasks 1, 2, 4, and 6 under stationary conditions at 30°C. Incubate flasks 3, 5, and 7 at 30°C on a shaker set at 175 rpm.
6. Observe flasks on a regular basis, noting any color changes in broth.
7. After one week of incubation, open the Brewer anaerobic jar and examine plates for the appearance of black colonies. Perform Gram stains on organisms from these colonies.

Laboratory Report 19

Name _______________ Date _______ Section _______

Initial SRB Isolation

			Change in Appearance or Color (days)						
			Tube						
Tube	Soil	Oxyrase	1	2	3	4	5	6	7
1	0.5 g	200 μl							
2	0.5 g	200 μl							
3	0.5 g	—							
4	1.0 g	200 μl							
5	1.0 g	200 μl							
6	1.0 g	—							
7	—	200 μl							

Secondary Enrichment

Plates in Anaerobic Jar

Sketch appearance of plates:

Sketch Gram stains of cells from isolated colonies:

SRB Broth Flasks

Flask	Oxyrase	Blackened Broth	Change in Appearance
1	2.0 ml	1.0 ml	
2	—	1.0 ml	
3	—	1.0 ml	
4	2.0 ml	—	
5	2.0 ml	—	
6	—	—	
7	—	—	

Review Questions

1. Explain the change of appearance in the initial SRB isolation tubes. What are the purposes of tubes 3, 6, and 7 in the experiment?

2. Explain the different appearances of the secondary SRB isolation flasks. Why might there be differences between a culture incubated under shaking versus stationary conditions?

LABORATORY

20 Growth of *Thiobacillus thiooxidans*

Safety Issues

- Use a pipetting device for all transfers.

OBJECTIVES

1. Become familiar with the nutritional requirements and metabolism of sulfur-oxidizing, chemolithotrophic bacteria.
2. Be able to identify carbon and energy sources for chemolithotrophic bacteria.
3. Cultivate sulfur-oxidizing bacteria and measure the amount of sulfuric acid formed during the oxidation of sulfur.

Members of the genus *Thiobacillus* are polarly flagellated, gram-negative bacilli. *T. thiooxidans* grows chemolithotrophically using elemental sulfur. It is resistant to the sulfuric acid it produces and can thrive at low pH.

Materials (per group)

1. *Thiobacillus* medium (200 ml)
2. *Thiobacillus thiooxidans* ATCC #19377 broth culture (4.0 ml)
3. Elemental sulfur (1.0 g)
4. 250-ml flasks (2)
5. 0.1 N NaOH solution (50 ml)
6. Alcoholic bromocresol green indicator (5.0 ml)
7. 5.0-ml pipettes (1 canister)
8. pH Meter

Procedure

1. Transfer 100 ml of the *Thiobacillus* medium to two 250-ml flasks.
2. Add 0.5 g of elemental sulfur to each flask.
3. Using a 5.0-ml pipette, inoculate each flask with 2.0 ml of the *Thiobacillus* broth culture.
4. Remove 10.0 ml of the medium from each flask and measure the pH using a pH meter.
5. After reading the pH, add two drops of alcoholic bromocresol green (blue at pH 5.4) to the media samples. Titrate to the end point using 0.1 N NaOH.
6. Incubate one flask in an incubator shaker at 30°C, set at 150 to 175 rpm.
7. Incubate the second flask under stationary conditions at 30°C.

8. Determine the pH of the medium in each flask at 3, 7, and 10 days. Also determine the amount of acid formed by titrating with the 0.1 N NaOH solution.
9. After ten days, perform a Gram stain on the culture medium in each flask.
10. Calculate the percent sulfuric acid present in each flask after ten days.
11. Use the graph paper in the Report section to plot a graph of the amount of NaOH needed for neutralization (*y*-axis) versus time (*x*-axis) for both flasks.

Laboratory Report 20

Name Date Section

pH Data and Acid Production

	Shaker Flask			Stationary Flask		
Time (days)	pH	NaOH Added	% H_2SO_4	pH	NaOH Added	% H_2SO_4
0						
3						
7						
10						

Plot graphs of the NaOH added versus time for both flasks. (Use the graph paper provided on the following pages.)

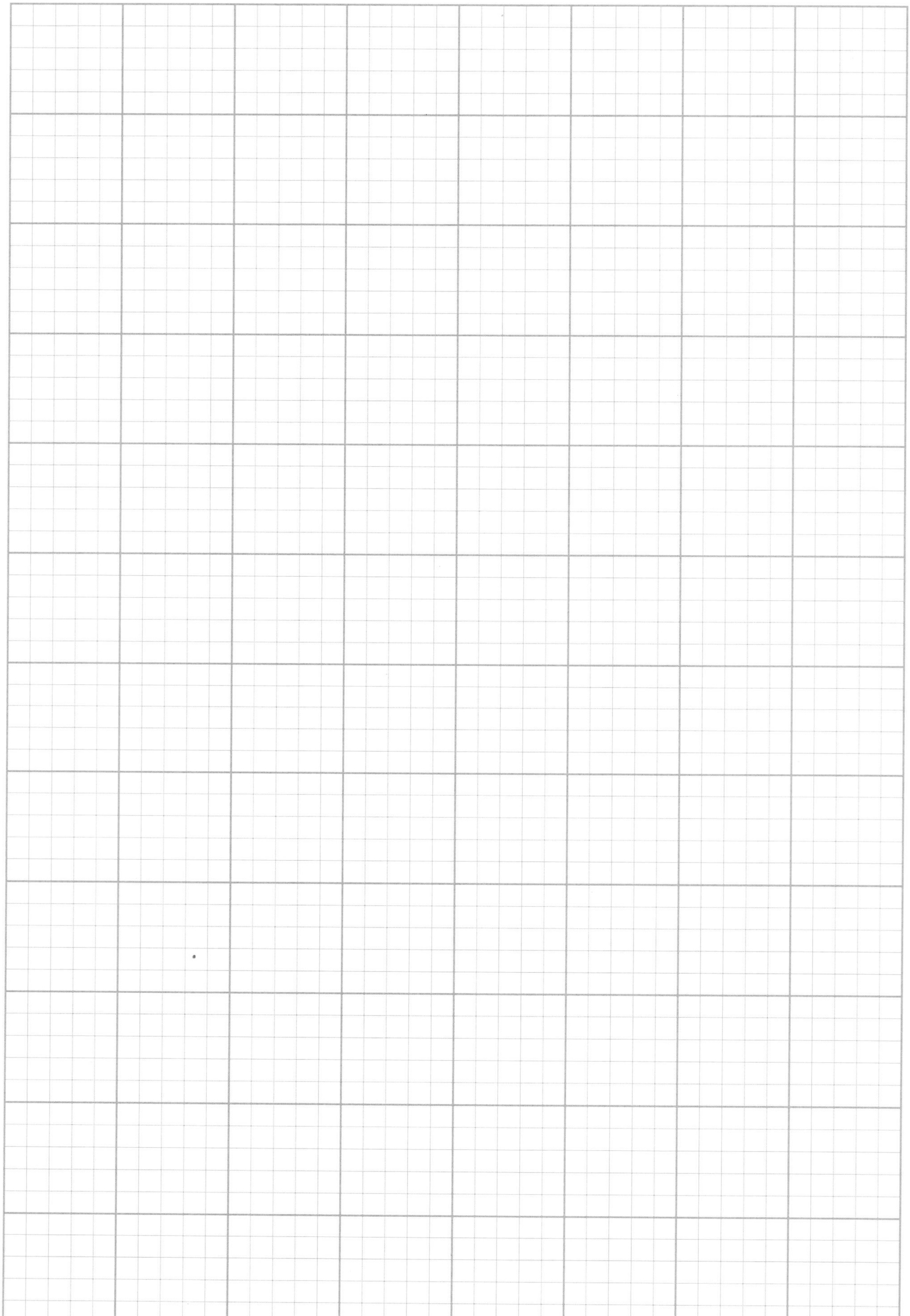

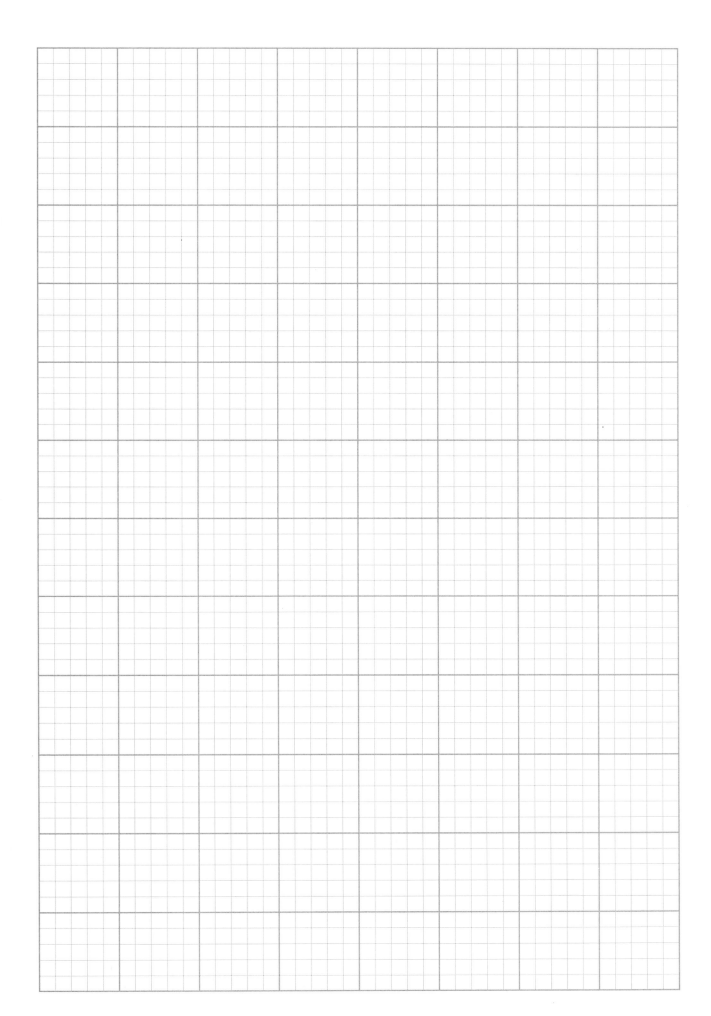

Review Questions

1. Why are aseptic conditions not required for this experiment?

2. What are the changes in the appearance and distribution of sulfur in the flasks? How can these changes be explained?

3. Write a balanced equation to describe the fate of the elemental sulfur.

4. What evidence is there to determine whether the oxidation of sulfur proceeds at different rates in the stationary versus the shaking flask?

5. What is the carbon source for the *Thiobacillus*?

6. What is the nutritional classification of the organism used in this experiment?

7. Where might you go in nature to isolate the type of microorganism used in this experiment?

UNIT V

Characterization and Identification of Experimental Microorganisms

Before beginning the characterization of any of the microorganisms isolated from any environment, one must be sure that each of the isolates is in **pure culture**. Verification is accomplished by careful examination of the colonial morphology on agar plates and by Gram staining. Further characterization of the organisms occurs by examining their growth on selective media, which aids in their identification.

After the completion of all biochemical tests, you should have accumulated enough information to be able to identify the experimental isolates. Use of the rapid identification systems should prove helpful. Also, refer to the bacterial identification schemes in Appendix G to assist in the identifications.

Biochemical tests are of two types:

1. Standard "tube" reactions, in which various types of media are inoculated with pure cultures of the microorganisms. The reactions are observed after incubation.
2. Rapid miniaturized identification techniques that use compartments of prepared media where reactions occur. Results are tabulated and identifications are made by comparing with a computerized bank of microorganisms.

The laboratories within this unit are organized by the following topic areas:

Extracellular Degradation
Fermentation of Carbohydrates
Tests Measuring Other Products of Metabolism
Biochemical Reactions Important in Types of Respiration
Tests Significant for Identification of Pathogens
Miniaturized Rapid Identification Systems

EXTRACELLULAR DEGRADATION

Every microorganism has different metabolic capabilities. By comparing these capabilities, one can group and identify the organisms being studied.

Microorganisms are metabolically quite diverse. Consequently, they are capable of degrading a variety of organic molecules. These molecules can be categorized by their structure and size. Some commonly utilized organic molecules are either macromolecules or constituents of macromolecules. The classes of organic macromolecules include proteins, carbohydrates, lipids or fats, and nucleic acids. Each of these macromolecules is composed of individual, smaller monomeric units. The proteins are composed of amino acids held together by peptide bonds. Monosaccharides in ether linkages comprise the building blocks of carbohydrates. The simple lipids or fats have glycerol and fatty acids in ester linkages as their basic structure. Nucleic acids are composed of individual nucleotides (sugars, bases, and phosphates). Key in the utilization of these molecules by microorganisms are catalytic proteins referred to as **enzymes**. Bacterial cells may produce hundreds of different enzymes. Each is unique in that each is capable of participating in and catalyzing a specific biochemical reaction. The metabolic diversity possessed by microorganisms is thus attributable to the variety of enzymes they have. Review your textbook for a more complete description of enzyme catalysis.

In addition to having to confront the various classes of molecules, bacteria must also deal with molecules that vary in size or molecular weight. The first task that confronts a microorganism when it is exposed to a potential nutrient is the task of transporting that molecule into the cytoplasm. The larger-molecular-weight molecules are incapable of permeating the cytoplasmic membrane. Therefore, to utilize large macromolecules such as proteins and carbohydrates as sources of energy, microorganisms must first release specific enzymes called **exoenzymes** to start the extracellular breakdown of these molecules. Since these reactions take place in the presence of water and involve the splitting of molecules, they are termed "hydrolytic reactions." The phenomenon is referred to as **hydrolysis**. Following the extracellular hydrolysis of macromolecules, transport proteins embedded in the cytoplasmic membrane import the simple monomers for the cell to metabolize.

In the procedures that follow, the activities of various exoenzymes that may be produced by isolated experimental microorganisms are investigated.

LABORATORY

21 Starch Hydrolysis

Safety Issues

- After flooding plates to detect the production of amylase, carefully pour off iodine into a discard beaker.

OBJECTIVES

1. Learn the metabolic products of starch hydrolysis.
2. Discern the ability of unknown isolates to hydrolyze starch.

Starch is a polymer made up of amylose, a polymer of glucose units and amylopectin. Amylopectin is a branched polymer containing phosphate groups (Figure 21.1). Bacteria capable of hydrolyzing starch produce the enzyme **amylase**. This enzymatic hydrolysis forms shorter polysaccharides called **dextrins**, which ultimately can be hydrolyzed to yield disaccharides and monosaccharides. These end products are transported into the cell and used in biosynthetic reactions (e.g., formation of cell wall components) or energy-yielding reactions (e.g., glycolysis or Kreb's cycle reactions).

Materials (per 20 students)

1. Starch agar plates (10)
2. Gram's iodine reagent (5 bottles)
3. Environmental isolates from Laboratory 13
4. *Bacillus subtilis* slant cultures (5)

Procedure

1. Conserve use of plates by streaking two organisms (isolate and known) onto each plate of starch agar (Figure 21.2).
2. Label the plates and incubate inverted at the appropriate temperature for 48 hr. The appropriate temperature should closely mimic the natural environments of your isolated organisms.
3. After 48 hr, flood the surface of the plates with Gram's iodine solution. Let the iodine react for 30 sec to 1 min, and then carefully pour off the excess iodine in a discard beaker. If starch is present in the agar, a blue-black color will be apparent. If the starch has been hydrolyzed by the production of amylase, a clear zone will appear surrounding the growth of organisms on the plates. This is one of those tests that provides a positive color reaction but actually represents a negative test for the presence of amylase, since the organism is not hydrolyzing the starch. The *Bacillus subtilis* control plate should give you a strong positive reaction (see Color Plate 11).

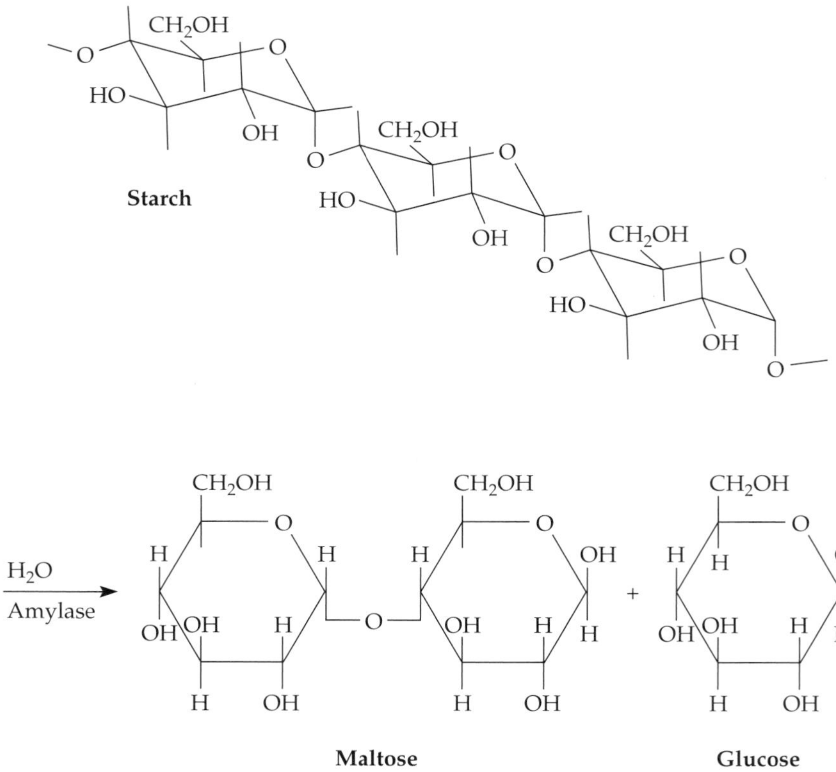

FIGURE 21.1 Starch hydrolysis products.

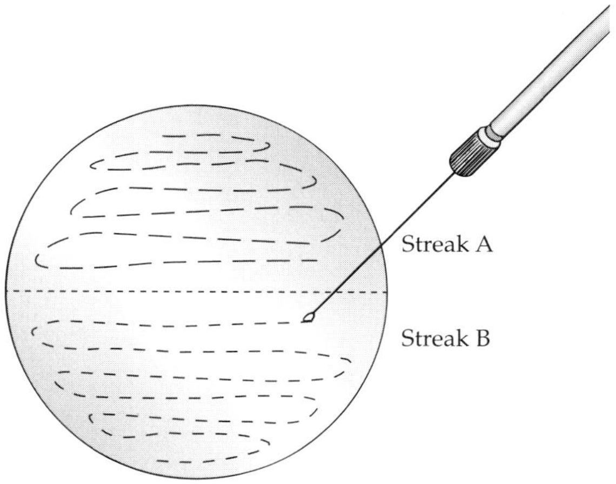

FIGURE 21.2 Inoculation of starch agar.

Laboratory Report 21

Name ____________ Date ____________ Section ____________

Starch Hydrolysis

Describe the appearance of plates after the addition of iodine.

Bacillus subtilis ____________	Reaction ____________
Culture # ____________	Reaction ____________
Culture # ____________	Reaction ____________
Culture # ____________	Reaction ____________

Review Questions

1. How can you have growth of an organism on a starch plate that is not metabolizing starch?

2. How is starch broken down when it is used as a sole carbon source by microorganisms?

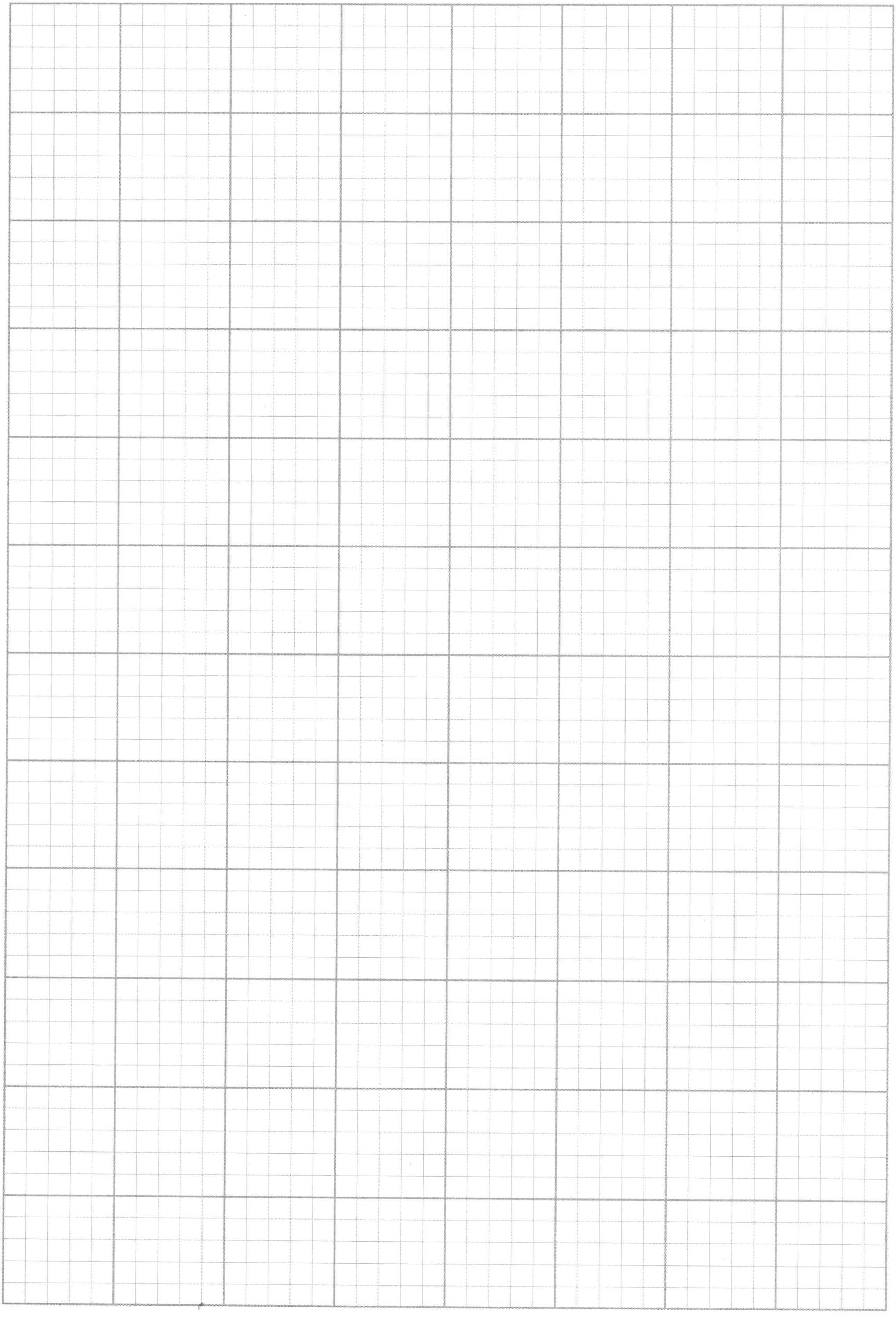

LABORATORY

22 Casein Hydrolysis

Safety Issues

- Be careful not to touch any of the growth on the plates when examining.

OBJECTIVE

Learn to detect caseinase enzyme activity in experimental isolates.

Casein is the large-molecular-weight protein found in milk. As with starch, this macromolecule must be hydrolyzed by exoenzymes into smaller molecules. In this case, amino acids are formed and are capable of being transported through the cell membrane. This hydrolysis is carried out by proteolytic exoenzymes known as **caseinase**.

Materials (per 20 students)

1. Skim milk (casein) agar plates (10)
2. Fresh cultures of environmental isolates from Laboratory 13
3. *Bacillus subtilis* slant cultures (5)

Procedure

1. With an inoculating loop, streak two of the unknown organisms per plate. Inoculate on separate sides of the plate as was done for starch hydrolysis.
2. Invert and incubate at the appropriate temperature for 24 to 48 hr.
3. After growth has appeared, examine the plates for a clear zone in the agar surrounding the growth of each of your test organisms. What does the clear area represent?

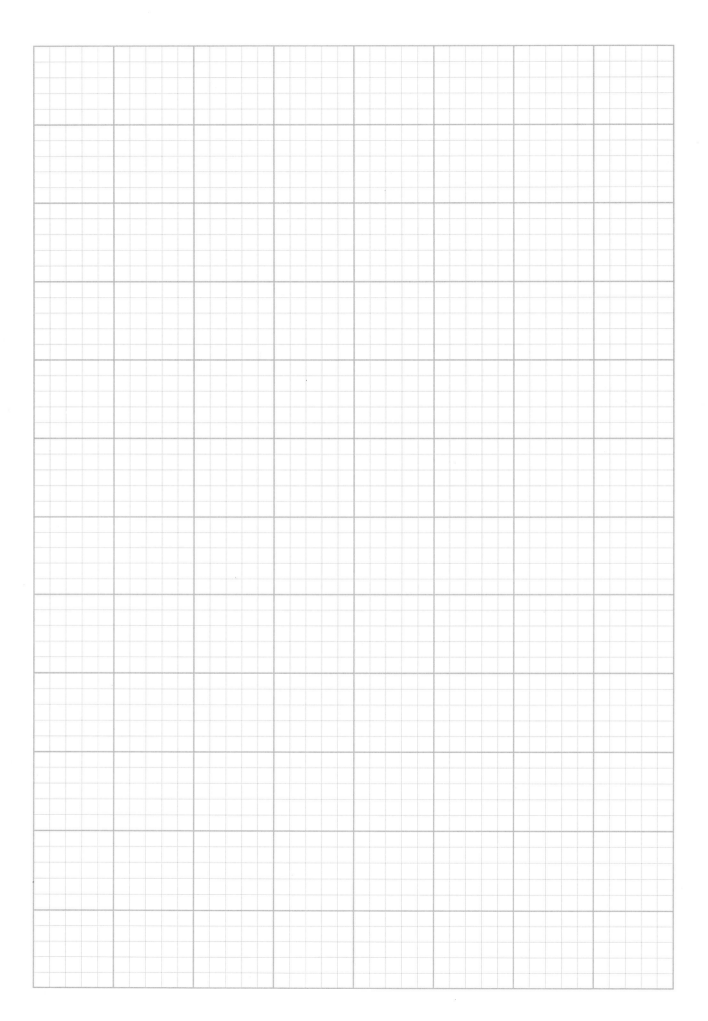

Laboratory Report 22

Name ______________________ Date ______________ Section ______________

Casein Hydrolysis

Sketch plates that have a positive caseinase reaction.

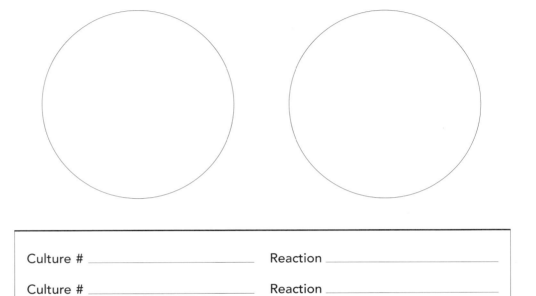

Culture # ______________	Reaction ______________
Culture # ______________	Reaction ______________
Culture # ______________	Reaction ______________

Review Questions

1. What type of experiment could you perform to demonstrate that the caseinase enzyme is actually an exoenzyme?

2. What property of milk allows for the detection of caseinase activity in a solid medium such as agar?

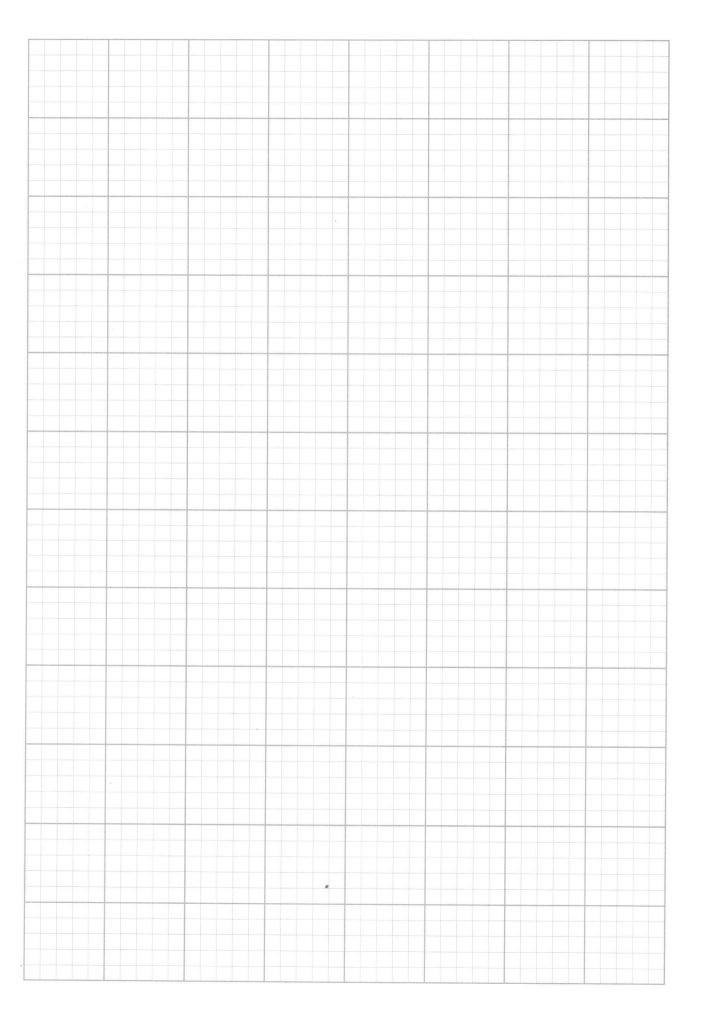

LABORATORY

23 Gelatin Hydrolysis

Safety Issues

- To prevent splattering, allow inoculating needle to cool after sterilization before transferring organisms.

OBJECTIVES

1. Learn straight-line stab inoculation technique.
2. Detect gelatinase enzyme activity in experimental isolates.

Upon boiling collagen, a major connective tissue, the product **gelatin** is formed. Certain bacteria are capable of producing a proteolytic exoenzyme called **gelatinase**, which hydrolyzes the protein to amino acids. These amino acids can then be transported into the cell for further metabolism. At temperatures below 25°C, gelatin will remain a gel, but if the temperature rises about 25°C, the gelatin will be liquid (remember what happens to Jello-O at room temperature). If you cool the liquefied gelatin, it will resolidify. Gelatin hydrolysis has been correlated with pathogenicity of some microorganisms. It is thought that pathogenic bacteria may break down tissue and spread to adjacent tissues.

Materials (per 20 students)

1. Tubes of solidified nutrient gelatin (20)
2. Cultures of experimental isolates from Laboratory 13.
3. Ice bath or refrigerator (needed for next session)
4. Slants of gelatinase positive organisms:
   Bacillus cereus (5)
   Proteus vulgaris (5)

Procedure

1. Using the inoculating needle, transfer some of the isolates via a straight-line inoculation stab to tubes of nutrient gelatin (Figure 23.1).
2. Incubate the tubes for 48 hr at the appropriate temperature.
3. After 48 hr, place the tubes in an ice bath or a refrigerator to determine whether the hydrolysis of gelatin (due to production of gelatinase) has occurred. If the gelatin has been hydrolyzed, the medium will remain liq-

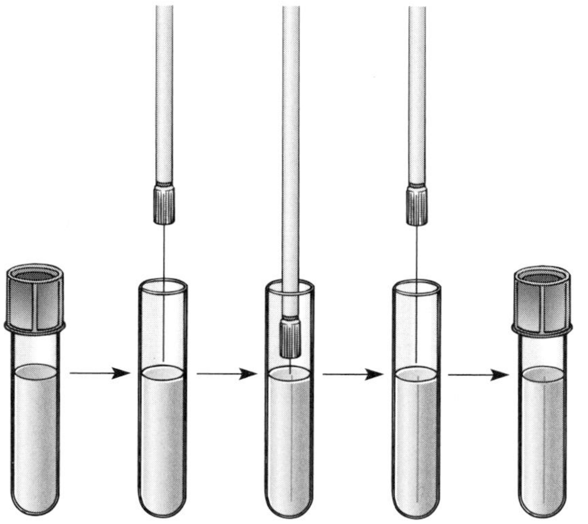

FIGURE 23.1 Straight-line inoculation technique.

uid after cooling. If gelatin has not been hydrolyzed, the medium will resolidify. Care must be taken to distinguish between **melting** of the gelatin and true irreversible **liquefaction** due to the production of the exoenzyme.

Laboratory Report

Name ____________ Date ____________ Section ____________

23

Gelatin Hydrolysis

Indicate with a sketch the degree of liquefaction in any tube.

Culture # ____________ Reaction ____________

Culture # ____________ Reaction ____________

Culture # ____________ Reaction ____________

Review Questions

1. Why do you suppose the manufacturers of gelatin desserts recommend against adding fresh pineapple to their products?

2. Why has agar replaced gelatin as a solidifying agent in laboratory microbiology?

3. Why is it important to refrigerate the gelatin tubes in order to detect the presence of gelatinase enzyme in any of the organisms tested?

4. What would happen if you accidentally refrigerate the gelatin tubes prior to incubation?

LABORATORY

24 Fat (Lipid) Hydrolysis

Safety Issues

- Cool inoculating loop before making any transfers.

OBJECTIVE

Observe lipase activity in experimental isolates and compare to that of known cultures.

The ability of certain microorganisms to degrade lipids such as triglycerides occurs through the action of exoenzymes called **lipases**. These enzymes cleave the ester bonds in the molecules, leading to the formation of glycerol and fatty acids (Figure 24.1). These degradation products can be taken up by microorganisms and metabolized aerobically for energy, or can be used in the biosynthesis of other fats or cellular components.

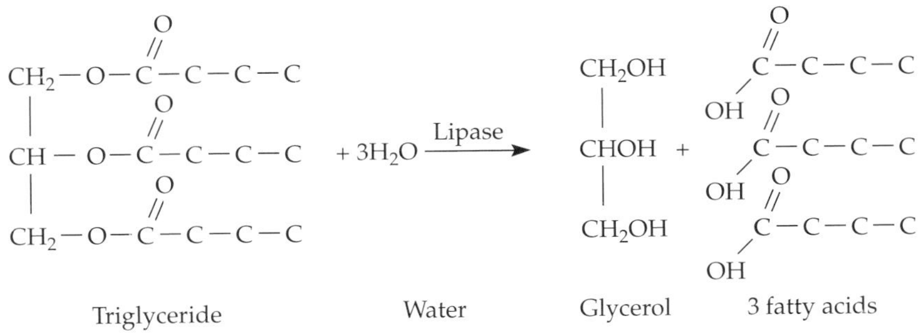

FIGURE 24.1 Products of lipid hydrolysis.

Materials (per 20 students)

1. Spirit blue agar plates (10)
2. Fresh cultures of experimental isolates from Laboratory 13
3. Slants of *Bacillus subtilis* (5)

Procedure

1. With an inoculating loop, use a straight-line inoculation method to apply two unknowns per plate of spirit blue agar.

2. Inoculate the second plate with the *Bacillus* culture.
3. Incubate the plates at the appropriate temperature for 24 to 48 hr.
4. Spirit blue agar contains a vegetable oil which, when hydrolyzed to produce fatty acids, lowers the pH of the medium and produces a dark blue precipitate. Any of the plates showing such a dark blue precipitate surrounding the bacterial growth should be scored as positive; that is, the organism produces the **lipase** enzyme.

Laboratory Report 24

Name ______ Date ______ Section ______

Fat Hydrolysis

Culture # ______	Reaction ______
Culture # ______	Reaction ______
Culture # ______	Reaction ______

Review Question

1. Why have organisms possessing lipase enzyme been implicated in food spoilage? Given the manner of metabolism of fatty acids, under what environmental conditions might growth of lipolytic organisms actually occur?

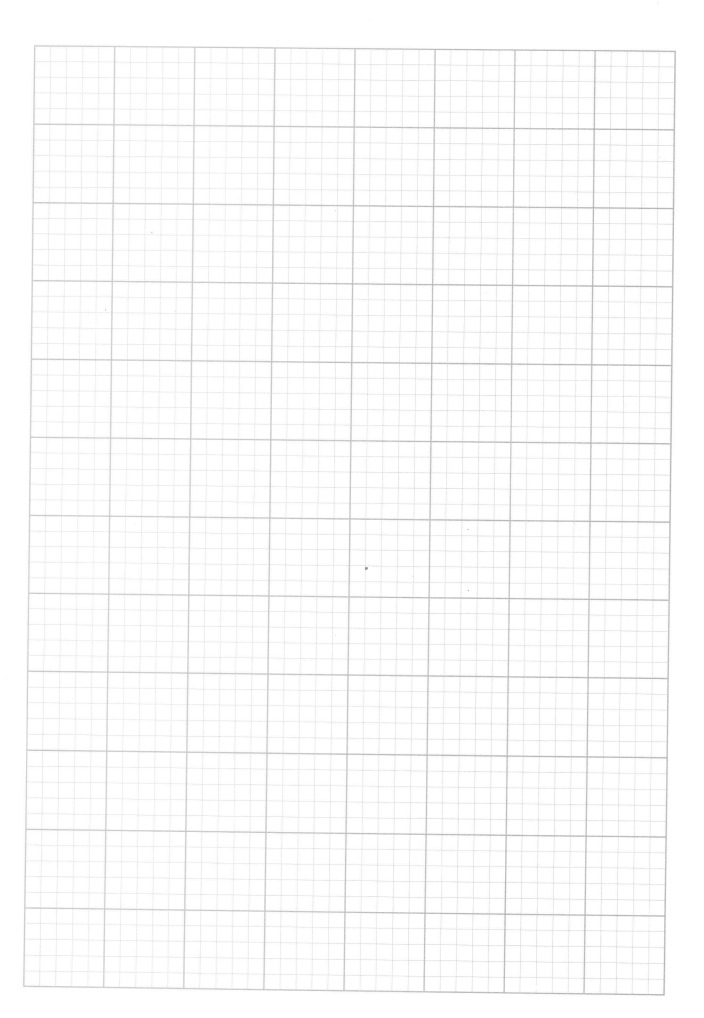

LABORATORY

25 Carbohydrate Fermentation

Safety Issues

- To prevent spillage, all fermentation tubes should be incubated upright and kept in a rack when being observed.

OBJECTIVE

To discern the ability of unknown isolates to ferment the monosaccharide glucose and the disaccharides lactose and sucrose by use of phenol red broth fermentation media.

FERMENTATION OF CARBOHYDRATES

Microorganisms can utilize carbohydrates for energy by using a complement of enzymes arranged in specific pathways. Some of these reactions require oxygen (aerobic), and others can take place in the absence of oxygen (anaerobic). Some microorganisms can metabolize the carbohydrates in either the presence or the absence of oxygen. This type of metabolism is referred to as facultative, and the organisms are usually called **facultative anaerobes**. In order to determine the utilization of carbohydrates, the carbohydrate source must be provided in the medium along with other non-carbohydrate nutrients required for growth. A medium referred to as phenol red broth base, which has been supplemented with the test carbohydrate to yield a final concentration of 1.0%, is the test medium used.

In fermentation, carbohydrates can be converted into a variety of end products, usually gases, acids, or alcohols. A number of different strategies can be used to detect the nature of the end products. In order to detect the production of gases, a small inverted tube called a **Durham tube** is placed in the fermentation broth. If gas is produced during the fermentation, it will displace the medium within the inverted tube and will appear as a bubble. If acids are produced during the metabolism of the carbohydrates, this can be detected by a pH change in the medium. To simplify this process, a pH indicator, phenol red, is incorporated into the medium. The phenol red indicator is red at neutral pH (7.0), and becomes yellow at a slightly acidic pH (6.8). If the medium becomes yellow, we surmise that acid has been produced as an end product in the breakdown of the carbohydrate (sugar) being used (see Color Plate 12). The reactions are scored in the following manner:

If acid is produced: A
If gas is produced: G
If both acid and gas are produced: AG

What if there is no color change in the tube? The first step is to determine whether growth has occurred. How may that be determined? What if there is growth and there is no color change or the color change is to deep red or purple? (*Hint*: Check the composition of the phenol red broth base medium.) What are the gases that may be formed? Can gas form in the absence of acids?

Materials (per 20 students)

1. Tubes of fermentation broth with glucose, lactose, and sucrose. Glucose tubes contain Durham tubes. (25 of each)
2. Cultures of experimental organisms from Laboratory 13
3. Known cultures of *Escherichia coli* and *Enterobacter aerogenes* (5 of each)

Procedure

1. Using a sterile inoculating loop, inoculate each of the sugar fermentation broth tubes with the unknowns to be identified. Gently twirl the loop as the tubes are inoculated. Do not shake the glucose tubes, as a bubble may accidentally form in the Durham tube. This would make the interpretation of the reaction difficult. Known cultures of *E. coli* and *Enterobacter* will be inoculated for the class and made available for interpretation.
2. Incubate the tubes at the appropriate temperature for seven days, being careful to note the reactions on a regular basis, the 24- and 48-hr readings being the most important. The tubes are kept for seven days, for some organisms carry out what is called the "alkaline swing": They ferment the carbohydrate to acids and then work on the protein in the medium, forming ammonia, which in solution becomes ammonium hydroxide. This produces an alkaline environment and a corresponding change in the phenol red indicator.

Laboratory Report

Name Date Section

25

CARBOHYDRATE UTILIZATION

	Appearance of Tubes		
	Glucose + Durham Tube	**Lactose**	**Sucrose**
Unknown 1			
Unknown 2			
E. coli			
Enterobacter			

Review Questions

1. In the carbohydrate fermentation reactions, is it possible to have accumulation of gas in the Durham tubes without changes in the color of the pH indicator?

2. What kinds of reactions are occurring when an organism undergoes an alkaline swing in a sugar-fermentation reaction?

3. Differentiate fermentation, respiration, and anaerobic respiration.

4. How does the Durham tube get filled with media?

LABORATORY

26 Mixed Acid Fermentation (Methyl Red Test) and the Butanediol Fermentation (Voges Proskauer Test)

Safety Issues

- Barritt's solution A contains alpha naphthol, which is toxic. Wear gloves when adding this reagent to tubes.
- Barritt's solution B contains 40% KOH, a strong base. Exercise caution when handling. Wear gloves when dispensing this reagent to tubes.
- Do not mouth pipet any material. Use the pipetting devices provided.

OBJECTIVES

1. Learn the types of fermentative end products formed by bacteria.
2. Detect the presence of acid and alcoholic end products of fermentation.

All enteric (intestinal) bacteria can utilize sugars for their energy demands. Mixed acid fermenters such as *Escherichia coli* ferment glucose to produce large amounts of acetic, formic, and succinic acids as end products. The large amount of acid produced lowers the pH of the medium to below 5.0. By using the indicator methyl red (MR), the production of these acids as an end product of fermentation can be monitored. If the pH drops below 4.5, the color of the methyl red indicator will be red. If the pH is above 6.0, the color will be yellow/orange. It is important not to confuse these color reactions with those of the phenol red indicator that was used previously.

⚠ Care should be taken when handling many of the chemical reagents used. In this case, Barritt's reagent contains potassium hydroxide and alpha naphthol.

Some bacteria, rather than producing abundant acid in the fermentation of glucose, produce other products such as alcohols. All species of *Enterobacter* and *Serratia* will form products such as ethanol and 2,3-butanediol rather than the large amount of acid, as does *Escherichia coli* (Figure 26.1). These reactions can be used to differentiate some common gram-negative rod-shaped organisms. A test for acetylmethylcarbinol (a precursor of 2,3-butanediol that appears in the growth medium) is performed. If Barritt's reagent is added, a pink color developing after a few minutes indicates the presence of acetylmethylcarbinol, and the test is positive (see Color Plate 13).

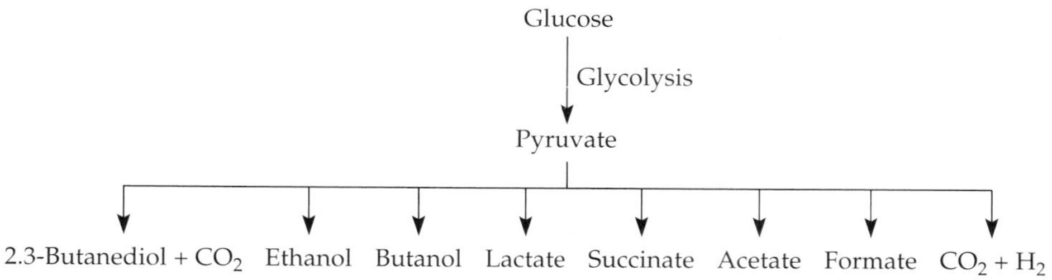

FIGURE 26.1 Fermentation end products.

Materials (per 20 students)

1. Tubes of MRVP Medium (a buffered glucose broth containing peptone and dipotassium phosphate) are used to perform both the methyl red (MR) and Voges Proskauer (VP) tests (20)
2. Experimental cultures (gram-negative rods) from Laboratory 13
3. Clean test tubes (20)
4. Pipetting device for reagents
5. Known cultures of *E. coli*, *Enterobacter*, and *Serratia*.
6. Methyl red indicator (60 ml)
7. Barritt's solution A (60 ml)
8. Barritt's solution B (60 ml)

Procedure

1. Using a sterile inoculating loop, inoculate the unknown culture into a tube of MRVP broth. Controls of the known cultures will be inoculated for you as controls.
2. Incubate the tubes at the appropriate temperature for 24 to 48 hr.
3. After it has become apparent that growth has occurred, transfer half of the contents of the broth tube to a clean test tube. There is no need to maintain aseptic conditions here, for tests are discarded after testing and reading. One half of the medium is used to perform the MR test, and the other half to perform the VP test.
4. To one tube add three to four drops of the methyl red reagent. A reddish color is indicative of a positive methyl red test.
5. To the second tube add 0.5 ml of Barritt's solution A (alpha-naphthol) and 0.5 ml of Barritt's B solution (KOH). Shake vigorously, being careful not to spill the contents of the tube. Let the tube stand for 1 to 2 hr. A positive VP test will produce a pink to red color in the medium.

Laboratory Report 26

Name ______________________ Date ______________ Section ______________

Organism	Methyl Red Reaction	VP Reaction
Escherichia		
Enterobacter		
Serratia		
Unknowns		

Review Question

1. *Escherichia coli* and *Enterobacter* are both capable of carbohydrate fermentation. Explain why *E. coli* is methyl-red positive and *Enterobacter* is methyl-red negative.

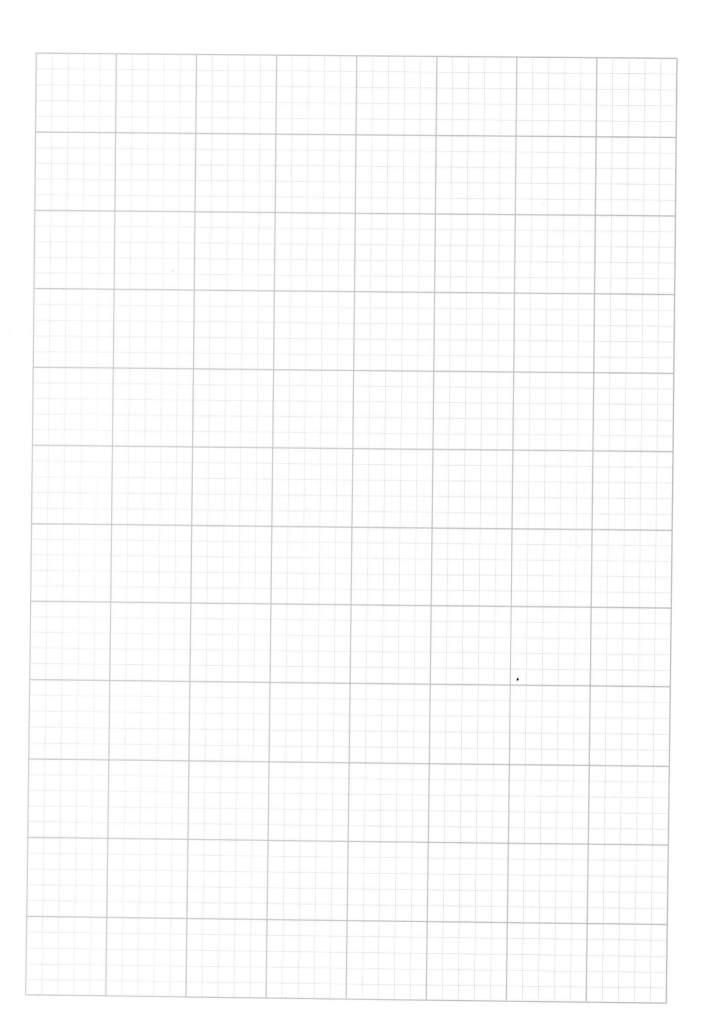

LABORATORY

27 Citrate Utilization Test

Safety Issues

- Incubate all tubes in an upright position.
- Cool inoculating loop after sterilization before attempting transfers.

OBJECTIVE

Determine whether unknown isolates can utilize citrate as a sole source of carbon.

Some microorganisms can metabolize citrate (citric acid) as a sole carbon energy source. Like many of the nutrients already tested, citrate needs to be transported into the bacterial cells before it can be metabolized. This transport depends upon the enzyme citrate permease. Once inside the cell, the citrate is broken down to pyruvate (pyruvic acid) and carbon dioxide. The carbon dioxide produced combines with the sodium in the medium and water to form sodium carbonate, an alkaline product. The rise in pH is detected by the color change in the bromothymol blue indicator present in the medium from green to a deep blue. This is indicative of a positive result and verifies that the organism is utilizing citrate. You should be careful to note any growth at all on the Simmons' citrate slant, since certain organisms grow very slowly at the expense of citrate and may not form sufficient carbon dioxide to cause a color change in the medium. Score any growth over the original inoculum as also a positive reaction (see Color Plate 14).

Materials (per 20 students)

1. Simmons' citrate agar slants (20)
2. Environmental organisms from Laboratory 13
3. Known cultures:
   Enterobacter
   Proteus
   Serratia

Procedure

1. Using an inoculating needle, transfer a small amount of the environmental isolate to a Simmons' citrate agar slant. This should be done by first stabbing the butt of the slant and then streaking the surface of the slant.
2. The known cultures will be inoculated for you in a similar fashion.
3. All tubes should be incubated for 24 to 48 hr at the appropriate temperature.

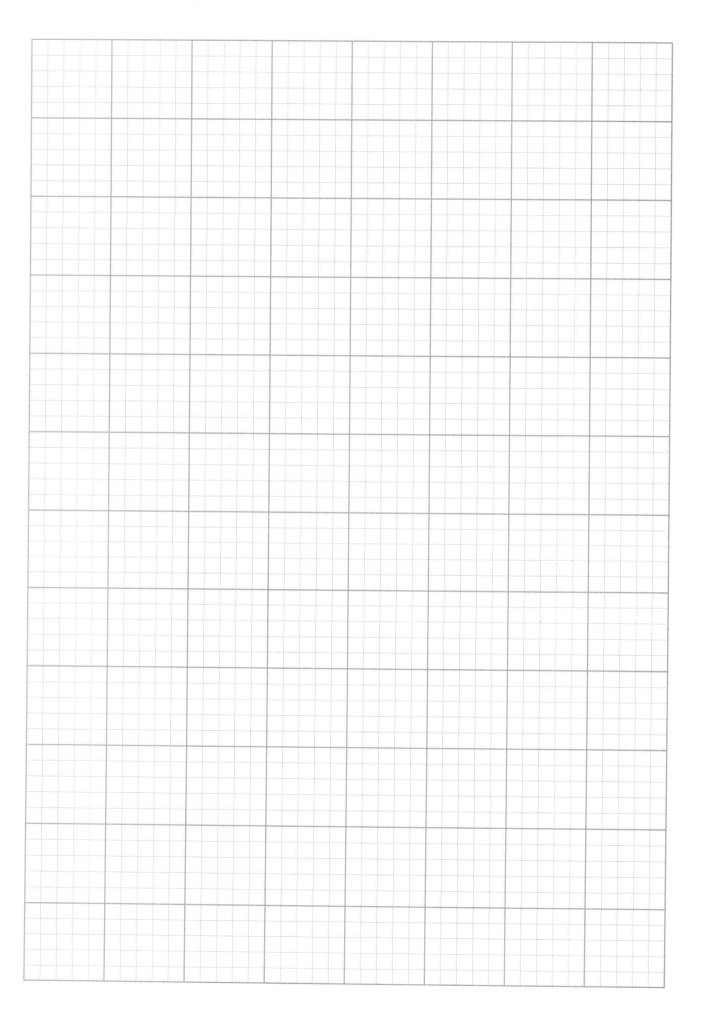

Laboratory Report 27

Name Date Section

Organism	Growth in Citrate Tube	Color Change in Citrate Medium
Enterobacter		
Unknown		

Review Questions

1. How can the change in the citrate agar be explained?

2. What are the metabolic products that are contributing to the change of color of the agar?

3. Draw the structure of citric acid. How is it being utilized as an energy source?

4. How do organisms that metabolize citric acid in the presence of inorganic nitrogen synthesize amino acids for proteins?

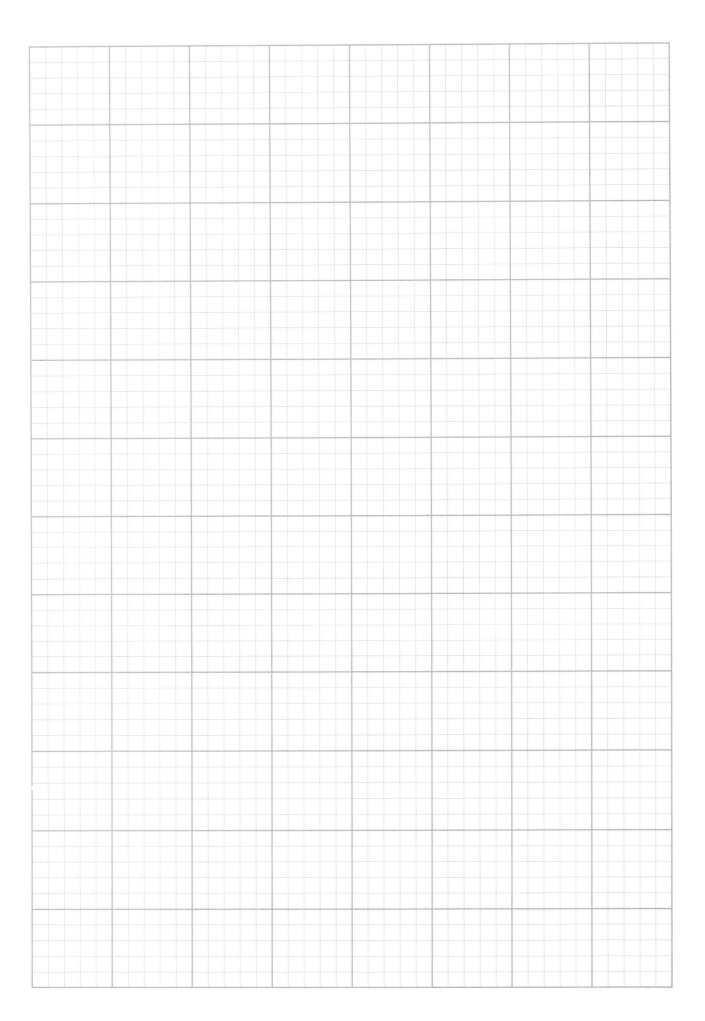

LABORATORY

28 Tryptophan Hydrolysis Test

Safety Issues

- Kovac's reagent is toxic and contains concentrated HCl and a carcinogen, para-dimethylamino-benzaldehyde. Wear gloves when dispensing the reagent and avoid inhaling any of the vapors.
- Use a pipetting device for all transfers.
- Dispose of Dry Slide Indole in a biohazard bag after completion of test.

OBJECTIVE

Detect the end product, indole, formed by the action of the enzyme tryptophanase.

Certain microorganisms can metabolize the amino acid tryptophan through the action of the enzyme tryptophanase. Once again, the activity of the enzyme is not measured directly, but rather, one of the end products is detected. The enzymatic degradation leads to the formation of pyruvic acid, indole, and ammonia. The pyruvic acid and ammonia may be utilized by some organisms but the indole produced accumulates in the medium. The presence of indole is detected by the addition of a reagent called Kovac's reagent. **Care should be taken in handling this reagent in that it contains two noxious chemicals: amyl alcohol and para-dimethylamino-benzaldehyde.** Cultures that exhibit a red layer after the addition of Kovac's reagent are indole-positive. The absence of color means that indole was not produced, and that the organism does not possess the tryptophanase enzyme.

Materials (per 20 students)

1. Tubes of tryptone broth (20)
2. Experimental isolates from Laboratory 13
3. Known culture of *Escherichia coli*
4. Kovac's reagent (60 ml)

Kovac's reagent is toxic and contains concentrated HCl and a carcinogen, para-dimethylamino-benzaldehyde. Wear gloves when dispensing the reagent and avoid inhaling any of the vapors.

Procedure

1. Inoculate a tube of tryptone broth using a sterile inoculating loop with the experimental unknown. The *E. coli* known cultures will be inoculated for you.
2. Incubate the tubes at the appropriate temperature for 24 to 48 hr.
3. After the incubation interval, add 0.5 ml (10 drops) of the Kovac's reagent to the tube, and shake the tube gently. A deep red color in the top alcohol layer indicates the presence of indole.

The previous four tests comprise a series of important determinations that are collectively called the IMViC series of reactions (I=indole; M=methyl red; V=Voges Proskauer; and C=citrate). The i is added for ease of pronunciation. For public health reasons, tests are conducted on water and food samples to detect the presence of intestinal (enteric) bacteria. Many of these intestinal organisms found in humans and lower mammals belong to the family *Enterobacteriaceae*. They all have the characteristics of being short, nonsporeforming, gram-negative rods. Some members of the family are considered to be pathogens: *Salmonella* and *Shigella*; others such as *Klebsiella* and *Proteus* are occasional pathogens; while others are considered to be normal intestinal residents (flora), such as *Escherichia* and *Enterobacter*. It should always be kept in mind that, while occasionally these distinctions of pathogenicity are made, many organisms considered to be nonpathogenic can cause serious problems in the appropriate host environment. The IMViC series of reactions allows for the differentiation of the various members of this family.

An alternative method for indole production by microorganisms is the use of Dry Slide Indole slides (Difco Laboratories). These slides have filter paper reaction areas containing para-dimethylamino-benzaldehyde (DMABA). If indole is produced by microorganisms, it will react with DMABA to form a pink-colored complex. The use of Dry Slide Indole eliminates the handling of the hazardous Kovac's reagent containing DMABA.

Procedure for Dry Slide Technique

1. Allow the Dry Slide Indole to equilibrate to room temperature before use.
2. With a sterile inoculating loop, obtain an isolated colony from an agar plate and smear it on the reaction area of the Dry Slide Indole.
3. Examine the area streaked for the appearance of a pink color within 30 sec. Indole-positive organisms change the color of the reaction area from buff-yellow to pink. Indole-negative organisms produce no change.

Laboratory Report 28

Name ______________________ Date __________ Section __________

Microorganism	Reaction in Tube
E. coli	
Unknown	

Review Question

1. Why are the IMViC series of reactions so important in the identification of gram-negative enteric microorganisms?

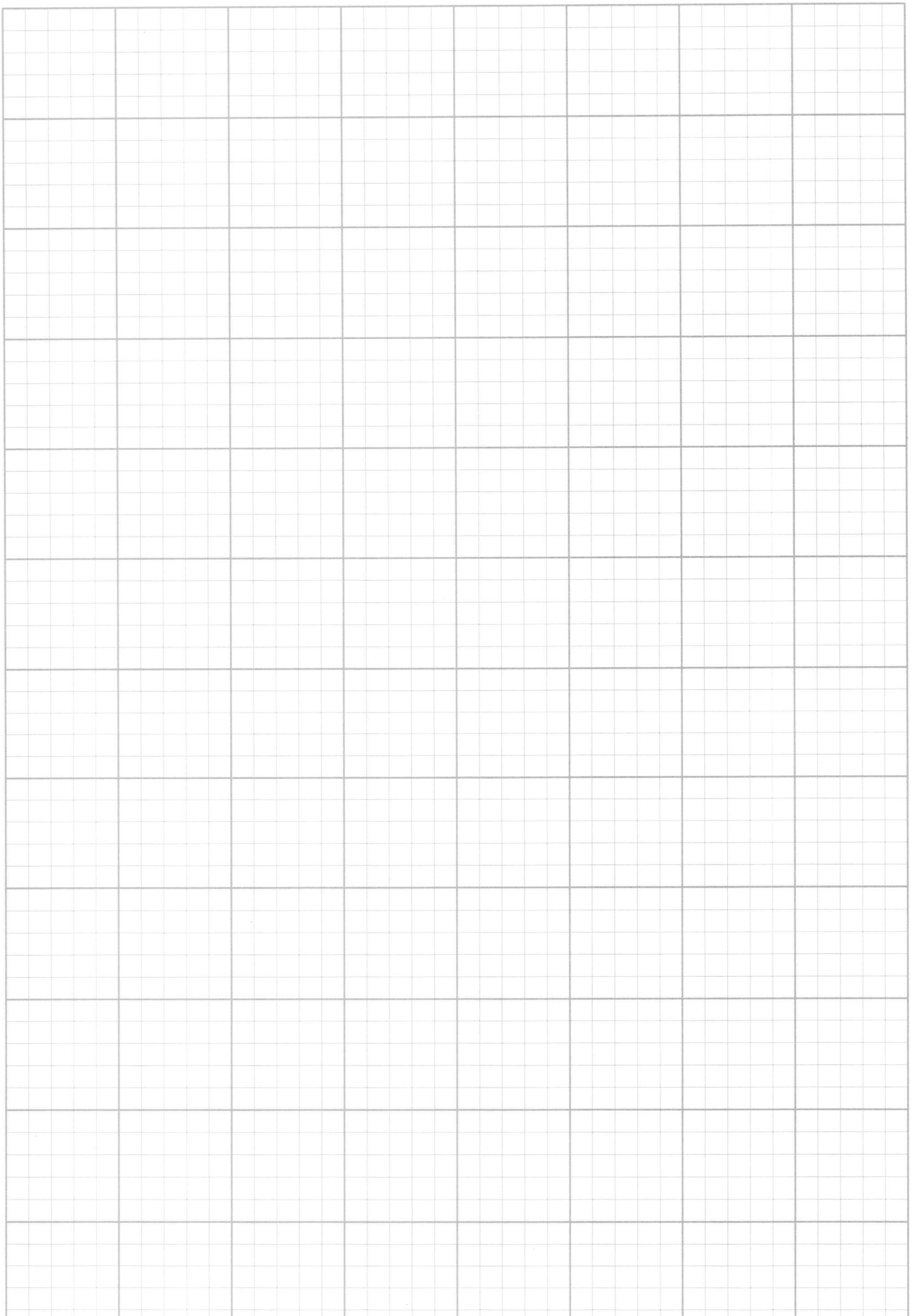

LABORATORY

29 Hydrogen Sulfide Test

Safety Issues

- All culture tubes should be kept upright in a rack to prevent spillage.
- Allow any heated inoculating loop time to cool before removing growth from a colony or liquid from a broth tube. Contact with a hot loop may cause splattering of material and the aerosol produced may be harmful to you.
- Exercise caution when handling the Kovac's reagent. Wear gloves.

OBJECTIVE

Learn use of SIM medium to detect formation of H_2S, indole, and motility in unknown isolates.

In this lab, three different tests are accomplished because of the nature of the medium. The medium is called SIM for sulfide, indole, and motility medium. Some microorganisms are motile by means of flagella. This motility can be measured by inoculating the microorganisms into this semi-solid (3 g/L of agar rather than 15 g/L) medium. A steady hand is necessary to perform the motility test accurately. The medium should be stabbed with an inoculating needle. Be careful to enter and exit the medium along the same line of inoculation. Upon incubation, motile bacteria will migrate in the semi-solid medium away from the line of inoculation and cause a cloudiness or turbidity throughout the tube. Nonmotile bacteria will grow along the line of inoculation. Therefore, it is critical that there be only one line of inoculation. If there is more than one line, it might be difficult to determine whether there is true motility or just growth surrounding all lines of inoculation.

Gaseous hydrogen sulfide may be formed from inorganic or organic molecules containing sulfur. Organic sulfur present in sulfur-containing amino acids, such as cysteine, can be released in the form of hydrogen sulfide by the action of the enzyme cysteine desulfurase. Gaseous hydrogen sulfide may also be produced by the reduction of inorganic sulfur compounds such as sulfates, sulfites, and thiosulfates. The SIM medium contains peptone as a source of amino acids (some of which contain sulfur) and sodium thiosulfate. Ferrous ammonium sulfate is also added and acts as the hydrogen-sulfide indicator. If hydrogen sulfide is produced, it combines with the ferrous ammonium sulfate to produce ferrous sulfide, a black insoluble precipitate. This black deposit can be seen along the line of inoculation if the organism is nonmotile, or throughout the medium if the organism is motile (see Color Plate 15).

Materials (per 20 students)

1. Tubes of SIM Medium (40)
2. Experimental organism from Laboratory 13
3. Known culture of *Proteus*

Procedure

⚠ Exercise caution when handling the Kovac's reagent. Wear gloves.

1. Using a sterile inoculating needle, inoculate tubes of SIM medium by a straight single-line inoculation with the unknown isolate and the known *Proteus* culture. Remember to exit the tube along the same line of inoculation entry.
2. Incubate the tubes at the appropriate temperature for 24 to 48 hr.
3. In reading the results of the reactions in this medium, it usually is wise to check for H_2S production first. Next, determine the motility. It is best to compare the inoculated tube with an uninoculated control, while holding both up to a light source. The last step should involve the addition of the Kovac's reagent to determine whether indole has been produced.

Laboratory Report 29

Name Date Section

REACTIONS

Organism	Motility	Sulfide Production	Indole Production
Proteus			
Unknown 1			
Unknown 2			

Review Questions

1. How is the production of H_2S observed in SIM medium?

2. What is an alternative to SIM medium in determining motility in bacteria?

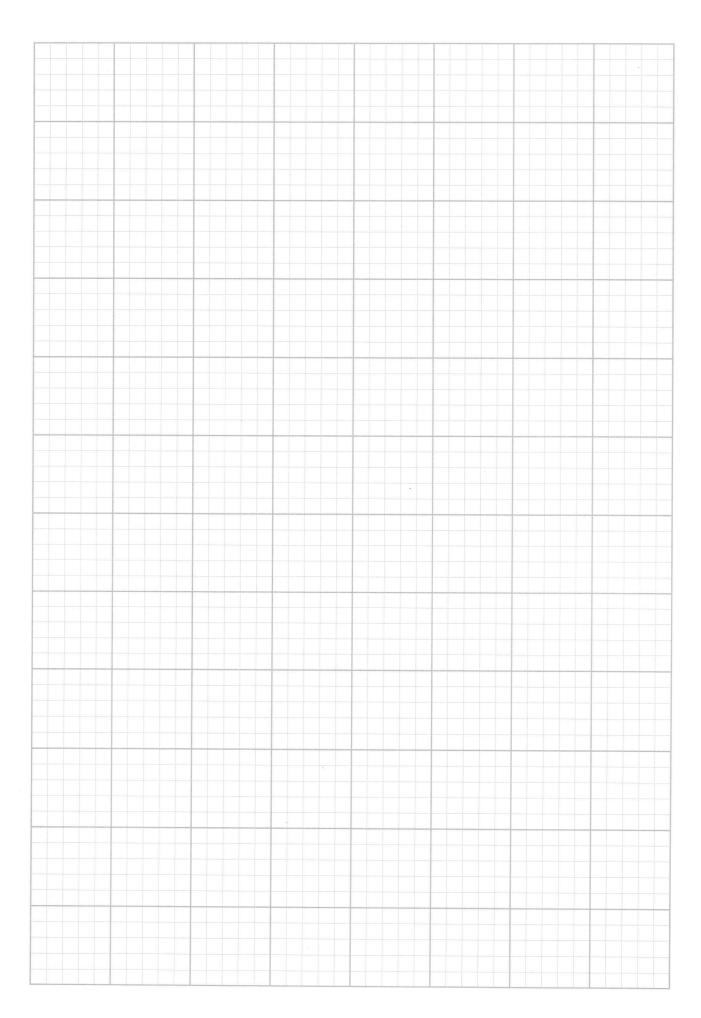

LABORATORY

30 Urea Hydrolysis

Safety Issues

- All culture tubes should be kept upright in a rack to prevent spillage.
- Always use a pipetting device when pipetting samples. Never attempt to mouth pipet any material.

OBJECTIVE

Detection of urease enzyme in unknown isolates.

Urease is an enzyme that catalyzes the conversion of urea to carbon dioxide and ammonia (Figure 30.1). The urease test is particularly useful in identifying members of the genus *Proteus*. Since other enteric organisms do not hydrolyze urea or do it slowly, the test is helpful in differentiating *Proteus* from other lactose-nonfermenting enteric microorganisms. The urea broth used for the test contains yeast extract, urea, and the indicator phenol red in a buffered solution. Care should be taken in the preparation of the medium, as urea is unstable at high temperatures and will break down if autoclaved. For this reason, the medium is normally prepared by membrane filtration. In the presence of urease, urea is split and ammonia and carbon dioxide are produced. Ammonia combines with water to produce ammonium hydroxide, a strong base which raises the pH of the medium. This rise in the pH causes the phenol red indicator to turn a deep pink. This is indicative of a positive reaction for urease.

Materials (per 20 students)

1. Tubes of filter-sterilized urea broth (2.0 ml). If not aliquoted, use aseptic conditions and deliver 2.0 ml to sterile tubes. (20)
2. Experimental isolate (use gram-negative rod) from Laboratory 13
3. Known culture of *Proteus*

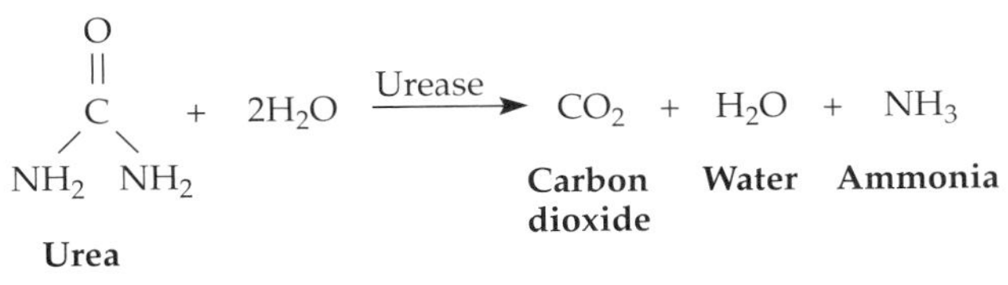

FIGURE 30.1 Breakdown of urea.

Procedure

1. Using a sterile inoculating loop, inoculate a tube of urea broth with the unknown isolate. A known positive culture of *Proteus* will be inoculated for you for comparison.
2. Incubate at the appropriate temperature for 4 hr.
3. Most members of the genus *Proteus* will show a positive reaction after 4 hr. If reaction is negative after this, incubate all tubes for 24 hr before discarding.

Laboratory Report 30

Name Date Section

Organism	Color of Urea Broth
Proteus	
Unknown	

Review Questions

1. Write the equation for the breakdown of urea by *Proteus*.

2. Explain the change in pH in the reaction as evidenced by the change in the color of the indicator dye.

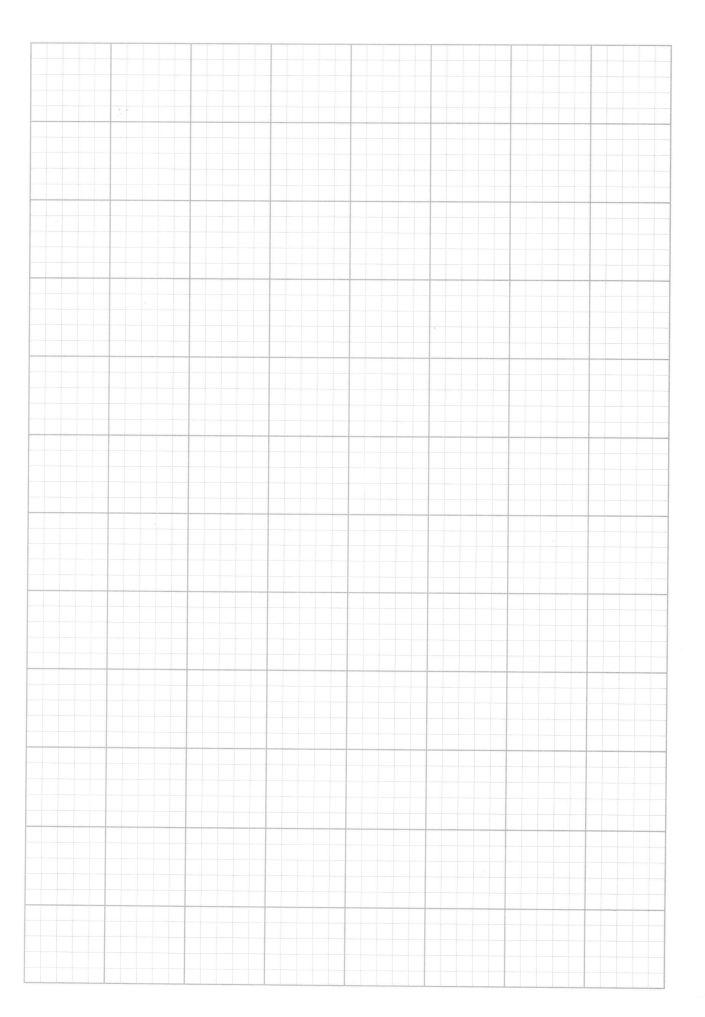

LABORATORY

31 Litmus Milk Reactions

Safety Issues

- Maintain all culture tubes in an upright position to prevent spillage.

OBJECTIVE

Examine the various ways microorganisms can attack the sugar and proteins in milk.

There are many nutrients in milk that can serve as substrates for bacterial growth. Milk contains the proteins casein, lactoalbumin, and lactoglobulin. The sugar that is present is lactose. There are also abundant vitamins present. By analyzing the end products formed, it is possible to determine which of the components of the litmus milk are being used by bacteria as an energy source. To assist in differentiating the metabolic changes that occur, the pH indicator litmus is also in the medium.

The following are the reactions we want to note in the litmus milk tubes (see Color Plate 16):

1. Lactose fermentation
2. Litmus reduction
3. Gas formation
4. Curd formation
5. Proteolysis
6. Alkaline reaction

Organisms possessing the enzyme β-galactosidase are capable of fermenting lactose to lactic acid. The lactic acid can be detected as the litmus indicator turns pink when the pH of the medium falls to about 4.

Litmus also serves as an artificial electron acceptor. The excess hydrogen ions produced during biooxidations are accepted by the litmus present in the medium. In the oxidized state, litmus is purple; in the reduced state, it turns white or the color of the milk.

The gases produced during the fermentation of lactose are carbon dioxide and hydrogen. These gases may be visualized as cracks, fissures, or bubbles in any curd present.

Two types of curds may be formed in the litmus milk tubes. Curds are identified as either acid or rennet curds. An acid curd is an insoluble clot formed when the acids precipitate the casein in the milk. To detect whether the curd is an acid curd, invert the tube and see if the clot remains in place. A rennet curd is a softer curd, which has a semi-solid consistency such as that of yogurt. The rennet curd will flow freely if the tube is tilted.

Some microorganisms may attack the protein in the milk. These organisms use proteolytic enzymes to hydrolyze the casein in the milk to its constituent amino acids. The tube will take on the appearance of a brownish straw-colored fluid. Large quantities of ammonia may also be released, causing the litmus to take on a deep purple color near the top of the tube.

An alkaline reaction occurs when there is no change in the litmus or the change is to a deeper color of blue. This represents a partial decarboxylation or deamination of the amino acids and the release of alkaline end products.

Materials (per 20 students)

1. Litmus milk tubes (60)
2. Experimental organisms from Laboratory 13
3. Known cultures:
   Pseudomonas aeruginosa (5)
   E. coli (5)

Procedure

1. Inoculate the litmus milk tubes with the unknown and known cultures.
2. Incubate at the appropriate temperature and observe regularly for seven days.
3. Record all reactions.

Laboratory Report 31

Name Date Section

TUBE REACTIONS

Organism	Curd Formation	Acid	Alkaline	Proteolysis

Review Questions

1. How can you differentiate curd formation and proteolysis?

2. Is litmus milk a selective or differential medium?

3. Contrast the pH indicators that are used in (a) carbohydrate fermentation, (b) methyl red test, (c) citrate test, (d) urease test, and (e) litmus milk test.

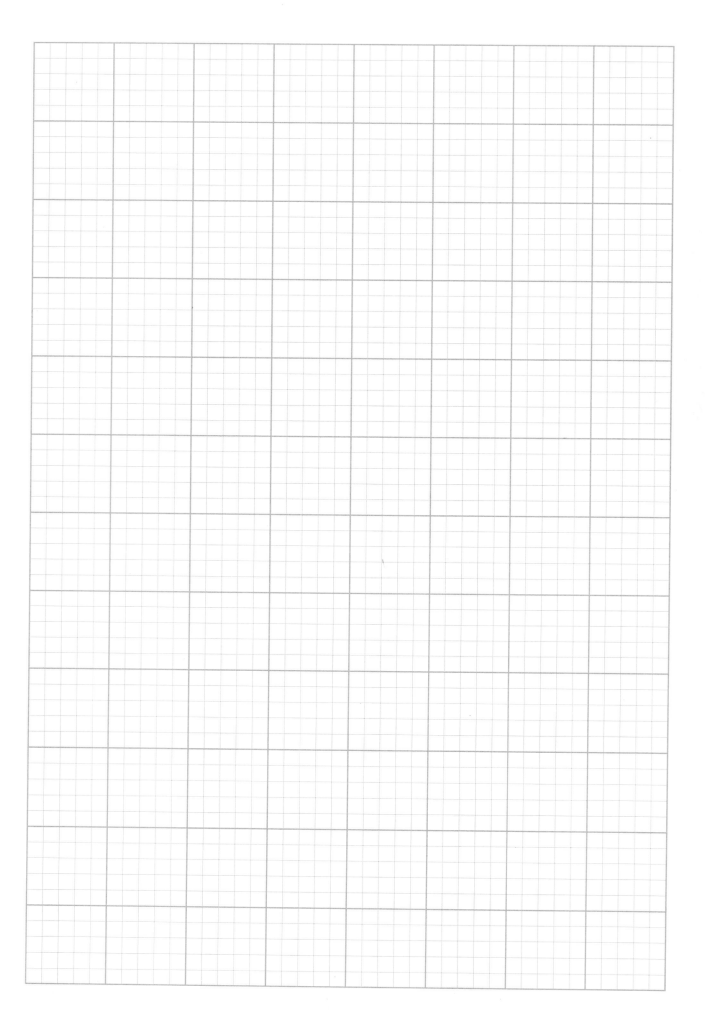

LABORATORY

32 Nitrate Reduction

Safety Issues

- Reagent A contains sulfanilic acid which is caustic. Wear gloves when dispensing the reagent.
- Reagent B contains dimethyl-alpha-naphthylamine, a suspected carcinogen. Wear gloves to avoid skin contact.
- Do not mouth pipet any material. Use the pipetting devices provided.
- Exercise caution when handling zinc. Do not inhale or allow to come into contact with skin.

OBJECTIVES

1. Become familiar with the ways in which microorganisms can reduce nitrates.
2. Detect the formation of nitrites by unknown isolates.

Many facultative microorganisms, when growing anaerobically, are able to use inorganic compounds rather than oxygen as their terminal electron acceptors. (Remember that electron acceptors become reduced when they gain electrons.) The process is referred to as anaerobic respiration. What is the process called if the terminal electron acceptor is oxygen? What if the terminal electron acceptor is an organic molecule?

During anaerobic respiration, some bacteria use nitrates as their terminal acceptors (i.e., nitrate becomes reduced), forming nitrites—ammonia or nitrogen gas (Figure 32.1).

$$\underset{\textbf{Nitrate}}{NO_3^-} + \underset{\textbf{Hydrogen electrons}}{2H^+ + 2e^-} \xrightarrow[\textbf{Reductase}]{\textbf{Nitrate}} \underset{\textbf{Nitrite}}{NO_2^-} + \underset{\textbf{Water}}{H_2O}$$

$$\underset{\textbf{Nitrite}}{NO_{2-}} \longrightarrow \underset{\textbf{Ammonia}}{NH_3}$$

$$\underset{\textbf{Nitrate}}{2NO_3^-} + 10e^- + 12H^+ \longrightarrow \underset{\textbf{Molecular nitrogen}}{N_2} + 6H_2O$$

FIGURE 32.1 Nitrate reduction products.

The broth used to detect nitrate reduction is a nutrient broth that has been supplemented with 0.1% potassium nitrate. Nitrate reduction is determined by the formation of nitrite and is detected by the addition of reagents A and B.

Materials (per 20 students)

1. Tubes of nitrate broth (60)
2. Unknown experimental isolates from Laboratory 13
3. Sulfanilic acid reagent (reagent A) (40 ml)
4. Dimethyl-alpha-naphthylamine (reagent B) (40 ml) **(This reagent is carcinogenic so avoid contact with skin.)**
5. Known bacterial cultures:
   Pseudomonas aeruginosa
   Escherichia coli

Procedure

⚠ **Reagent A contains sulfanilic acid, which is caustic. Wear gloves when dispensing the reagent. Reagent B contains dimethyl-alpha-naphthylamine, a suspected carcinogen. Wear gloves to avoid skin contact.**

1. Inoculate nitrate broth tubes with your unknown culture and the known organism.
2. Incubate at the appropriate temperature for 24 to 48 hr.
3. Add three to four drops of reagent A and an equal amount of reagent B to the broth tubes to check for the presence of nitrite. If nitrite is present (nitrates have been reduced), then a red color should appear almost immediately. This constitutes a positive test result for nitrate reduction. The test could be negative for one of two reasons: (a) nitrates have not been reduced and remain in the broth, or (b) nitrates have been completely reduced to nitrous oxide or nitrogen.
4. Add a small amount of zinc to broth tubes that are negative for nitrite. Zinc is an inorganic catalyst that reduces nitrates; therefore, a red color indicates that nitrites are being formed from nitrates that remain in the tube. This would constitute a negative nitrate-reduction test result. If no red color develops after adding zinc to the tubes, no nitrate is present; the nitrate has been completely reduced to nitrous oxide or nitrogen. This constitutes a positive test result for nitrate reduction.

Laboratory Report 32

Name Date Section

Organism	Nitrite	Nitrate	Nitrogen Gas

Review Question

1. In the nitrate reduction test, what happens if a negative reaction occurs after the addition of reagents A and B?

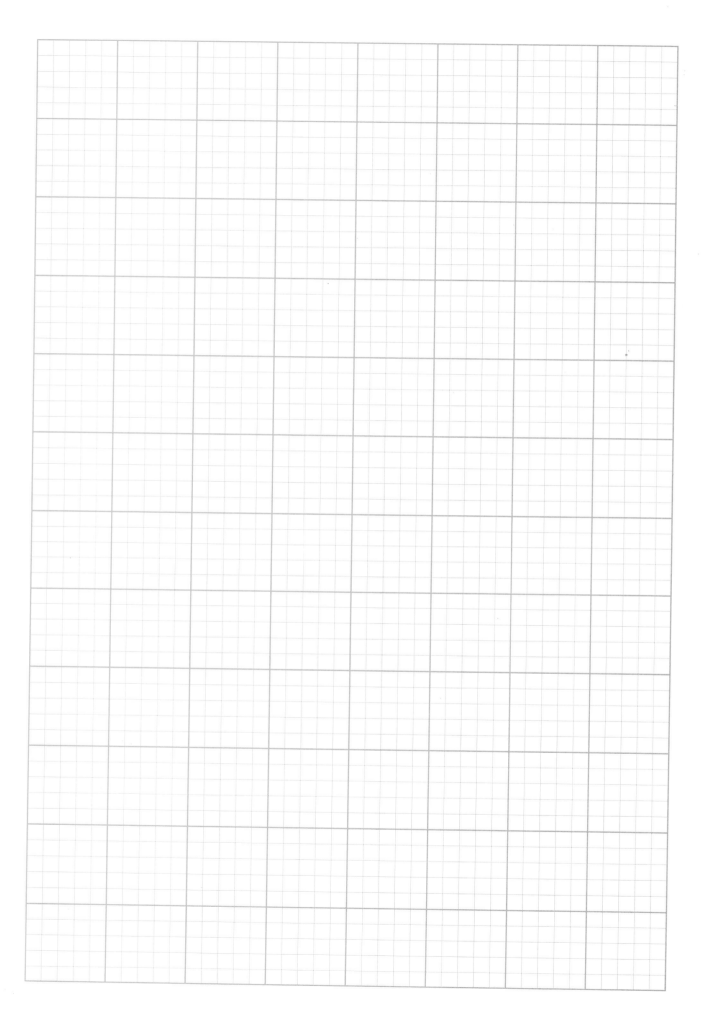

LABORATORY

33 Catalase Test

Safety Issues

- Make sure to dispose of the slide containing the hydrogen peroxide and the test organism in the tray of disinfectant provided.
- Do not pipet any material by mouth.
- Hydrogen peroxide is caustic. Avoid contact with skin.

OBJECTIVE

Determine presence of catalase enzyme in unknown isolates using the slide hydrogen peroxide test.

The presence of oxygen can exert a variety of effects on obligate anaerobic bacteria. Many obligate anaerobes can withstand oxygen for some time, but the growth of others, such as the methanogens, is inhibited by brief exposures to oxygen. The oxygen kills them because toxic products of oxygen reduction accumulate in the organisms. The toxic products are superoxide radical (O_2^-), hydroxyl radical (OH), and hydrogen peroxide (H_2O_2). Some electron carriers such as flavoproteins can reduce oxygen directly and form the superoxide radical.

$$O_2 + e^- \rightarrow O_2^-$$

The presence of two enzymes, superoxide dismutase and catalase, allows anaerobes to resist the toxic effects of exposure to oxygen. Superoxide dismutase catalyzes the conversion of the O_2^- anion to hydrogen peroxide and O_2.

$$O_2^- + O_2^- + 2\,H^+ \rightarrow H_2O_2 + O_2$$

One superoxide radical transfers its extra electron to another radical, which becomes reduced to hydrogen peroxide. Catalase catalyzes the breakdown of hydrogen peroxide to water and oxygen.

$$H_2O_2 + H_2O_2 \rightarrow 2\,H_2O + O_2$$

Two electrons are transferred from one hydrogen peroxide molecule to a second hydrogen peroxide molecule. The first molecule is oxidized to oxygen, and the second molecule is reduced to two molecules of water.

Aerobic and aerotolerant microorganisms produce superoxide dismutase and, as a result, do not accumulate superoxide radicals. Most strict anaerobes lack superoxide dismutase and catalase and therefore cannot tolerate the presence of oxygen.

Materials (per 20 students)

1. 3% Hydrogen peroxide reagent (three 40-ml bottles)
2. Clean glass slides
3. Fresh cultures of experimental isolates from Laboratory 13.
4. Know cultures of *Staphylococcus aureus*

Procedure

1. Place a few drops of hydrogen peroxide on a clean slide.
2. Using a sterile inoculating loop, mix some of the unknown in the hydrogen peroxide.
3. The vigorous production of bubbles indicates a positive reaction (see Color Plate 17).
4. Repeat the procedure with the known culture.

Laboratory Report 33

Name Date Section

Organism	Bubble Formation

Review Questions

1. Write the equation for the catalase reaction.

2. Why is the catalase enzyme not found in obligate anaerobic bacteria?

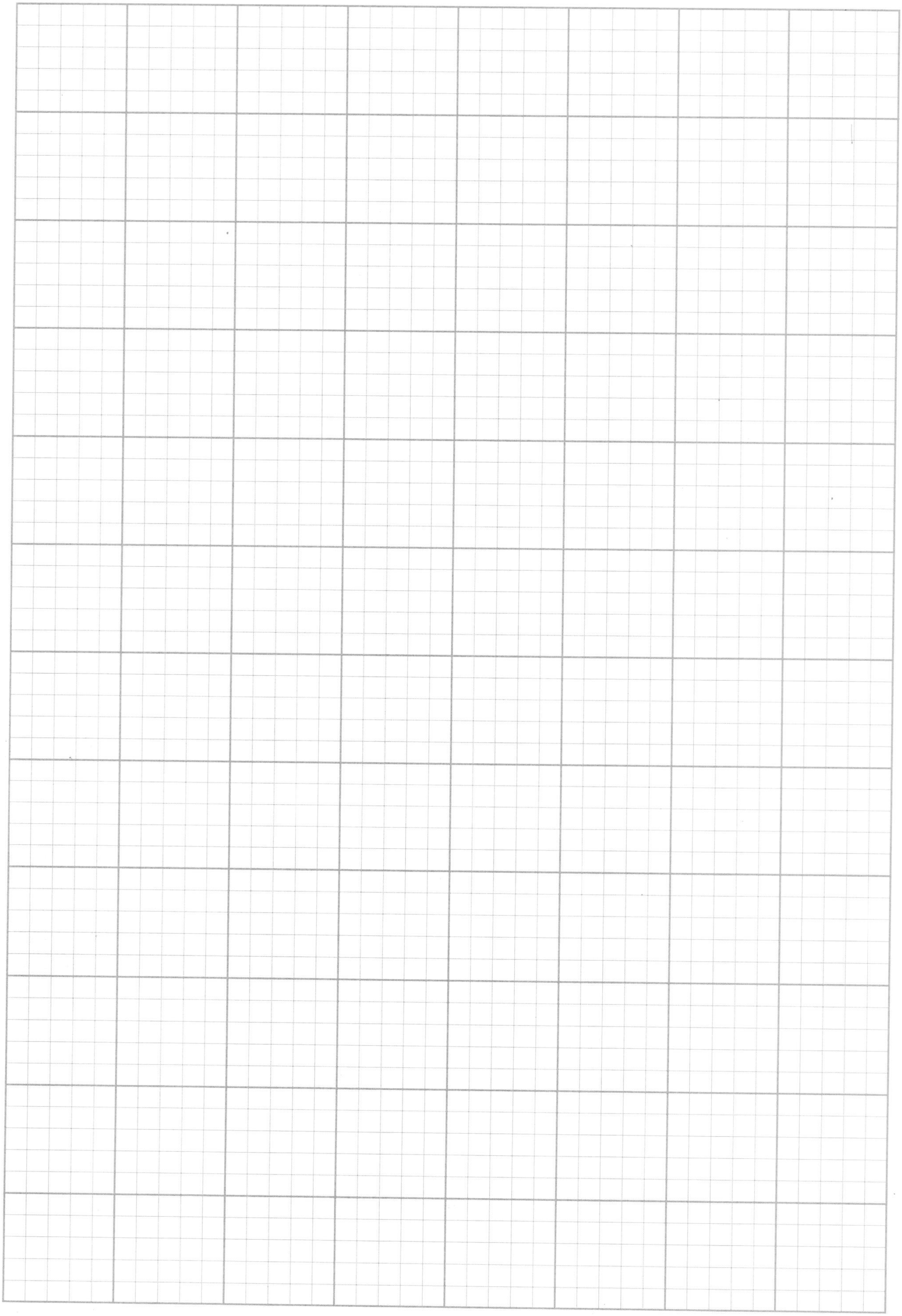

LABORATORY

34 Oxidase Test

Safety Issues

- Dispose of test panels by placing them in a biohazard bag.
- Allow any heated inoculating loop time to cool before removing growth from a colony or liquid from a broth tube. Contact with a hot loop may cause splattering of material and the aerosol produced may be harmful.

OBJECTIVE

Detection of the presence of cytochrome oxidase in isolates by using Difco Dry Slide oxidase technique.

The oxidase test determines the presence of cytochrome c in bacteria. Recall that in aerobic bacteria the cytochromes serve as electron carriers during aerobic respiration. The detection of this cytochrome is extremely beneficial in differentiating many groups of bacteria. All of the members of the Family *Enterobacteriaceae* are oxidase-negative, while other gram-negative rods, such as *Pseudomonas*, are oxidase-positive. The oxidase test is based on the ability of organisms to turn a reagent (dimethyl-para-phenylenediamine hydrochloride) to a purple color. Since this reagent is very toxic, a packaged test kit is used to perform the test.

Materials (per 20 students)

1. Difco Dry Slide oxidase test kit (10)
2. Experimental isolates from Laboratory 13 plated on Mueller–Hinton agar.
3. Culture of *Pseudomonas* on Mueller–Hinton agar slants (5)

Procedure

1. Open a fresh package of the oxidase test.
2. Using a sterile inoculating loop, rub some of the unknown culture on the plastic film of the oxidase test card. A deep purple color should appear within 30 sec. The rapid color development is a positive oxidase result. Any test that shows a delayed color development (greater than 30 sec) should be scored as negative (see Color Plate 18).
3. As a control, rub a plain sterile wire loop over the plastic film. It should produce no color change. Some of the wire used in inoculating loops have been known to react with the oxidase reagent and thus give a false-positive result.
4. Rub some of the *Pseudomonas* culture on another section of the plastic film. Score the reaction.

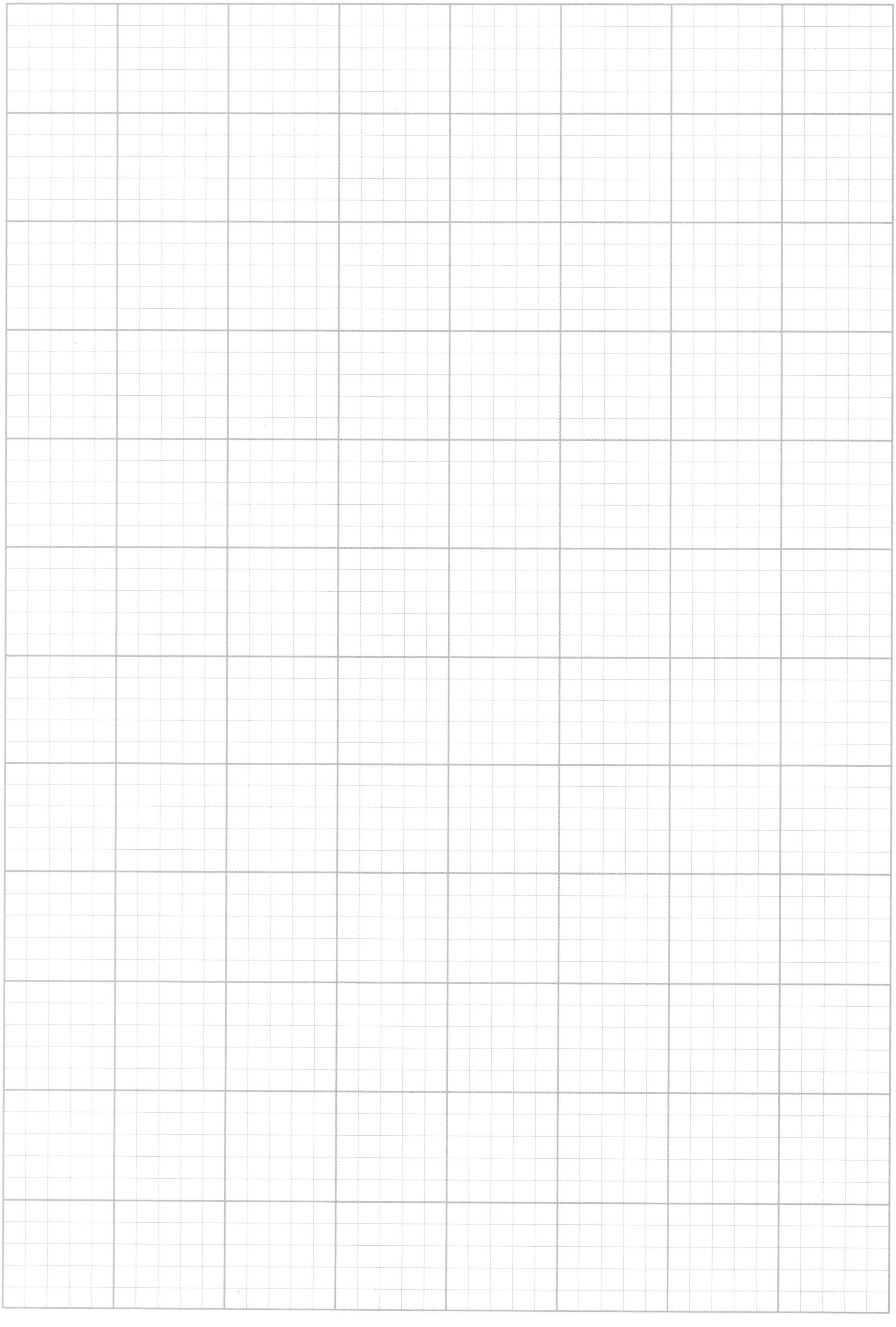

Laboratory Report 34

Name _______________ Date _______________ Section _______________

Organism	Color Development Within 20–60 sec
Pseudomonas	
Unknown	

Review Questions

1. Why is the oxidase test such an important test in helping to differentiate microorganisms?

2. How would inhibiting the cytochrome oxidase enzyme affect cellular metabolism?

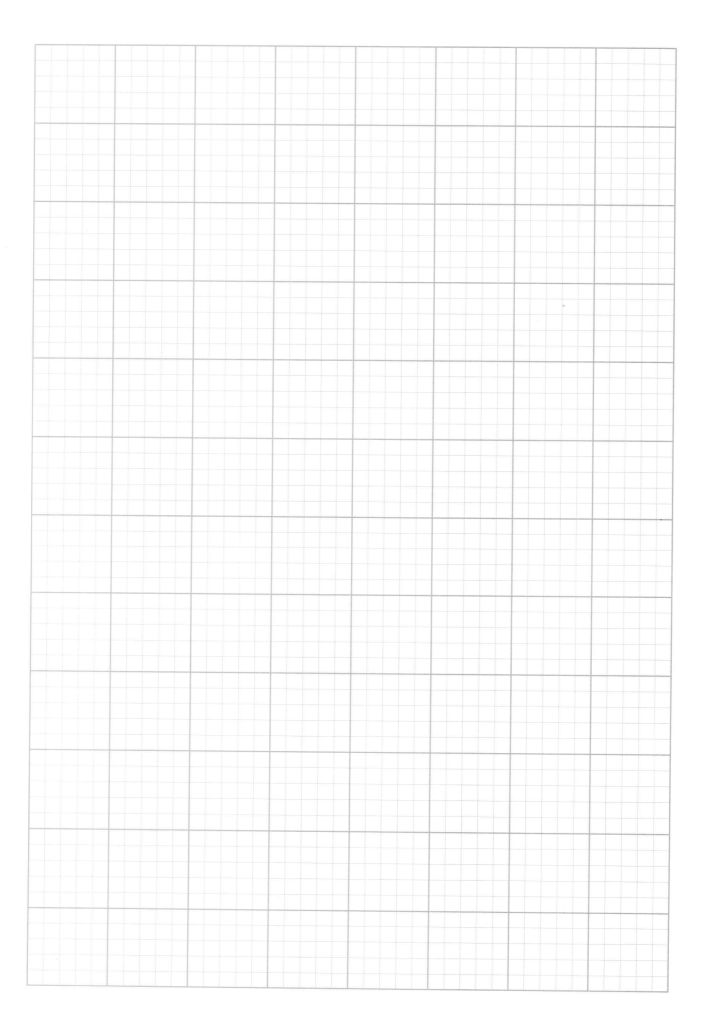

LABORATORY

35 Deoxyribonuclease

Safety Issues

- Be cautious when working with stains as clothing may become permanently stained.
- Do not pipet any materials by mouth.
- HCl can cause burns to skin and irritate mucous membranes. Avoid skin contact and do not inhale vapors.

OBJECTIVE

Detect DNase activity in experimental isolates and compare to activity in *Staphylococcus aureus.*

Many pathogenic members of the genus *Staphylococcus* produce the enzyme **deoxyribonuclease** (DNase), which enables them to degrade DNA.

Materials (per 20 students)

1. DNA agar plates (20)
2. Experimental isolates from Laboratory 13
3. 1 N HCl or 0.005% toluidine blue O (5 bottles)
4. Slants of *Staphylococcus aureus* (5)

Procedure

1. With an inoculating loop, inoculate two of the unknown organisms per plate of DNA agar. Use a circular motion to create an area of inoculation about the size of a dime.
2. Incubate at the appropriate temperature for 24 to 48 hr. After incubation, observe DNase activity by adding either 1 N HCl or 0.005% toluidine blue to the agar surface. If using HCl, a zone of clearing indicates a positive DNase test. The zone represents the absence of DNA. The medium around colonies not producing DNase remains opaque, which is a reflection of the precipitation of DNA by the added acid. With toluidine blue, positive DNase cultures will show a rose-pink halo around their area of growth on the plate. Toluidine blue appears blue when bound to intact DNA. The breakdown of DNA by DNase changes the structure of the dye and causes it to reflect a different wavelength of light, leading to the red or pink color instead of blue.

An alternative method for the detection of DNase involves the incorporation of 0.1% methyl green into the DNA agar. Inoculate in the same manner, and note zones of color change surrounding the colonies as an indication of the production of DNase and breakdown of DNA.

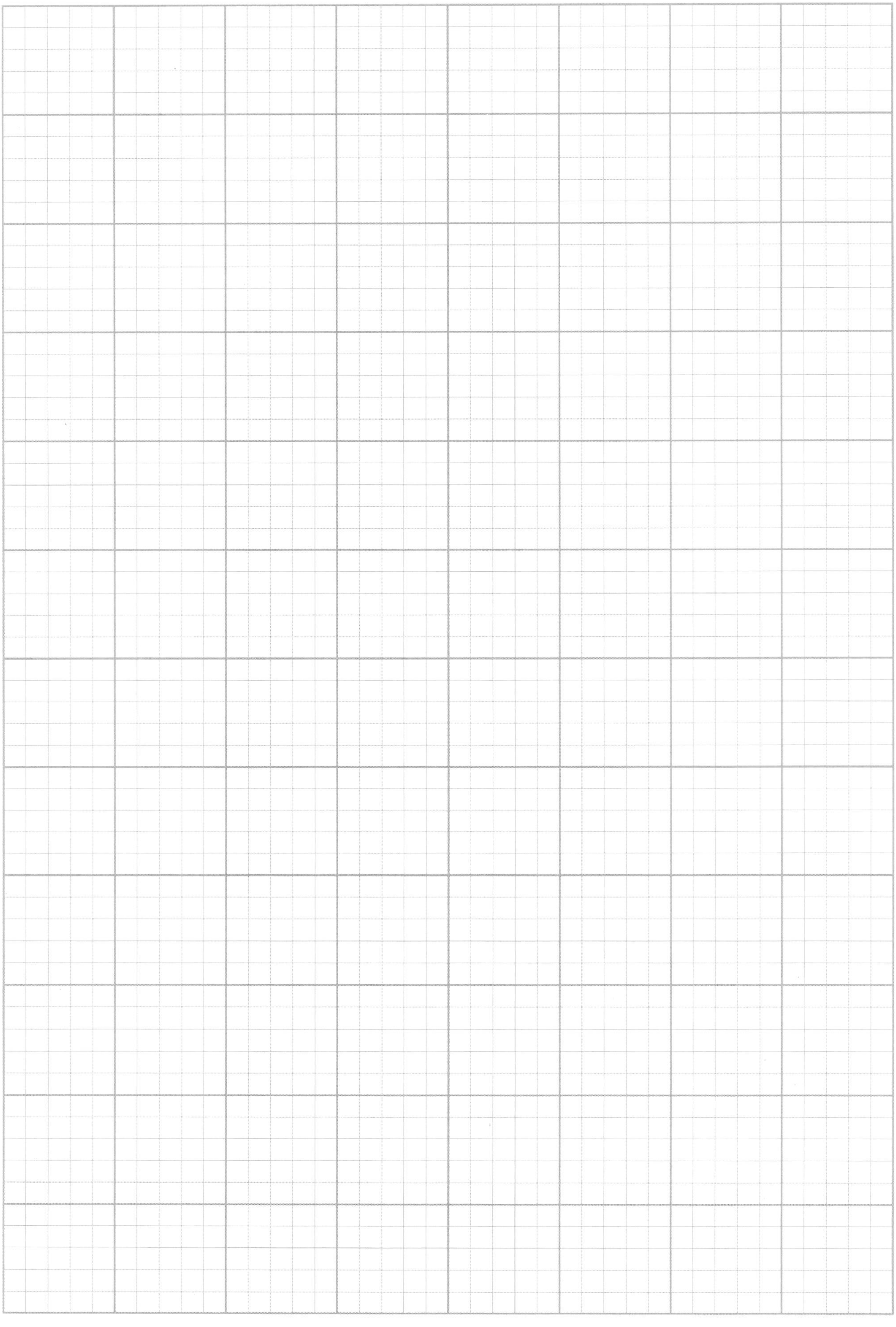

Laboratory Report 35

Name ______ Date ______ Section ______

Deoxyribonuclease

Indicate plates that are positive for the presence of DNase.

Culture # ______ Reaction ______

Culture # ______ Reaction ______

Culture # ______ Reaction ______

Review Questions

1. What are the likely metabolic products formed by the action of deoxyribonuclease?

2. What role does the enzyme play in most microorganisms?

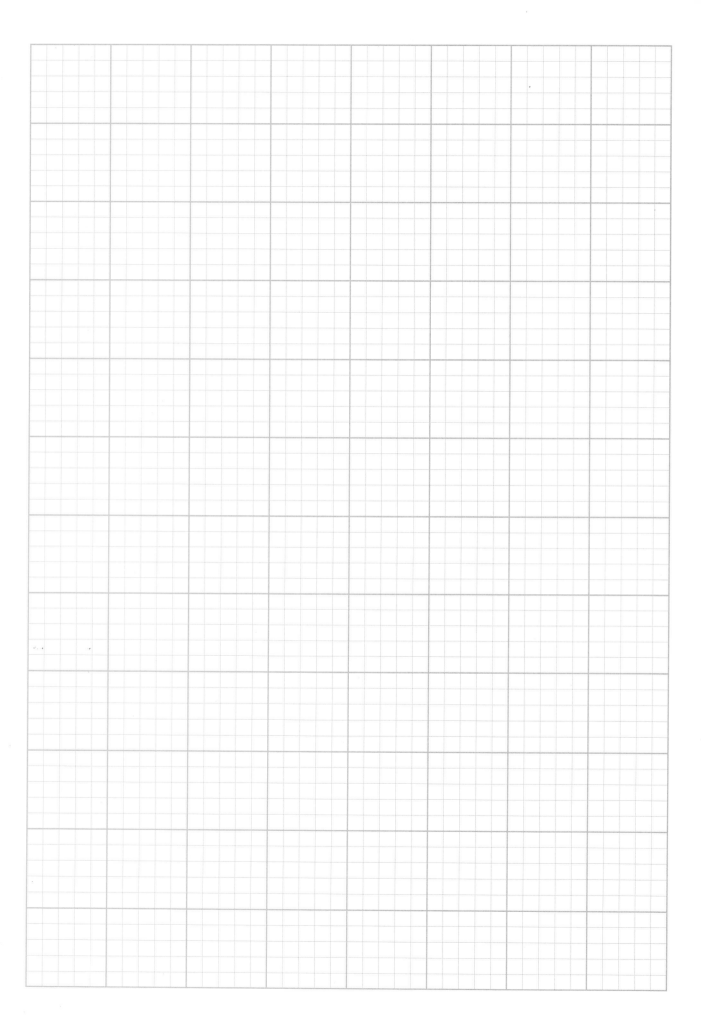

LABORATORY

36 Coagulase Test

Safety Issues

- Caution should be exercised when transferring the coagulase-positive *Staphylococcus* as the organism is pathogenic.

OBJECTIVE

Learn the use of plasma to differentiate coagulase-positive from coagulase-negative staphylococci.

Blood plasma may be clotted by **coagulase** produced by pathogenic staphylococci. The enzyme acts by converting host tissue fibrinogen to fibrin. It is thought that coagulase-positive staphylococci may avoid host defense mechanisms by forming this fibrin clot around them. Such pathogens have a positive-coagulase test (see Color Plate 19).

Materials (per 20 students)

1. Citrated human or rabbit plasma (20 ml)
2. Fresh cultures of environmental organisms from Laboratory 13
3. 37°C water bath
4. Coagulase-positive *Staphylococcus aureus* (ATCC #29213)
5. 13 × 100 mm sterile tubes (40)

Procedure

Caution should be exercised when transferring the coagulase-positive *Staphylococcus* as the organism is pathogenic.

1. Add 0.5 ml of the citrated plasma to two 13 × 100 mm tubes.
2. Inoculate one tube with a loop of fresh culture of a gram-positive environmental isolate and the second tube with the *Staphylococcus* culture.
3. Incubate the tube in a 37°C water bath for 1 to 4 hr.

Any degree of clotting during the time of incubation should be interpreted as a positive result. Since some gram-negative organisms may cause false positives, the coagulase test should be performed only on gram-positive isolates.

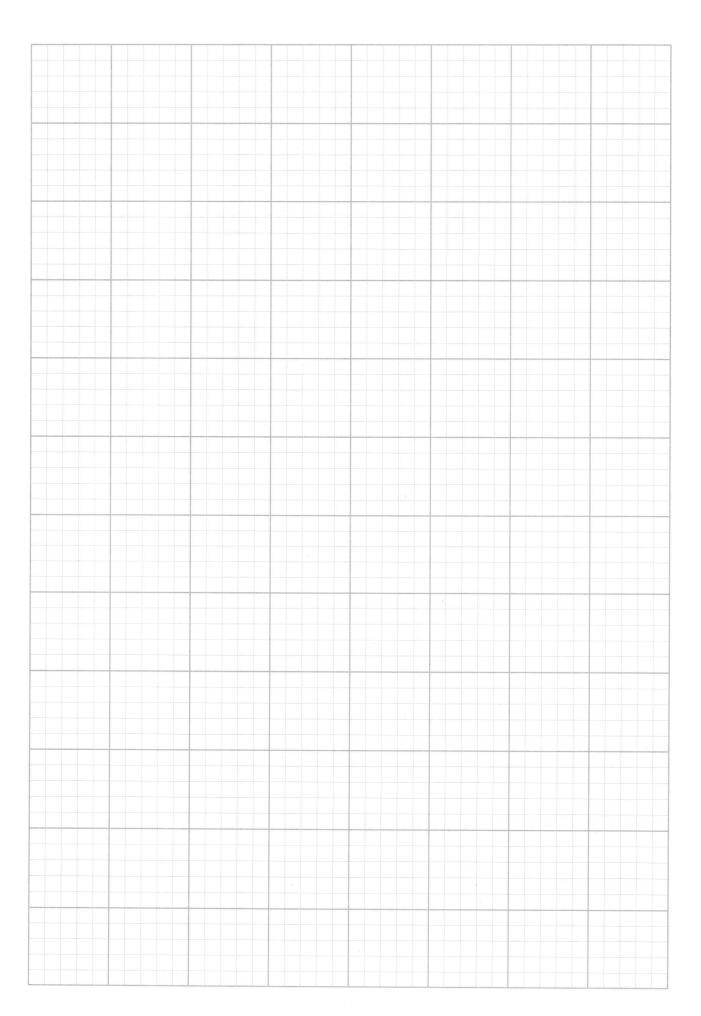

Name Date Section

COAGULASE

Sketch a positive and negative coagulase test.

Culture # ______________ Reaction ______________

Culture # ______________ Reaction ______________

Culture # ______________ Reaction ______________

Review Questions

1. How is the production of the coagulase enzyme related to an organism's pathogenicity? Can coagulase-negative organisms be pathogenic?

2. What role does the coagulase enzyme play in the pathogenicity of a microorganism?

3. What is the appropriate control for this procedure?

LABORATORY

37 Hemolysin Production

Safety Issues

- Beta hemolytic streptococci are known to be pathogenic, so extreme care needs to be exercised when handling these organisms.

OBJECTIVE

Differentiate types of hemolysis present in known and unknown cultures.

Certain members of the genus *Streptococcus* produce exoenzymes that can lead to the destruction of red blood cells. These enzymes are referred to as hemolysins and the destruction is called hemolysis. Pathogenic streptococci produce β-hemolysins, which causes the complete destruction of red blood cells and is exhibited by a clear zone on a blood agar plate. Five-percent sheep red blood is most commonly used, as it is the most sensitive to hemolytic reactions and produces consistent hemolysis. Sheep blood is used because it will not support the growth of *Haemophilus haemolyticus*, whose β-hemolytic reaction could be confused with the β-hemolysis of streptococci.

Materials (per 20 students)

1. Blood agar plates (40)
2. Experimental isolates from Laboratory 13
3. β-hemolytic *Streptococcus*

Procedure

Exercise extreme caution when transferring any β-hemolytic organism, due to its potential pathogenicity.

1. Streak the entire surface of one blood agar plate with a gram-positive experimental isolate.
2. Streak a second plate with the β-hemolytic *Streptococcus*.
3. Incubate the plates at 37°C for 48 hr.
4. Any plate showing a clear area surrounding any colony should be labeled as **β-hemolysis** and scored as a positive reaction (see Color Plate 20).

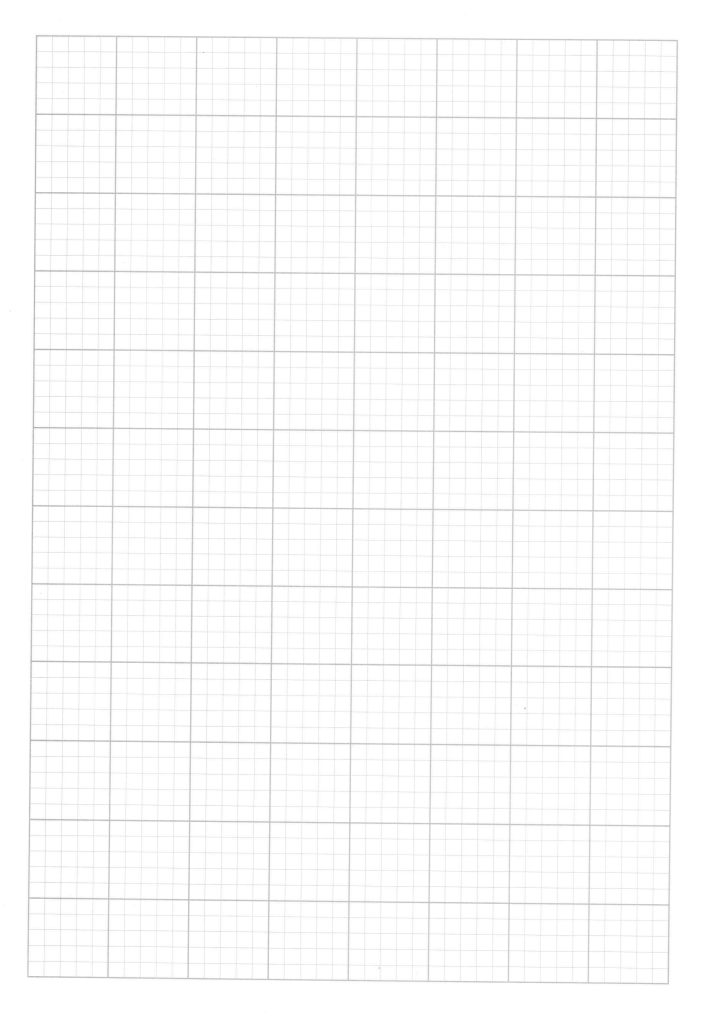

Laboratory Report 37

Name ______ Date ______ Section ______

Hemolysis

Indicate which cultures show positive hemolysis.

Culture # ______ Reaction ______

Culture # ______ Reaction ______

Culture # ______ Reaction ______

Review Questions

1. Is the blood agar used for the hemolysin test a differential or selective medium?

2. How does the differentiation of α and β hemolysis in members of the genus *Streptococcus* aid in determining their likely pathogenicity?

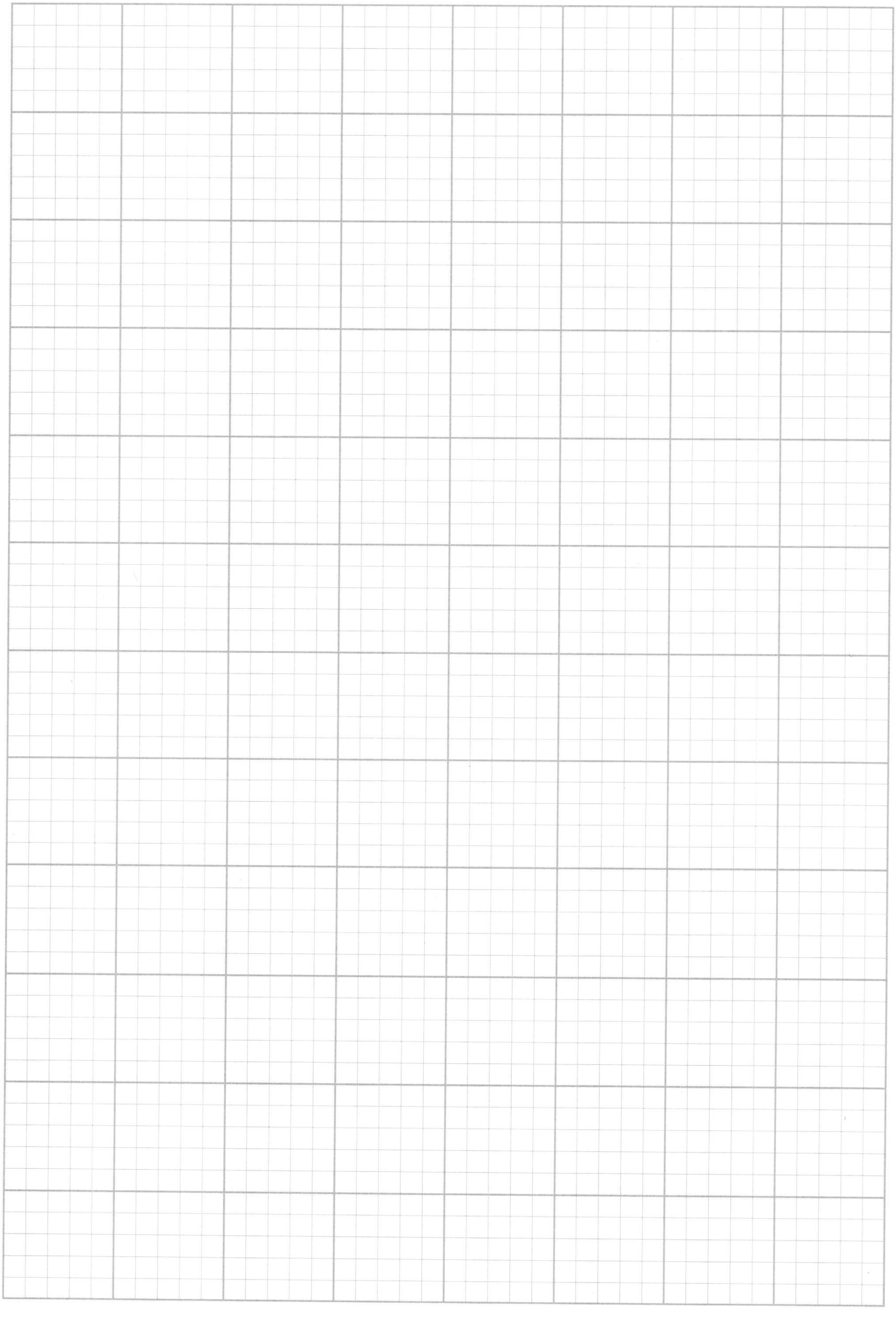

LABORATORY

38 API-20E

Safety Issues

- Many of the reagents are either caustic or suspected carcinogens. Wear gloves when dispensing reagents for nitrate, indole, and VP tests.
- Dispose of plastic strips in a biohazard bag when experiment is completed.

OBJECTIVE

Become familiar with miniaturized strip techniques for the identification of oxidase-negative, gram-negative bacilli.

This test is designed to help identify members of the Family *Enterobacteriaceae*. In it we use a plastic strip with 20 separate compartments containing cupules of dehydrated medium. Upon inoculation of a suspension of an unknown organism, the media become hydrated.

Care needs to be taken in the preparation of the saline suspension of the organisms to be tested. After incubation the reactions are recorded, test reagents added, and all results tabulated. Once all results are recorded, a Profile Number (a seven- or nine-digit number) is obtained. Using a profile recognition system codebook supplied by the manufacturer, one can attempt to determine the identity of the unknown organism.

Materials (per 20 students)

1. Plate or slant of gram-negative, oxidase-negative unknown organism from Laboratory 13
2. Tubes containing 5.0 ml of sterile saline (20)
3. API test strip (20)
4. API incubation chamber (20)
5. Sterile mineral oil (3 bottles)
6. Pasteur pipettes or disposable 1.0-ml pipettes
7. 10% Ferric chloride (40 ml)
8. Barritt's reagents A and B (40 ml)
9. Nitrite test reagents A and B (40 ml)
10. Hydrogen peroxide (40 ml)
11. Zinc dust (5 g)
12. API-20E Analytical Profile Index
13. Result sheet

Procedure

For further detailed information, consult the manufacturer's brochure located in the laboratory. Information about the setup and interpretation of API-20E can also be found in Appendix J.

1. Be sure the test organism is oxidase-negative. The API-20E is designed to identify oxidase-negative rod-shaped bacteria. The next test (API-NFT) will be used to identify the oxidase-positive gram-negative rods.
2. Once it has been verified that the organism is oxidase-negative, prepare a suspension of the organism in the tube of sterile saline by carefully removing a well isolated colony (2–3 mm or larger in diameter) with a sterile inoculating loop and mixing it with the saline.
3. Place the API-20E test strip in the bottom corrugated tray.
4. Using a sterile Pasteur pipette or a disposable 1.0-ml pipette, deposit the suspension of the test organism into the cupules of the test strip. This is best done by holding the tray at an angle and pipetting the suspension down the side wall of the cupule (Figure 38.1). The idea is to add the suspension to the cupules without introducing bubbles. Remember to make sure that the dehydrated media in the bottom of the cupules are hydrated. Slightly underfill the LDC, ADH, ODC, H_2S, and URE compartments (these are shown by underlining on the test strip).
5. Completely fill the cupules marked CIT, VP, and GEL. These cupules are shown in brackets on the test strip. Place tray flat on the lab bench when filling the remainder of the cupules.
6. Add sterile mineral oil to completely fill the ADH, LDC, ODC, H_2S, and URE compartments.
7. Label the flap of the incubation tray with your name and the number of the test organism. Add about 5 ml of tap water to the corrugated bottom tray, being careful not to spill any of the tap water into the inoculated compartments.

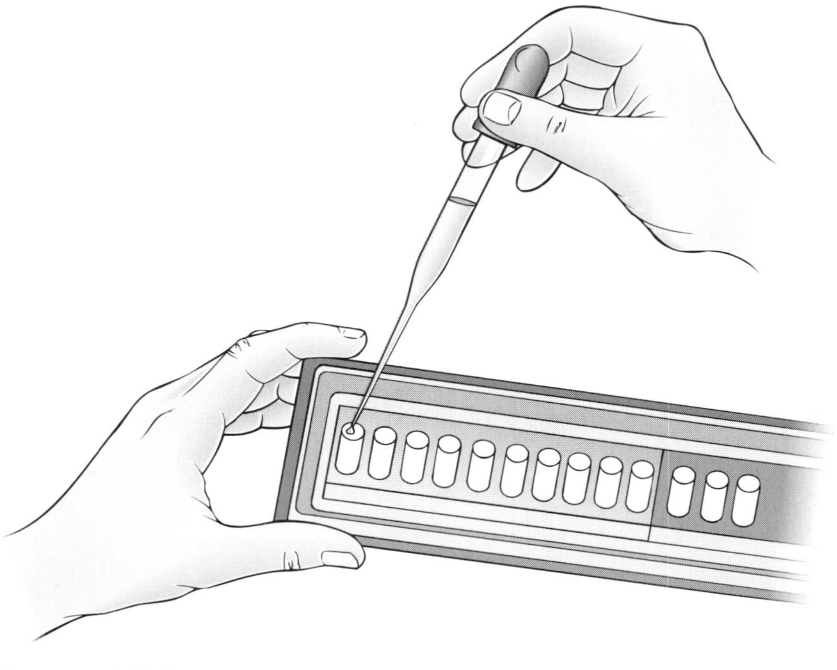

FIGURE 38.1 Inoculation of API-20E strips.

⚠ **Many of the reagents are either caustic or suspected carcinogens. Wear gloves when dispensing reagents for nitrate, indole, and VP tests.**

8. Place the lid on the incubation tray and incubate at 37°C for 18 to 24 hr.
9. Using the manufacturer's instructions, read all reactions that do not require the addition of a test reagent. Record your observations.
10. Add the test reagents to the appropriate compartments:
    (a) Add one drop of Kovac's reagent to IND. Look for a red positive ring. This reaction should occur within a few minutes.
    (b) Add one drop of Barritt's A and B to VP. A positive reaction here may take up to 10 min. A positive test is dark pink or red.
    (c) Add one drop of 10% ferric chloride to TDA.
    (d) Examine the GLU cupule for presence of bubbles. This indicates the reduction of nitrate to nitrogen gas. Add two drops of NIT 1 and NIT 2 GLU. A positive test is the development of a red color within 2 to 3 min. If the test is negative, add a small amount of zinc dust. A pink-orange color after 10 min indicates that nitrate is present and therefore reduction did not occur.
    (e) Add one drop of hydrogen peroxide to MAN, INO, and SOR cupules. If gas bubbles appear immediately, the presence of catalase is indicated. Note all color changes and compare with negative and positive test results shown in Color Plate 21.
11. Record results on the chart provided and determine the seven- or nine-digit profile index number.
12. Identify the unknown by looking up the profile number in the API-20E Analytical Profile Index.

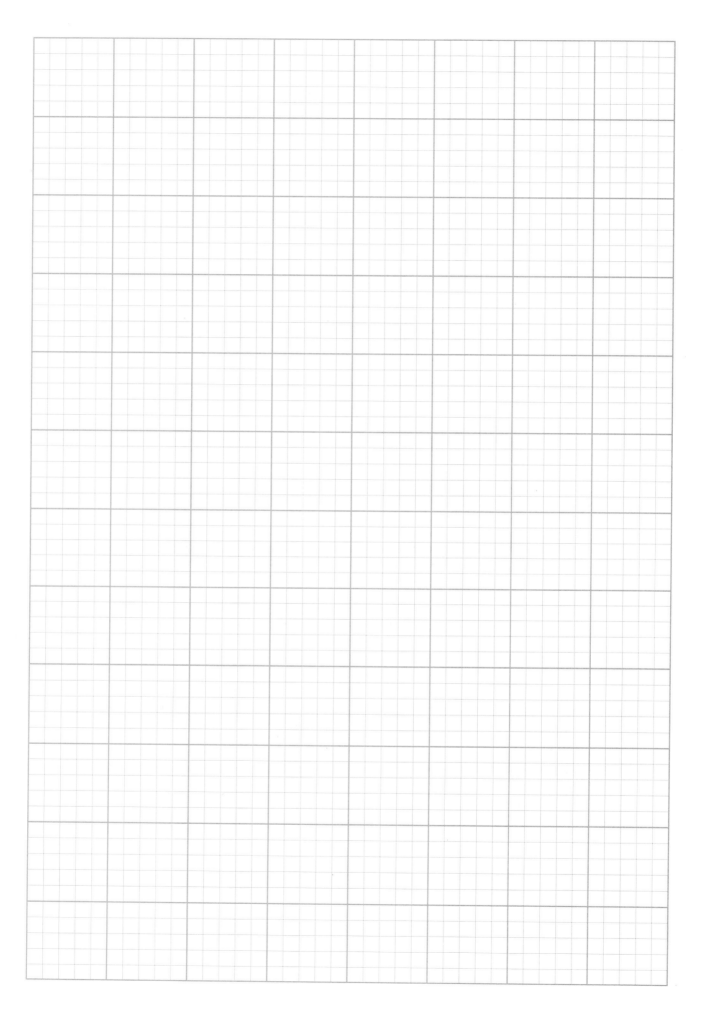

Laboratory Report 38

Name Date Section

API-20E Data Sheet

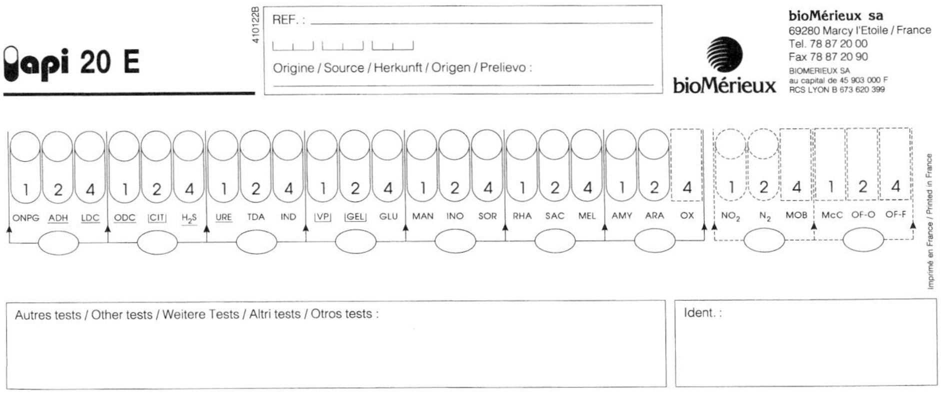

Review Questions

1. What is the function of the mineral oil in this test?

2. What is the purpose of the water in the tray?

3. What should be done if no positive reactions are visible after incubating the test strip for 24 hr?

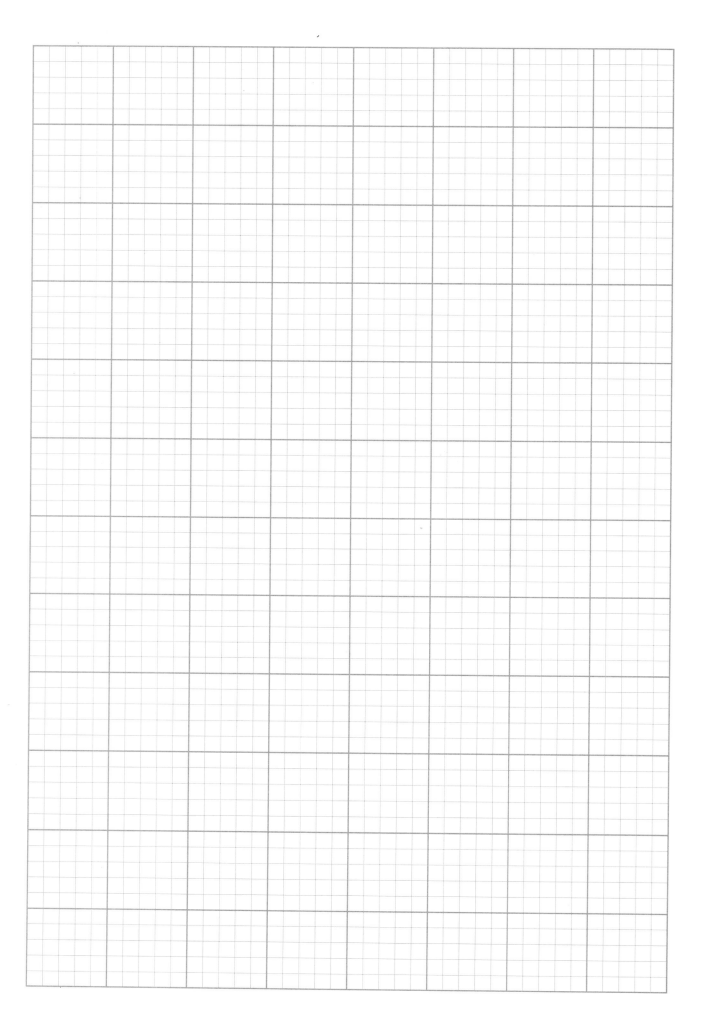

LABORATORY

39 API-NFT

Safety Issues

- NIT reagent contains caustic material. Wear gloves when dispensing.

OBJECTIVE

Learn the use of the 20-cupule technique for the identification of oxidase-positive, gram-negative, non-fermentative bacteria.

This test is designed to identify oxidase-positive, gram-negative, rod-shaped bacteria. It provides a miniaturized method consisting of 20 biochemical and assimilation tests for the identification of gram-negative, non-fermentative bacteria. The rapid NFT strip contains 20 cupules of dehydrated substrates for the demonstration of either enzymatic activity or assimilation of carbon sources (see Color Plate 22). Similar to the API-20E, the substrates are reconstituted with the addition of the bacterial suspension. The microorganisms metabolize the contents of the cupules and, for the enzymatic tests, yield metabolic end products that produce a color change. The assimilation tests are based upon the observation of microbial growth in the presence of single sources of carbon.

Materials (per 20 students)

1. Experimental organisms (oxidase-positive, gram-negative rods) from Laboratory 13
2. API-NFT test strips (20)
3. Incubation chamber with lid (20)
4. Ampules containing AUX medium (20)
5. Tubes containing 2 ml of sterile saline (20)
6. Sterile mineral oil (3 bottles)
7. NIT reagent 1 (40 ml)
8. NIT reagent 2 (40 ml)
9. TRP reagent (Light-sensitive reagent. Store in dark and do not leave on lab bench for an extended period of time.) (10 ml)
10. Zinc dust (5 g)
11. 0.5 McFarland turbidity standard
12. Rapid NFT Identification Codebook
13. Sterile Pasteur or disposable 1.0-ml pipettes (4 canisters)

Procedure

Information concerning interpretation of API-NFT can be found in Appendix K.

1. Pick one to four isolated colonies of a pure culture from an agar plate and emulsify in a tube of sterile saline. Carefully homogenize the sus-

pension and adjust the turbidity to a 0.5 McFarland barium sulfate turbidity standard.

2. Place the NFT strip in the tray provided.
3. Inoculate the enzymatic tests, that is, tubes NO_3 through PNPG (cupules with no colored lines), by delivering the suspension down the side wall of the cupule as you did in the API-20E procedure.
4. Open an ampule of AUX medium by applying pressure with your thumb on the plastic cap, and add four drops of the saline suspension of the organism. Mix well, avoiding the formation of bubbles.

⚠ **Make sure to use protective plastic cap when breaking ampule to prevent any cuts to your fingers.**

5. Inoculate the assimilation tests, that is, cupules GLU through PAC (cupules with colored lines), by holding the tray on an angle and pipetting the suspension down the side wall of the cupule. Lay the tray flat on the lab bench and fill the cupules entirely so that a flat liquid surface without a miniscus is obtained. The assimilation medium should fill the tube and the cupule. Tubes insufficiently filled or overfilled may be the source of inaccurate results.
6. Add sterile mineral oil to the GLU, ADH, and URE cupules until there is a convex meniscus on the oil.
7. Add approximately 5 ml of tap water to the bottom of the tray.
8. Cover the tray with the lid and incubate for 24 hr at 29 to 31°C.
9. After incubation, read the color changes in the cupules not requiring the addition of reagents, following the manufacturer's instructions.
10. Add test reagents to the following cupules:

⚠ **NIT reagent contains caustic material. Wear gloves when dispensing.**

    (a) Add 1 drop each of the NIT 1 and NIT 2 reagents to the NO_3 test cupule and within 5 min a positive test will be indicated by the formation of a red color. If the reaction is negative (colorless), it is important to determine whether nitrogen gas is being produced. Add 2 to 3 mg of zinc dust to the NO_3 cupule. Read the reaction within 5 min. If the reaction remains colorless, the reaction is positive. If it turns pink, it is negative. Recall what this means. The pink color is due to the remaining presence of unreduced nitrates.
    (b) Add 1 drop of the TRP reagent to the TRP cupule and within 5 min observe any color changes. A red color indicates the presence of indole being formed by the action of the enzyme tryptophanase.
11. Assimilation tests
    (a) Look for growth. If there is visible growth (opaque), the reaction is positive. If there is no visible growth (clear), the reaction is negative.
    (b) If the assimilation tests are very weak (clear) and a good identification is difficult (profile cannot be found in the codebook), the test strip may be reincubated for another 24 hr. Immediately cover the NO_3 and TRP cupules with sterile mineral oil until a convex meniscus is formed. Always note on the report sheet the NO_3, TRP, and GLU test results after 24 hr. Do not read after 48 hr.
    (c) Using the report sheet, calculate the seven-digit profile number. In each group of three tests, the first test is assigned a value of 1, the second test a value of 2, and the third a value of 4. The profile number is acquired by summing all of the values.
    (d) Identification is made by using the Identification Codebook. Remember that the oxidase test is the 21st test and needs to be included in your calculation of the profile number.

Laboratory Report 39

Name ____________________ Date ____________ Section ____________

API-NFT Data Sheet

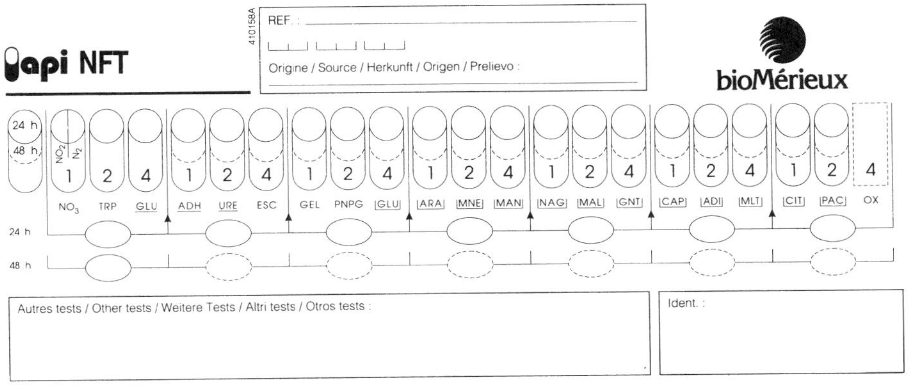

Review Question

How are API-20E and API-NFT alike? Different?

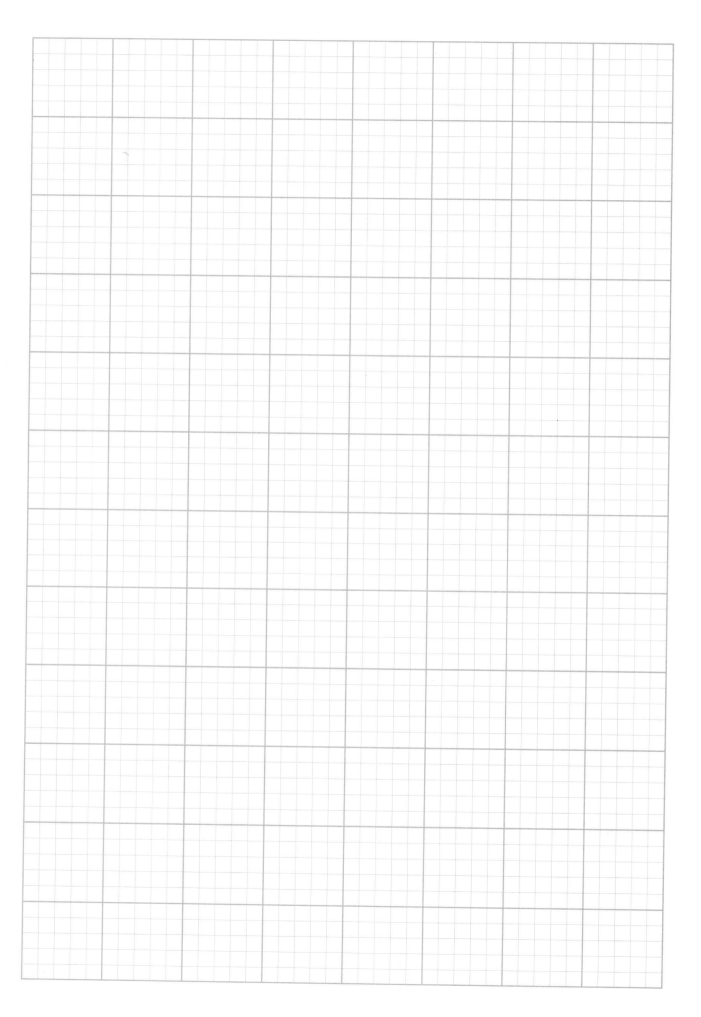

LABORATORY

40 Enterotube II*

Safety Issues

- Kovac's reagent and VP reagents contain either acids or toxic material. Wear gloves when dispensing.
- Exercise caution not to jab yourself with the needle of the syringe.
- After recording results dispose of Enterotube II in a biohazard bag.

OBJECTIVE

Become proficient in use and interpretation of the miniaturized test designed for identification of oxidase-negative, gram-negative bacilli.

This system was developed as a miniaturized test system consisting of 12 different media compartments for the identification of gram-negative, oxidase-negative rods belonging to the Family *Enterobacteriaceae.* If an oxidase-positive rod needs to be identified, use the Oxi/Ferm II test in Laboratory 41. Examine the compartments of the Enterotube. Many of the same biochemical tests should be familiar at this point. Repeating these tests enables one to compare the conventional tube tests with a number of the computerized, miniaturized systems available. Observe that a number of the compartments have openings for air access. These reactions are primarily aerobic. Others have a layer of paraffin over the media and these are anaerobic reactions. An inoculating wire extends through all of the compartments. To inoculate the compartments, one merely picks a portion of a colony on the wire from an agar plate and pulls the wire through each of the compartments (Figure 40.1). This single-step procedure is one of the easiest of the miniaturized systems. *Note:* Since you are required to obtain the organisms on the end of the wire, it behooves you to prepare your culture on an agar plate. It will be difficult to obtain microorganisms from an agar slant. What can you do if you have a fresh slant culture and do not wish to reinoculate a culture for another day? How could you still perform the experiment?

Materials (per 20 students)

1. Experimental organism (gram-negative, oxidase-negative rod) from Laboratory 13
2. Enterotube II (20)
3. Kovac's reagent (40 ml)
4. Barritt's 1 and 2 reagents (40 ml)

*Courtesy of Becton Dickinson and Company

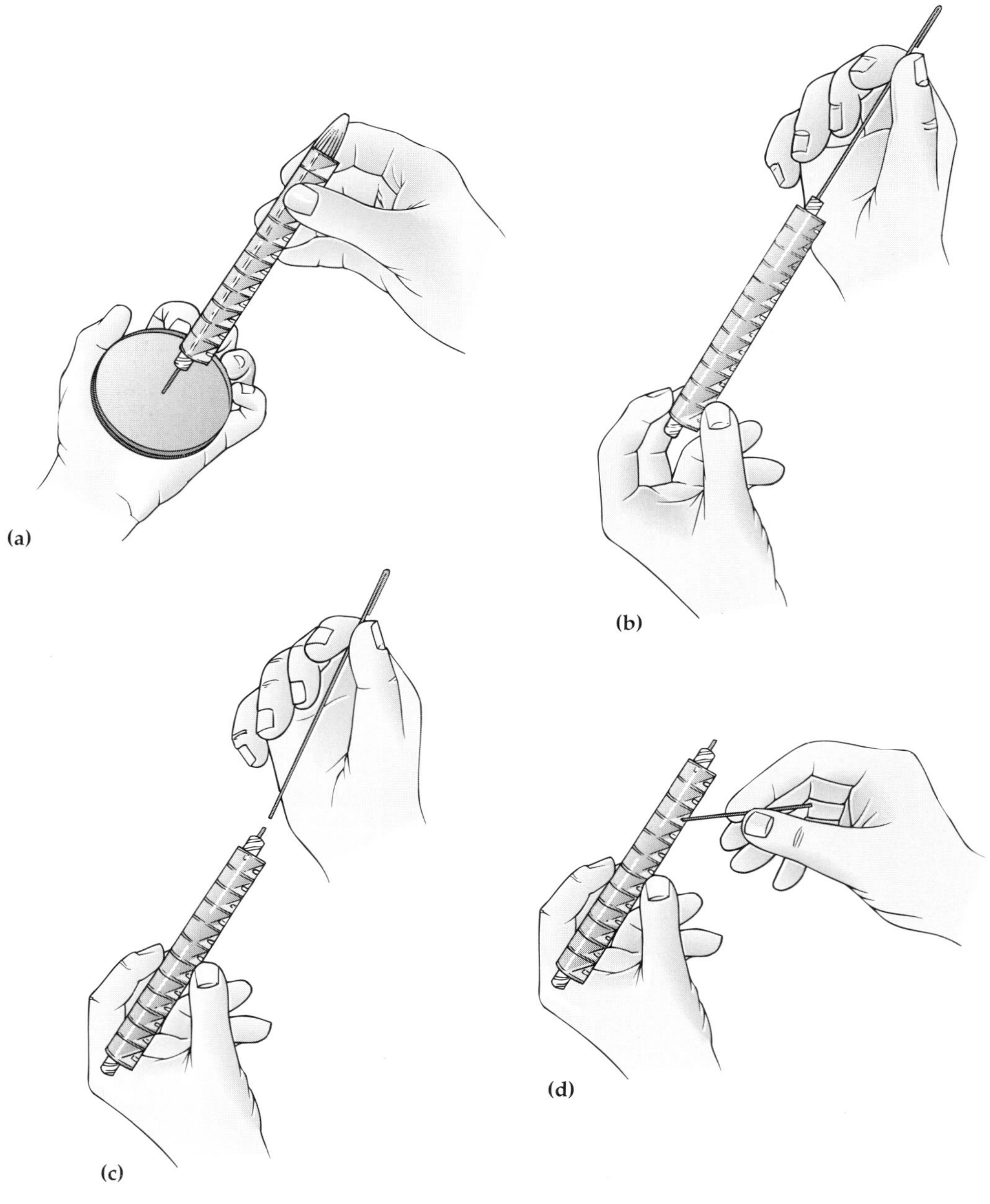

FIGURE 40.1 Inoculation of Enterotube II.

Procedure

Interpretative information on Enterotube II is found in Appendix L.

1. Write your name or initials and the culture number on the tube.
2. Unscrew both caps of the Enterotube II.
3. Grasp the wire-handle end and insert the wire end into a colony on an agar plate. A visible inoculum should be seen at the tip and the side of the wire. Do not touch agar with the wire.
4. Transfer the organisms by pulling the wire in a rotating fashion through all 12 compartments. **Be careful not to pull the wire out of the tube completely.**
5. Reinsert the wire by using a turning motion through all 12 compartments, until the notch on the wire is aligned with the opening of the

tube. Break the wire by giving it a gentle bend. The task is made easier since a notch exists in the wire where you are to bend it. This helps to maintain an anaerobic environment. The tip of the wire should be seen in the citrate compartment.

6. Punch holes with the broken portion of the wire in the covered air inlets in the last eight compartments.
7. Replace both caps.
8. Incubate the Enterotube II at 37°C for 18 to 24 hr. Be sure the tube is placed with its flat side down in the incubator. Interpret and record all reactions except for the indole and VP tests (see Color Plate 23).

⚠ Kovac's reagent and VP reagents contain either acids or toxic material. Wear gloves when dispensing.

9. Perform the indole test by adding Kovac's reagent to the H_2S/indole compartment. This can be accomplished by using a syringe and needle. A positive reaction is indicated by the development of a red color within 10 sec.
10. For the VP test, use a syringe and needles to add two drops of 20% KOH containing 0.3% creatine and three drops of alpha-naphthol/ethanol to the VP compartment. A positive test is indicated by the formation of a red color within 20 min.

⚠ Exercise caution when using a needle to dispense the toxic VP reagents into the Enterotube.

11. After reading, discard the Enterotube into a biohazard bag for autoclaving.

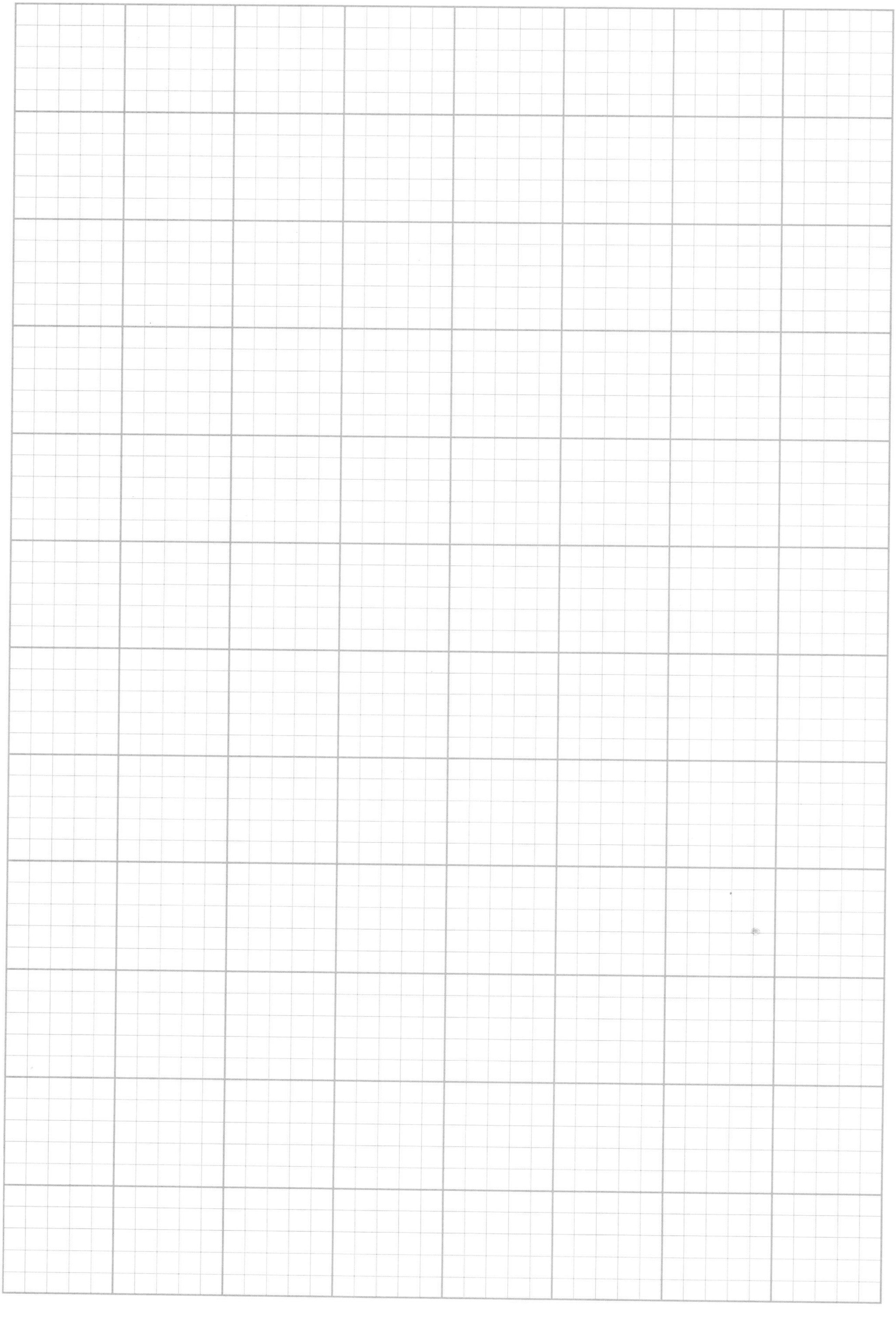

Laboratory Report 40

Name   Date   Section

Enterotube II Data Sheet

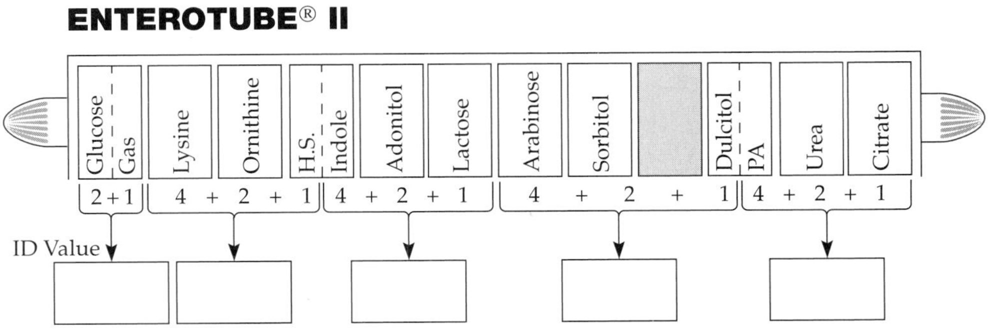

Culture Number or Patient Name   Date   Organism Identified

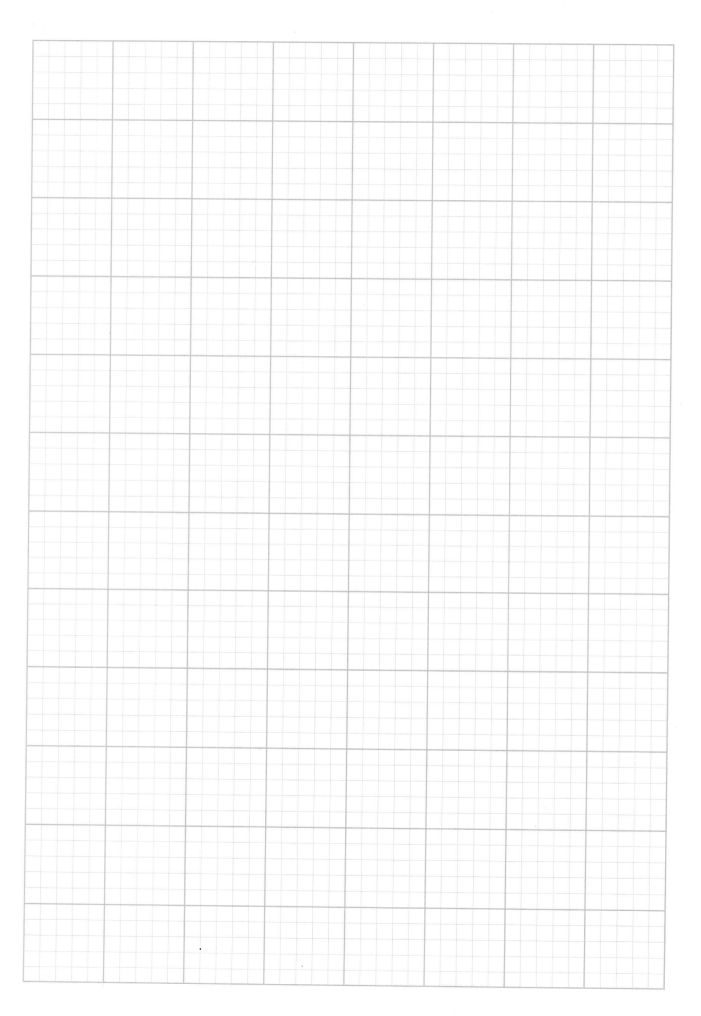

LABORATORY

41 Oxi/Ferm Tube II*

Safety Issues

- Kovac's reagent contains concentrated HCl and dimethyl-naphthylamine, a suspected carcinogen. Wear gloves when dispensing.
- Be careful not to stick your hands or fingers when using the syringe and needle to inject compartments.

OBJECTIVE

Learn the use of a 20-media compartment tube for the rapid identification of oxidase-negative, gram-negative bacilli.

Oxi/Ferm II is a self-contained, sterile, compartmented tube containing 12 different media. It operates in a fashion similar to that of the Enterotube II system, except that it is used to identify the oxidase-positive, oxidative-fermentative gram-negative rods.

Materials (per 20 students)

1. Oxidase-positive gram-negative rods from Laboratory 13
2. Oxi/Ferm Tube II (20)
3. Kovac's reagent (40 ml)

Procedure

1. Remove both caps from the tube.
2. The tip of the needle is under the white cap. Pick a well-isolated colony from an agar plate onto the tip of the needle.
3. Inoculate the Oxi/Ferm tube by twisting the needle and drawing it through the compartments, using a turning motion.
4. Reinsert the needle until the notch on the wire is aligned with the opening of the tube. The tip of the needle should be seen in the citrate compartment.
5. Break the needle and replace the end caps.
6. With the broken portion of the needle, punch holes through the foil covering the air inlets.
7. Incubate at 35 to 37°C for 48 hr with the Oxi/Ferm tube lying on its flat surface.

*Courtesy of Becton Dickinson and Company

8. Interpret the color reactions according to the manufacturer's instructions.
9. Perform the indole test by holding the Oxi/Ferm tube with the anaerobic glucose compartment pointing downward. Add three to four drops of Kovac's reagent through the plastic film of the sucrose/indole compartment, using a needle and syringe. A positive test is indicated by the development of a red color in the added reagent.
10. Calculate the five-digit identification code. Use the Oxi/Ferm II Biocode Manual to identify your unknown.

Laboratory Report

Name Date Section

41

Oxi/Ferm II Data Sheet

<Oxi/Ferm Tube> II Roche

Trade Mark

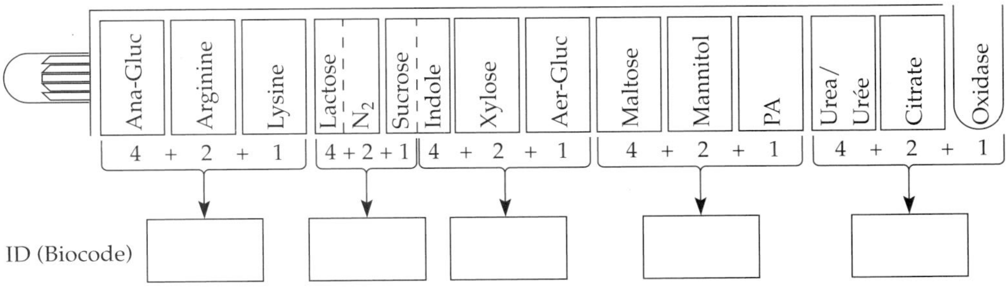

Patient Date

Review Question

How do the results of the Oxi/Ferm test compare with the results from the API-NFT test (Laboratory 39)?

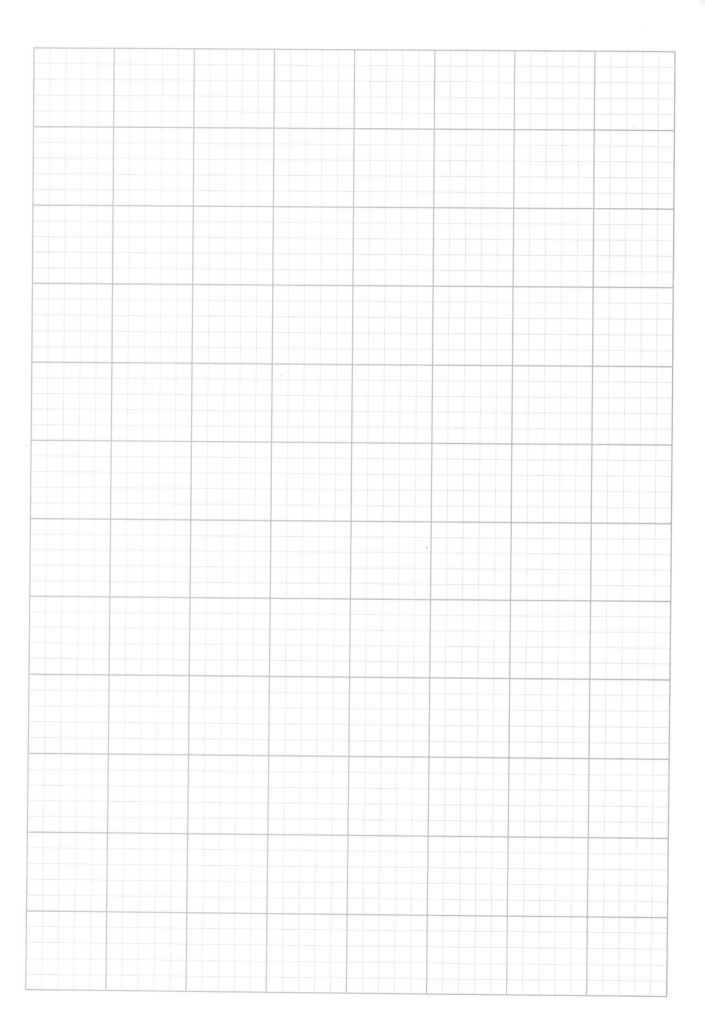

LABORATORY

42 Biolog GN

Safety Issues

- If drops of any culture fall on the lab bench, treat with disinfectant solution.
- Eject pipette tips into jar of disinfectant after use.
- After reading, dispose of panel in a biohazard bag.

OBJECTIVES

1. Learn the use of a micropipette.
2. Learn the procedure for the inoculation and interpretation of a 96-well microplate identification method.

The Biolog identification system uses 96-well microtiter plates with 95 wells containing different carbon sources. One well (A-1) contains no carbon source and serves as a negative control or reference well (see Color Plate 24). All of the necessary nutrients are prefilled and dried into the 96 wells of the plates. The test measures the ability of a microorganism to utilize (oxidize) the 95 carbon sources. The MicroPlates utilize Biolog's patented redox chemistry to colorimetrically indicate the respiration of live-cell suspensions. All of the wells start out colorless when inoculated. In wells that contain a chemical that is oxidized, there is a burst of respiration and the cells reduce a tetrazolium dye, forming a purple color. Other wells remain colorless, as does the reference well with no carbon source. The Biolog Corporation currently markets a number of identification systems. In this experiment you will run the system used to identify gram-negative bacteria.

Materials (per 20 students)

1. Unknown gram-negative organism from Laboratory 13
2. Biolog GN MicroPlates (20)
3. 25 ml of sterile saline (0.85%) in a sterile disposable plastic tube (20)
4. Sterile cotton or Dacron swabs (20)
5. Spectrophotometer or colorimeter
6. Micropipette with sterile disposable tips (5)

Procedure

1. Grow an organism to be identified on a nutrient medium that promotes vigorous growth. The cells should be in an active phase of growth.

2. Remove the MicroPlates from the refrigerator and allow to warm to room or incubator temperature.
3. Use saline to blank a spectrophotometer set at 590 nm.
4. Care should be taken to gently remove cells from the agar plate. Wet a sterile swab with sterile saline and roll it over the colonies on the plate. It is most important not to remove any nutrients from the surface of the growth medium. You do not wish to carry over any nutrients to your microtiter identification plate.
5. Immerse the swab in the saline solution and gently remove the organisms. Mix gently to obtain a uniform suspension. If clumps are evident, let the tube stand for several minutes and allow them to settle out.
6. Using the spectrophotometer, adjust the cell suspension to an absorbance of 0.127 to 0.148. ***Note:* This adjustment is critical.**
7. Inoculate the MicroPlates within 10 min as cells may lose metabolic activity if held in the saline solution for a long period of time.
8. Label the MicroPlate with your name and the number of the organism.
9. Using a micropipette, deliver **exactly** 150 microliters to each well. If you have a multichannel micropipette available, the inoculation process will be speeded up immeasurably. If not available, be patient.
10. Place the lid on the microplate and incubate at the appropriate temperature for your organism. Try 28 to 30°C for most experimental isolates.
11. Incubate plates for 4 hr or 16 to 24 hr.
12. Many microorganisms give results in 4 hr and some become unreadable at 24 hr, so carefully observe your plates after 4 hr.
13. Read the color development in your plate and score results on test sheet.

Laboratory Report 42

Name Date Section

Biolog GN Report Sheet

	1	2	3	4	5	6	7	8	9	10	11	12
A												
B												
C												
D												
E												
F												
G												
H												

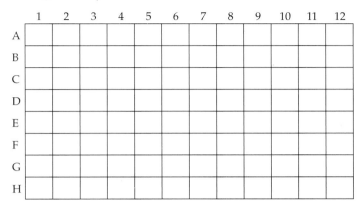

Review Question

What is the significance of a positive reaction in the A-1 well in the Biolog test?

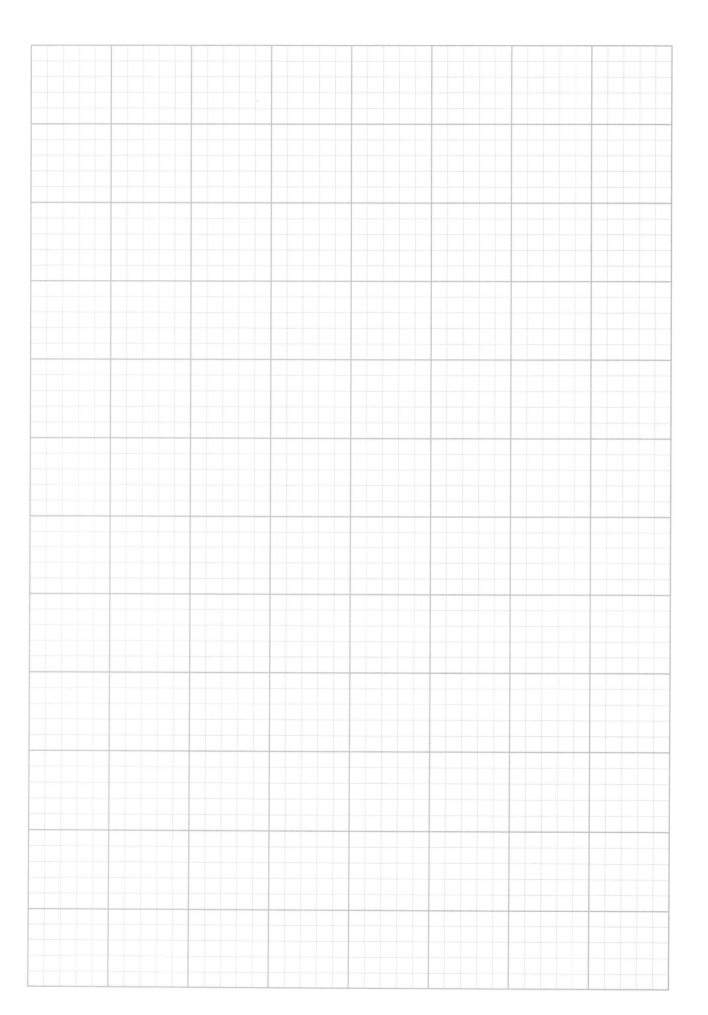

LABORATORY

43 Biolog GP

Safety Issues

- If drops of any culture fall on the lab bench, treat with disinfectant solution.
- Eject pipette tips into jar of disinfectant after use.
- After reading, dispose of panel in a biohazard bag.

OBJECTIVE

Become proficient in using a 96-well microtiter plate identification method for gram-positive bacteria.

The Biolog GP identification system is similar to the Biolog GN system in that it uses a microtiter plate with 96 wells prefilled with nutrients and biochemicals. The preparation of the bacterial inoculum differs slightly from the GN technique.

Materials (per 20 students)

1. Unknown gram-positive organisms from Laboratory 13
2. 25 ml of sterile saline (0.85%) in sterile disposable tube (prewarmed to 28–35°C).
3. Biolog GP MicroPlates (20)
4. Sterile cotton-tipped swabs (20)
5. Spectrophotometer or colorimeter
6. Micropipettes (4)
7. Sterile disposable micropipette tips
8. BUGM-based culture plates (20)

Procedure

1. Perform a Gram stain on the experimental isolate to verify the gram-positivity of the organism and its morphology.
2. The choice of growth medium is important when performing the GP test. Most of the gram-positive organisms can be identified by growing the organisms on Biolog Universal Growth Medium (BUGM) with 5% sheep blood (BUGM + B). Spore-forming gram-positive rods are prone to sporulation unless they are grown with an excess of nutrient. These rods should be cultured on BUGM + glucose (1%) (BUGM + G).
3. The inoculum should be freshly grown. Usually the inoculum should be grown for 4 to 18 hr on the appropriate medium.

4. Remove MicroPlates from the refrigerator and warm to 28 to 35°C.
5. Use saline to blank a spectrophotometer set at 590 nm.
6. Remove cells from the agar growth medium by rolling a saline-wetted cotton swab over the colonies. It is important not to carry over any nutrients from the agar plate into the cell suspension. Gently roll the swab, do not streak or slide it over the surface of the plate.
7. Immerse the swab in the sterile saline solution in the sterile disposable tube. Gently twirl the swab and press it against the side of the tube to break up clumps and to release cells into the saline.
8. Using the spectrophotometer, adjust the cell suspension to an absorbance of **0.227 to 0.273**. *Note:* This adjustment is critical. The approximate cell density should be about 4.5×10^8 cells/ml.
9. Inoculate the cell suspension into the GP MicroPlate within 10 min of preparing the cell suspension.
10. Label the MicroPlate with your name and the number of the culture tested.
11. Using a micropipette, deliver **exactly** 150 μl to each well. (This task is made easier if an eight-channel repeating pipette is available.) Be sure not to carry over liquids from one well to another or to splash chemicals from one well to another.
12. Cover the MicroPlate with its lid and incubate at the appropriate temperature. If the experimental isolate is an environmental isolate, incubate at 30°C. If of human, veterinary, or food origin, incubate at 35°C.
13. Incubate the plates for 4 to 6 hr or overnight (16–24 hr).
14. It is advisable to provide a source of moisture in the incubator to prevent the drying of the outer wells of the MicroPlate. This can be accomplished by placing a pan of distilled water on the bottom shelf of the incubator.
15. Read the plates after 4 hr. Many gram-positive organisms give an adequate pattern after 4 to 6 hr, and a few species become unreadable if incubated overnight.
16. The development of a positive (purple) color is referenced against the negative control well, A-1. All wells visually resembling the A-1 well should be scored as "negative" (−), and all wells with a purple color should be scored as "positive" (+). Wells with extremely faint color are best scored as "borderline" (\).
17. Use the Microlog computer software program to interpret the readings.

Laboratory Report 43

Name Date Section

Biolog GP Data Sheet

	1	2	3	4	5	6	7	8	9	10	11	12
A												
B												
C												
D												
E												
F												
G												
H												

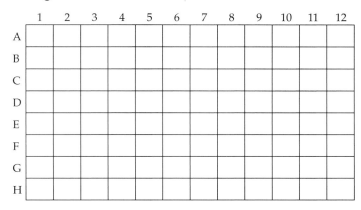

Review Question

How do your results compare when attempting to make an identification of one of your gram-negative environmental isolates using (a) API-20E, (b) Enterotube II, and (c) Biolog. How do you explain any discrepancies in the results?

UNIT VI

Effects of Chemical and Physical Agents

Microorganisms can be inhibited or killed by a variety of chemical agents and by changes in their physical environment. The experiments in this section examine the effectiveness of various antibiotics on bacterial growth, the use of disinfectants and antiseptics to control bacterial populations, and the lethal effect of ultraviolet light.

The control of microorganisms is critical in the treatment of infectious disease. Identification of an effective antibiotic and its proper dosage is of utmost clinical significance.

Evaluation of disinfectants used to kill or inhibit microorganisms on inanimate objects is important in order to prevent accidental infection from a contaminated instrument or surface. Antiseptics are in widespread use to prevent the proliferation of microorganisms on tissue surfaces. These chemicals can either kill or inhibit microorganisms, but generally are not as toxic as disinfectants since they are normally applied to tissue surfaces. Biocides are chemical agents used to kill bacteria in commercial products (e.g., cosmetics) or in industrial systems (e.g., cooling towers). They are evaluated in a manner similar to that used for disinfectants.

Ultraviolet light is lethal to microorganisms, but the main drawback is that it is not able to penetrate some substances, such as glass and turbid water. UV radiation is used in biological safety cabinets, and in rooms to sterilize air. Commercial UV units are installed for water treatment to kill microorganisms that may be present in water. Small units are available to purify drinking water and are usually mounted under sinks. Industrial UV systems are used to treat large volumes of water that have become contaminated in manufacturing processes. The intensity and wavelength of light, time of exposure, and thickness of water are important considerations in any application where UV light is used to control the presence of microorganisms in water.

LABORATORY

44 Antimicrobial Testing. The Kirby–Bauer Method (Filter-Paper Disk Method)

Safety Issues

- After inoculation of plates, the swabs should be placed in a tray or beaker of disinfectant.

OBJECTIVE

Understand the use of the Kirby–Bauer disk technique for measurement of antibiotic sensitivities.

Chemicals used in the treatment of disease are called chemotherapeutic agents. Antibiotics historically were defined as chemicals produced by specific microorganisms that would inhibit or kill other microorganisms. Antibiotics are now produced in a number of ways:

1. Naturally, through the action of a bacterial or fungal species.
2. Semi-synthetically, by a combination of natural products with added chemical synthesis.
3. Synthetically, through chemical synthesis alone.

Most antibiotics are produced or synthesized through the action of a small group of bacterial and fungal genera. *Penicillium* and *Cephalosporium* are the main fungal genera, with *Bacillus* and *Streptomyces* being the predominant bacterial genera involved.

Before administering an antibiotic, it is essential to check the susceptibility of the microorganism to the agent prescribed. There has been an emergence of significant antibiotic resistance among microorganisms. To minimize the selection and spread of resistant strains, care must be taken in the use and administration of antibiotics.

Several methods are employed to measure the sensitivity of bacteria to antibiotics; the one most commonly used is the filter-paper disk method. In this method an organism is distributed evenly over the surface of an agar plate. Then filter-paper disks impregnated with the appropriate antibiotics are placed on the agar surface. The impregnated disks are commercially available and greatly facilitate testing.

1. Saline suspensions of the test organisms, adjusted to a 0.5 McFarland standard (Absorbence of 0.10 measured at 600 nm) (5)

TABLE 44.1 Code Sheet for Antibiotics

BBL					
AM-10	Ampicillin	E-15	Erythromycin	OA-2	Oxolinic Acid
AN-10	Amikacin	FM-300	Nitrofurantoin	OL-15	Oleandomycin
B-10	Bacitracin	GM-10	Gentamycin	P-10	Penicillin G
C-30	Chloramphenicol	K-30	Kanamycin	PB-300	Polymyxin B
CB-50	Carbenicillin	L-2	Lincomycin	RA-5	Rifampicin
CB-100	Carbenicillin	N-30	Neomycin	S-10	Streptomycin
CC-2	Clindamycin	NA-30	Nalidixic acid	SSS-25	Triple sulfonamides
CF-30	Cephalothin	NB-30	Novobiocin	Te-30	Tetracycline
CL-10	Colistin	NF-1	Nafcillin	Va-30	Vancomycin
DP-5	Methicillin	NN-10	Tobramycin		
Difco					
AM10	Ampicillin	GP2	Penicillin	OFX5	Ofloxacin
B10	Bacitracin	GP10	Penicillin	PB300	Polymyxin B
CB100	Carbenicillin	K30	Kanamycin	RA5	Rifampicin
C30	Chloramphenicol	L2	Lincomycin	S10	Streptomycin
CC2	Clindamycin	LOR30	Loracarbef	SAM20	Ampicillin/Sulbactam
CEC	Cefactor	ME5	Methicillin	SD300	Sulfadiazine
CID30	Cefonicid	MOX30	Moxalactam	SSS300	Triple Sulfa
CIP5	Ciprofloxacin	N5	Neomycin	SXT25	SxT
CR30	Cephalothin	NA30	Nalidixic Acid	T30	Oxytetracycline
CRO30	Ceftriaxone	NB30	Novobiocin	TE30	Tetracycline
DO30	Doxycycline	NF1	Nafcillin	TM10	Tobramycin
E15	Erythromycin	NY100	Nystatin	TMP5	Trimethoprim
GM10	Gentamicin	OX1	Oxacillin	VA30	Vancomycin

2. Mueller–Hinton agar plates (20)
3. Antimicrobial sensitivity disks loaded in Difco or BBL dispensers (see Color Plate 25) (2)
4. Rulers (20)
5. Sterile cotton or Dacron swabs (20)

Procedure

1. Label the agar plate with the organism to be tested.
2. Inoculate the Mueller–Hinton agar plate by wetting a swab with the test organism. Remove any excess fluid from the swab by pressing it against the side of the tube, and then swab the entire surface of the plate. Be sure to swab in a number of different directions so that the plate is completely covered.
3. Allow the plates to dry for approximately 5 min.
4. Remove the lid of the Petri plate and position the automatic dispenser over the surface of the agar plate. Push down on the plunger (Difco), or slide the dispensing bar (BBL) to release the antibiotic disks onto the surface of the plates.
5. Gently press each disk down with alcohol-sterilized forceps. In the event that a disk has fallen errantly on the plate, it can be moved with sterilized forceps. Be sure **not** to press the disks down into the agar. Note that each disk has an identification code. Refer to the manufacturer's table given to identify the antibiotic (Table 44.1).
6. Incubate the plates in an inverted position at the appropriate temperature for 18 to 24 hr.
7. After the incubation interval, measure the zone of inhibition (in mm) with a ruler (see Figure 44.1 and Color Plate 26).
8. Record observations. Determine whether the bacteria are resistant or sensitive by using Table 44.2.

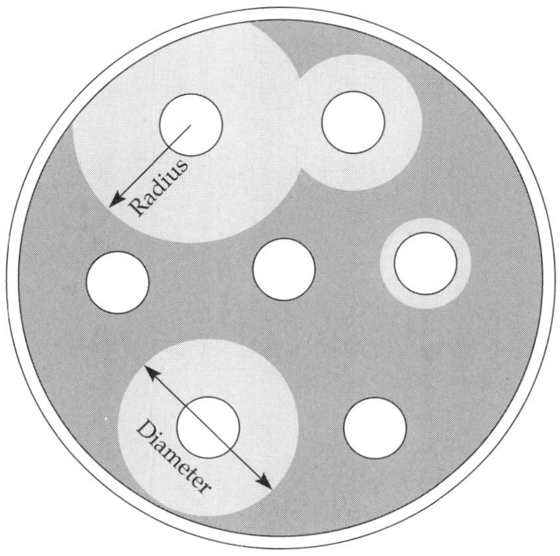

FIGURE 44.1 Measuring zones of inhibition.

TABLE 44.2 INTERPRETATION OF ANTIBIOTIC INHIBITION ZONES[1]

Antimicrobial Agent	Disk Content (μg)	Diameter of Zone of Inhibition (mm)		
		Resistant	**Intermediate**	**Susceptible**
Ampicillin[2] when testing Gram negative enteric organisms	10	≤13	14–16	≥17
Ampicillin when testing staphylococci[3,4]	10	≤28	—	≥29
Ampicillin when testing enterococci	10	≤16	—	≥ 7
Ampicillin when testing *Listeria*	10	≤19	—	≥20
Carbenicillin[3] when testing *Pseudomonas*	100	≤13	14–16	≥17
Carbenicillin when testing other Gram-negative organisms	100	≤19	20–22	≥23
Cefoxitin[4]	30	≤14	15–17	≥18
Cephalothin[4,5]	30	≤14	15–17	≥18
Chloramphenicol	30	≤12	13–17	≥18
Clindamycin	2	≤14	15–20	≥21
Erythromycin	15	≤13	14–22	≥23
Gentamycin	10	≤12	13–14	≥15
Kanamycin	30	≤13	14–17	≥18
Methicillin when testing staphylococci[6]	5	≤ 9	10–13	≥14
Nitrofurantoin	300	≤14	15–16	≥17
Penicillin when testing staphylococci[2,3]	10 units	≤28	—	≥29

TABLE 44.2 (continued)

Antimicrobial Agent	Disk Content (μg)	Diameter of Zone of Inhibition (mm)		
		Resistant	**Intermediate**	**Susceptible**
Penicillin when testing enterococci	10 units	≤14	—	≥15
Rifampicin	5	≤16	17–19	≥20
Streptomycin	10	≤11	12–14	≥15
Tetracycline	30	≤14	15–18	≥19
Trimethoprim	5	≤10	11–15	≥16
Vancomycin when testing enterococci	30	≤14	15–16	≥17
Vancomycin when testing other Gram-positive organisms	30	≤9	10–11	≥12

[1]Permission to use portions of M100-S6 (Performance Standards for Antimicrobial Susceptibility Testing. Sixth Informational Supplement) has been granted by the National Committee for Clinical Laboratory Standards (NCCLS). MS100-S6 updates M2-A5 (Performance Standards for Antimicrobial Disk Susceptibility Tests-Fifth Edition. Approved Standard). The intrepretive data are only valid only if the methodology in M2-A5 is followed. NCCLS frequently updates the M2 tables through new editions of the standard and supplements. Users should refer to the most recent edition. The current standard may be obtained from NCCLS, 940 West Valley Road, Suite 1400, Wayne, PA 19087, U.S.A.

[2]Class disk for ampicillin, amoxicillin, bacampicillin, cyclacillin, and hetacillin.

[3]Resistant strains of *Staphylococcus aureus* produce β-lactamase and the testing of the 10-unit penicillin disk is preferred. Penicillin should be used to test the susceptibility of all penicillinase-sensitive penicillins, such as ampicillin, amoxicillin, azlocillin, bacampicillin, hetacillin, carbenicillin, mezlocillin, piperacillin, and ticarcillin. Results may also be applied to phenoxymethyl penicillin or phenethicillin.

[4]Staphylococci exhibiting resistance to methicillin, oxacillin, or nafcillin should be reported as also resistant to other penicillins, cephalosporins, carbacephems, and β-lactamase inhibitor combinations despite apparent *in vitro* susceptibility of some strains to the latter agents. This is because infections with methicillin-resistant staphylococci have not responded favorably to therapy with β-lactam antibiotics.

[5]Cephalothin can be used to represent cephalothin, cepapirin, cephradine, cephalexin, cefactor, and cefadroxil. Cefazolin, cefuroxime, cefpodoxime, cefproxil, and loracebef (urinary isolates only) may be tested individually because they may be active when cephalothin is not.

[6]Methicillin/oxacillin resistance in coagulase-negative staphylococci is most reliably detected using oxacillin salt agar screening plates with 48 hours incubation at 35°C (see document M7).

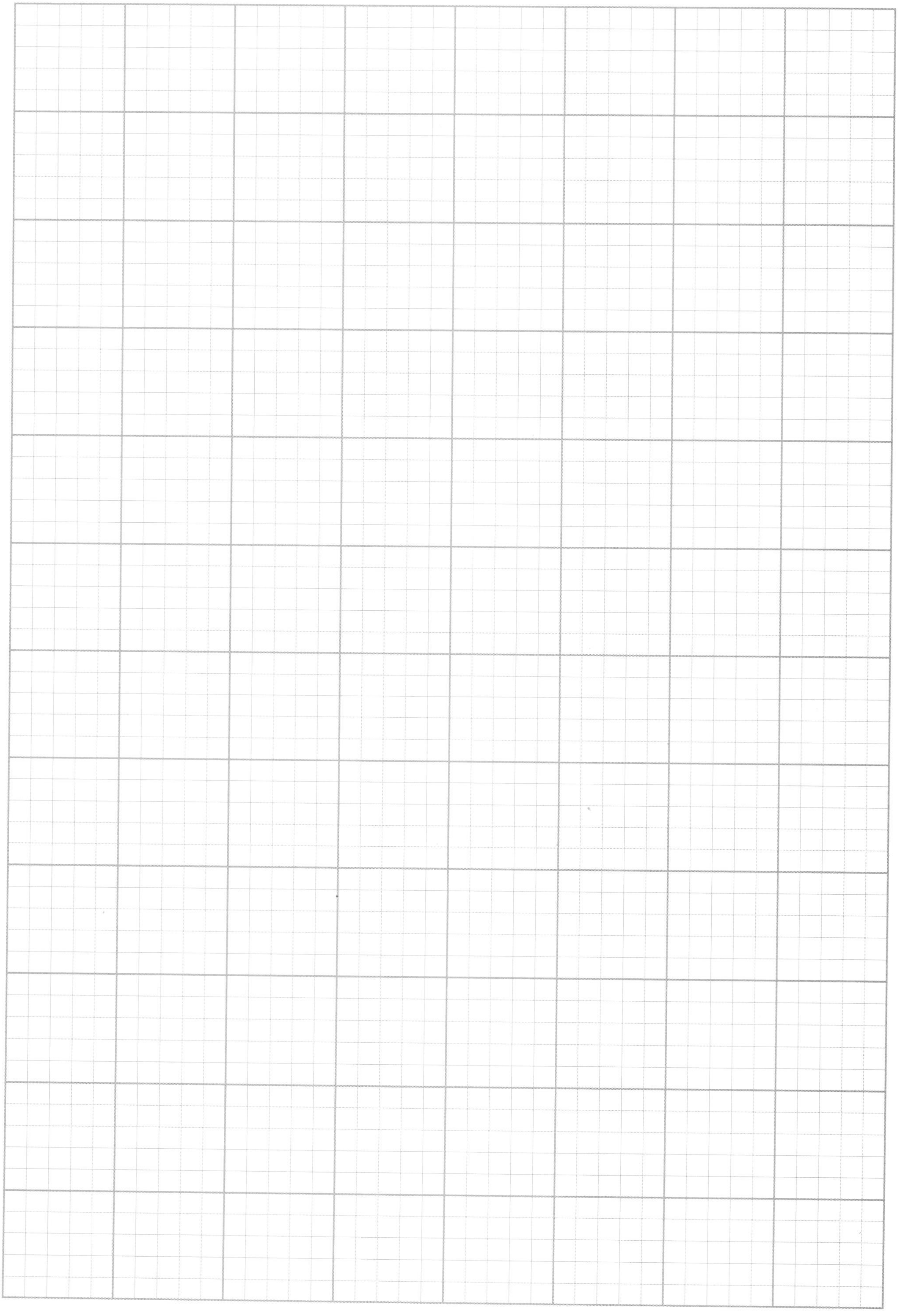

Laboratory Report 44

Name Date Section

ANTIMICROBIAL TESTING

Test Organism	Antibiotic Concentration	Zone of Inhibition	Resistance (+ or −)

Review Questions

1. How does a bacteriostatic effect of an antibiotic differ from a bactericidal effect?

2. What is meant by a broad-spectrum antibiotic? Is there an antibiotic that is effective against all of the tested organisms?

3. How do antibiotic-resistant organisms appear in the population?

LABORATORY

45 Effects of Commercially Available Disinfectants and Antiseptics on Bacteria

Safety Issues

- Since 95% ethanol is used to disinfect forceps in this experiment, caution should be exercised to keep the ethanol away from the open flame.
- Caution needs to be taken not to allow an open flame to come into contact with any aerosol from cans of deodorant or disinfectant, as the aerosol may ignite. Many liquid disinfectants are alcohol-based and will ignite if exposed to an open flame.

OBJECTIVE

Measure the effectiveness of various disinfectants and antiseptics.

Chemical agents have been used for centuries to control the growth of unwanted bacteria. Chemicals used to disinfect inanimate objects such as laboratory benches, restroom plumbing, or other large pieces of equipment are called **disinfectants**. These agents are usually too toxic to be used on human tissue. The agents that are safe to use on living tissue are referred to as **antiseptics**. Many commercially available agents are advertised with promotions such as "kills germs on contact," and "get rid of unwanted germs." Can these claims be substantiated? There is some terminology that has to be clarified before proceeding further. Some chemicals are capable of killing "germs" and are sometimes referred to as germicides. Their action is called **bactericidal** or viricidal, since bacterial death and viral inactivation actually occur. Many antiseptics are **bacteriostatic**, that is, they do not kill bacteria directly, but prevent the further multiplication of organisms. There are many tests designed to measure the effectiveness of a bactericidal or bacteriostatic agent. In this experiment, procedures to measure both the effectiveness of antiseptics and disinfectants are used.

Materials (per 20 students)

1. Nutrient agar plates (40)
2. Saline suspensions of the test microorganisms (20)
3. Cotton or Dacron swabs (20)
4. Antiseptics to be evaluated (some will be available in the lab, but you are encouraged to bring some of your own agents to test) (6)

5. Forceps (20)
6. Beakers (10)
7. Filter-paper disks (300)

Procedure

1. Using aseptic technique, wet a cotton swab with the test microorganism and spread it uniformly over the surface of the agar plate.
2. Label the plate with the name of the test organism.
3. Select five to six antiseptics to be tested, and place a small amount of each in a beaker. If the material to be tested is a solid, such as toothpaste, prepare a slurry in a beaker using some of the agent and sterile distilled water.
4. Sterilize forceps by dipping into 95% alcohol and then passing through the flame of a Bunsen burner to burn off the alcohol.

⚠ Be careful not to ignite the beaker of alcohol. If this should occur, extinguish the flame by smothering.

5. Pick up a sterile filter-paper disk and dip it into the antiseptic solution to be tested. Remove the excess fluid from the disk by touching the disk to the side of the beaker. Avoid placing an excess of liquid on the disks.
6. Place the disk on the surface of the inoculated plate and gently pat the disk in place with the forceps.
7. Label the bottom of the plate below the disk to identify the agent being tested.
8. Repeat the procedure with the other antiseptics to be tested. Up to six disks can be placed on each plate. Remember to mark below each disk the name of the agent being tested.
9. Select a second microorganism and repeat the procedure by first swabbing a plate and then applying the wetted filter-paper disks.
10. Incubate the plates at the appropriate temperature for 18 to 24 hr.
11. Measure the diameter (mm) of the zone of inhibition of growth, as was done in the evaluation of antibiotic effectiveness.
12. Compare the effectiveness of each of the agents tested.
13. When interpreting the results, keep in mind that the agents tested may diffuse to a different degree in the agar medium used. Therefore, this type of testing, while simple, only provides relative effectiveness data.

Laboratory Report

Name Date Section

45

Antiseptic Evaluation

Test Organism	Antimicrobial Agent	Zone of Inhibition

Review Question

What factors must be taken into account when comparing the zones of inhibition of the tested antiseptics?

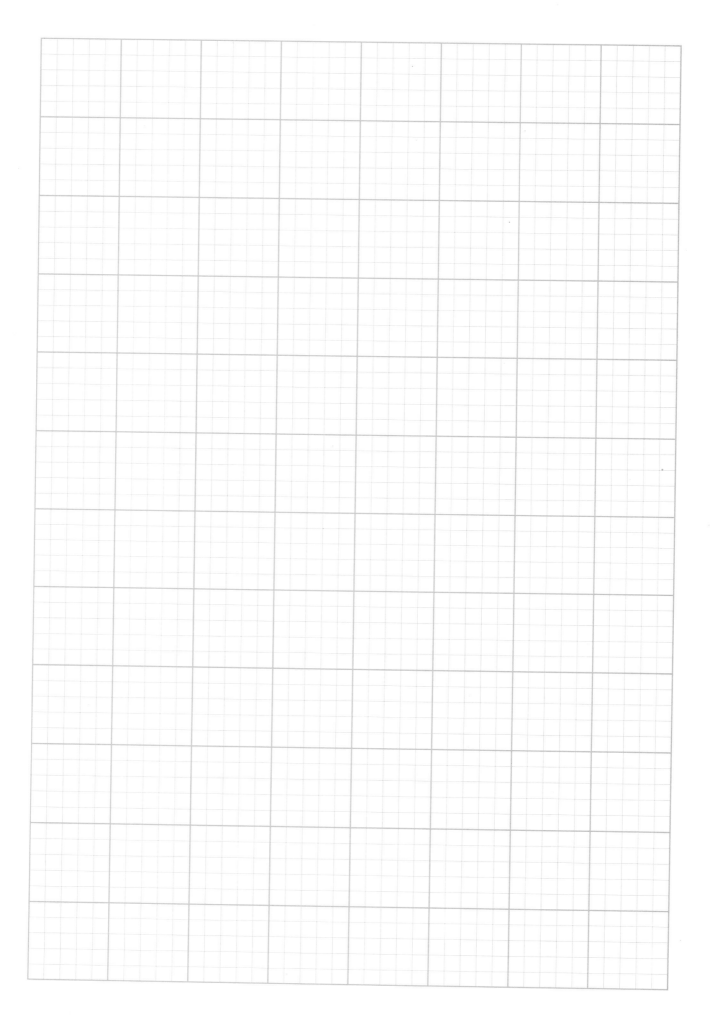

LABORATORY

46 Lethal Effects of Ultraviolet Light

Safety Issues

- Do not mouth pipet any material. Use the pipetting devices provided.
- Keep the alcohol away from open flame and do not place hot spreaders in the alcohol, as the alcohol may ignite.

OBJECTIVE

Determine the bactericidal effect of ultraviolet radiation.

By definition, ultraviolet light includes electromagnetic radiations whose wavelengths fall between 40 Angstroms (4 nm) and 4000 Å (400 nm). Wavelengths in the range of 210 to 300 nm are those capable of producing the greatest lethal effects. It has been demonstrated that the peak lethality for most organisms is at 265 nm. The lethal action of ultraviolet light is due to its action on the DNA molecule. The radiation causes the formation of thymine dimers, that is; the linking, via covalent bonds, of pyrimidine bases that lie next to one another on the same strand of DNA, and sometimes on opposite strands of the DNA molecule. The dimer which is formed cannot serve as a template for the DNA polymerase and, as a result, replication is prevented. The extent of the damage can be lethal for the organism unless repair takes place. There are a number of repair mechanisms that can occur. Dark repair mechanisms also exist. RecA protein is involved in an inducible repair system known as SOS repair, which acts rapidly, but is very error-prone and introduces mutations in the DNA. Where the damage is extensive, repair may not be possible, or else it is so error-prone that the exposure to ultraviolet radiation is lethal for the organism.

Materials (per group)

This experiment should be performed in groups, with two groups determining the original titer of microorganisms and the other groups determining the titer at various irradiation intervals.

1. Saline suspension of organism to be tested (1)
2. Saline dilution blanks (4.5 ml/tube) (5)
3. Sterile 1.0- and 10-ml pipettes (1 canister of each)
4. Sterile glass Petri plate (1)
5. Nutrient, tryptic soy agar, or plate-count agar plates (8)
6. Glass spreaders (2)
7. Alcohol (50 ml)
8. Turntables (1)

9. Ultraviolet lamp (1)
10. Ultraviolet-light protective goggles (2)

Procedure

Two groups should determine the titer of the suspension of microorganisms to be irradiated by performing a viable spread-plate technique.

A. Determination of Titer or Unirradiated Culture

1. With a sterile 1.0-ml pipette, transfer 0.5 ml of the culture to a 4.5-ml sterile saline dilution tube. Mix well.
2. With a fresh 1.0-ml pipette, transfer 0.5 ml to a second dilution tube marked 10^{-2}.
3. Repeat the procedure until dilutions have been prepared to 10^{-6}.
4. Pipette 0.1 ml each of the 10^{-4}, 10^{-5}, and 10^{-6} dilutions onto duplicate agar plates.
5. Using an ethanol-sterilized glass spreader, distribute the fluid on the surface until it is absorbed into the agar.
6. Invert the plates and incubate at the appropriate temperature for 48 hr.
7. After the incubation period, count the colonies on each of the plates and calculate the titer of the unirradiated culture.

B. Determination of Titer of Organisms after Irradiation

⚠ Ultraviolet radiation can cause damage to eyes. Wear safety goggles during any irradiation process.

1. Pipette 10 ml of the suspension of microorganisms into a sterile glass Petri plate.
2. Put on protective goggles to protect your eyes, and irradiate the cells for 30 sec. (Note the distance from the light source. Also make sure to remove the glass lid of the Petri plate as UV light does not penetrate glass!) (See Color Plate 27.)
3. Withdraw 0.5 ml of the irradiated culture and prepare serial 10-fold dilutions (10^{-1} to 10^{-4}) in the 4.5-ml saline tubes provided.
4. Spread plate 0.1 ml of the dilutions onto the surfaces of two agar plates.
5. Irradiate at regular intervals up to 2 min, and repeat the viable plate-count determination for each interval by withdrawing 0.5 ml of the irradiated culture and placing it into a tube of 4.5-ml saline (1:10 dilution). (Suggested times = 30 sec, 1 min, and 2 min.)
6. Incubate all plates in an inverted position for 48 hr at the appropriate temperature.
7. Count the colonies and determine the population of microorganisms after each of the irradiation intervals.
8. Calculate the percentage of organisms surviving at each interval by dividing the population at each interval by the original population and then multiplying by 100%.

Laboratory Report 46

Name Date Section

Bactericidal Effect of Ultraviolet Light

Time of Irradiation	Non-irradiated Culture		Irradiated Culture	
	Dilution	**Number of Colonies**	**Dilution**	**Number of Colonies**

Review Questions

1. What is the mechanism of action of the bactericidal effect of ultraviolet light on bacteria? How can the damage of UV light be reversed?

2. Many biological safety cabinets and laminar air-flow hoods have ultraviolet light for disinfection purposes. What precautions should be taken in using these pieces of equipment?

UNIT VII

Applied Microbiology

Microorganisms are ubiquitous, being found in water, soil, air, and in various food products. The presence of pathogenic bacteria in water can pose a significant risk to human health. The enumeration and identification of these organisms is important in establishing the source of the water contamination.

The first few inches of topsoil contain billions of microorganisms which are responsible for soil fertility and decomposition of organic material. Pour-plating techniques are normally used to enumerate and identify the number and types of bacteria in a soil sample.

Bacteria can be transmitted in airborne droplets and may cause infections or allergic reactions if inhaled. Concerns have been raised about the numbers of microorganisms in the air in tightly-sealed buildings and in aircraft where the percentage of fresh air circulating may be low.

Most foods contain some microorganisms. Many microorganisms carry out fermentations responsible for the production of a variety of foodstuffs. Food improperly stored or preserved is susceptible to spoilage, and food improperly cooked may harbor disease-causing bacteria.

The laboratories in this unit have been formulated to allow the demonstration of the diversity and activities of microorganisms in air, water, soil, and food.

LABORATORY

47 Water Microbiology

Safety Issues

- Use mechanical pipetting devices. Never attempt to pipette any material by mouth.
- Keep alcohol away from open flames. Burn alcohol from forceps by quickly and cautiously passing through the burner flame. Never place hot forceps into alcohol container as the ethanol may ignite.
- Wear gloves when handling any water sample that is suspected to contain sewage or sewage runoff. There may be human pathogens present in the sample.
- All filtered and unfiltered water samples need to be decontaminated at the conclusion of the laboratory. Pour all samples into beakers or flasks in preparation for autoclaving. Do not pour any of the water samples down the drain.

OBJECTIVE

Become familiar with techniques used in the microbiological analysis of water.

The quality of drinking (potable) water is certainly an extremely important matter. A variety of agents causing gastroenteritis are carried in water supplies, including typhoid and dysentery. The testing of large numbers of water samples for the variety of disease-causing agents would be very tedious and time-consuming. As well, the occurrence of disease-causing agents in small numbers would make detection very difficult. The primary contamination of water comes from sewage pollution, so a test that detects sewage pollution could provide a good indication of the presence of potentially pathogenic microorganisms. **Coliforms** are facultatively anaerobic, non-spore-forming, gram-negative rods that ferment lactose with gas formation within 48 hr at 35°C. An example of an organism that fits these criteria is *Escherichia coli*, a component of the normal flora of the gastrointestinal tract of warm-blooded vertebrates and, therefore, a component of feces. The coliforms present serve as **indicator** organisms, in that they indicate the presence of sewage pollution.

This laboratory presents two methods for detection of pollution of water with microorganisms: (a) membrane-filter method and (b) one-step method.

MEMBRANE-FILTER METHOD

This technique utilizes membrane filters that have a pore size of 0.45 μm. When the water sample is filtered via suction through the membrane, microorganisms larger than 0.45 μm are trapped on the surface of the filter. The filter is then transferred to a media-saturated pad in a small Petri plate. Nutrients can diffuse from the pad to the filter surface and may provide nutrients for the cell division or growth of any microbe trapped on the filter. An organism trapped on the filter can eventually give rise to a colony on the surface of the filter. If needed, a microscope can be used to count the number of colonies present on the filter. There are advantages and disadvantages in using this technique. The one obvious advantage is the simplicity of the technique. Other advantages include the capability of processing large sample volumes, and the speed of obtaining results. A major disadvantage is that any turbidity in the water sample will quickly clog the filter and prevent further filtration. Depending on the type of media placed on the absorbent pad or in the plate, the enumeration of a number of different organisms is possible. In this experiment, the following counts will be determined:

(a) **The total heterotrophic bacterial count**, using m-HPC as the nutrient.
(b) **The total coliform count (TCC)**, using m-endo medium.
(c) **The fecal coliform count (FCC)**, using m-FC broth.
(d) **The fecal streptococcus count (FSC)**, using m-enterococcus or KF agar.

The total heterotrophic count is determined by the number of colonies formed on the filter surface when incubated on an absorbent pad containing m-HPC agar.

Total coliforms are detected as colonies producing a green metallic sheen on filters exposed to endo broth saturated pads.

Fecal coliforms can be identified by their ability to ferment lactose at 44.5°C with the production of acid and gas. These are organisms that enter a water supply via the fecal discharge of warm-blooded animals and humans. *E. coli* is the most common fecal coliform species.

Fecal streptococci are native to the gut of warm-blooded animals, especially humans, and are considered typically nonpathogenic. Their presence may be more indicative of pollution in tropical climates and in brackish or marine waters.

Guidelines have been established for the allowable limits of microorganisms in various types of water. The maximum permissible coliform count is 1/100 ml for municipal drinking water, if less than 40 samples are processed per month. Recreational waters have a maximum permissible coliform count of 70/100 ml.

Materials (per group)

1. Water sample
2. Sterile filtration apparatus
3. 0.45-μm membrane filters and pads
4. Sterile 50-mm Petri plates (4)
5. Sterile distilled water (500 ml)
6. 99.0-ml sterile water dilution bottles
7. Vacuum source
8. 95% ethanol (50 ml)
9. Forceps
10. Sterile 10-ml pipettes (1 canister)
11. Stereomicroscope
12. Magnifying lens

13. Sterile m-HPC agar or R2A agar* (25 ml)
14. m-endo broth (prepared fresh as per instructions)
15. m-FC broth (prepared fresh as per instructions)
16. m-enterococcus (KF agar) (prepared fresh as per instructions)

If verification tests are performed:

17. Lauryl tryptose broth (5 tubes)
18. Brilliant-green lactose bile broth (5 tubes)
19. Brain-heart infusion agar plates (3)
20. Brain-heart infusion broth tubes (6)
21. 3% hydrogen peroxide in dropping bottles
22. Bile esculin agar plates (3)
23. Brain-heart infusion broth tubes with 6.5% NaCl (3)

Procedure

⚠ Be careful not to ignite the alcohol when sterilizing forceps. Keep the alcohol away from any open flame.

Each group should test one of the water samples. (See Figure 47.1.)

1. Attach the filter apparatus to the vacuum device.
2. With alcohol-sterilized forceps, place a membrane filter on the platform base of the filter apparatus. Make sure not to perforate or tear the filter. Also be sure that none of the colored filter separators have been loaded along with the filter. If there is difficulty with filtration, check to see whether one of the separators has been loaded.
3. Pour about 20 ml of the sterile distilled water into the filter funnel.
4. The type of water being evaluated will dictate the volume and dilution of the sample to be filtered. Drinking (potable) water should contain few organisms if properly treated, so larger volumes (100–500 ml) are usually filtered. Pond water or other samples, which are suspected to have a higher population of microorganisms, will need to be diluted (Appendix F) prior to filtration. **For total heterotrophic counts, try 10^{-2} to 10^{-5} dilutions. For total coliform, fecal coliform, and fecal strep, use undiluted to 10^{-3} dilutions.** Filter at least 25 ml of these water samples.

Standard Methods for the Treatment of Water and Wastewater, 18th edition, 1992, suggests that the following sample volumes be used for total coliform tests using membrane filtration:

Drinking water	100 ml
Swimming pools	100 ml
Wells	10–100 ml
Lakes	10–100 ml
Bathing beaches	0.1–10 ml
River water	0.001–1.0 ml
Raw sewage	0.0001–0.1 ml

*R2A can be used as an alternative. This lower nutrient agar gives a higher count than the high nutrient formulations.

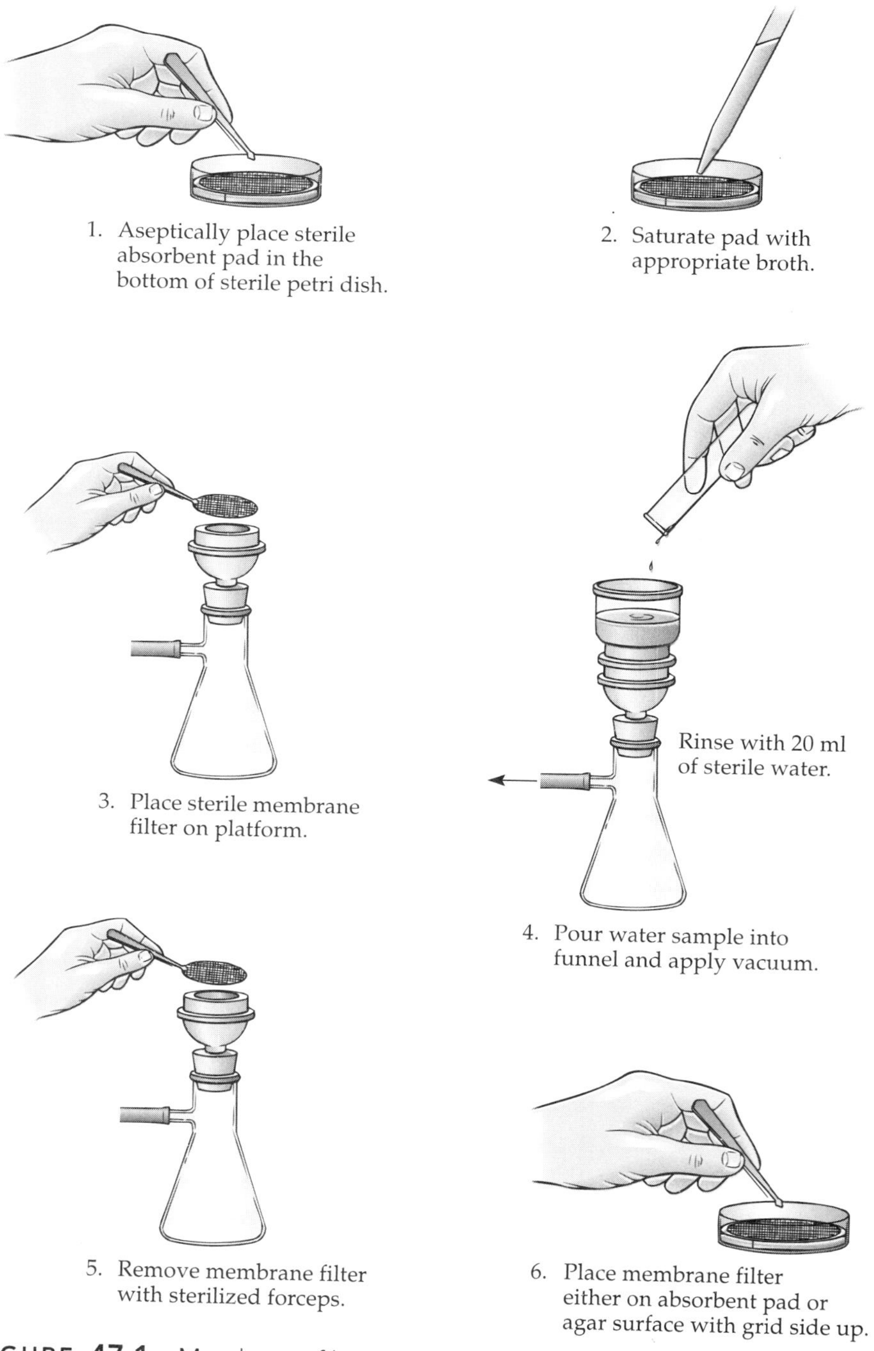

FIGURE 47.1 Membrane-filtration apparatus.

5. If using an undiluted sample, add the sample directly to the filter funnel. If using a diluted sample, start with the highest dilution and add to the filter funnel containing the sterile distilled water. Filter by applying a vacuum.
6. Turn off the vacuum and wash the walls of the filter funnel with at least 40 ml of sterile distilled water.
7. Turn on the vacuum to filter the wash. It is important to rinse to the filter surface any organisms that might be adhering to the walls of the filter funnel.

8. Place absorbent pads in the 50-mm Petri plates, and label appropriately.
9. Add 2.0 ml of the proper medium to each absorbent pad. In the case of the fecal strep test and the total heterotrophic count, no pads are necessary since an agar base is used.
10. Using sterile forceps, transfer the filter to the appropriate saturated pad in the Petri dish. For the fecal strep and the total heterotrophic count, place the filters directly onto the agar surface.
11. Incubate the heterotrophic-count plates at 28°C for two to five days. The total coliform plates should be incubated at 35°C for 22 to 24 hr. Fecal strep plates should be incubated at 35°C for 48 hr.
12. Fecal coliform plates need to be placed in a water-tight bag that has been weighted down, and then placed in a water-bath incubator set at 44.5°C for 24 hr.
13. After incubation, remove all filters from the plates and allow them to dry on absorbent paper for 1 hr. If necessary, use a dissecting microscope or hand lens to observe the colonies present on the filters.

The colonies on the **total heterotrophic-count** filters will have various pigments and will be of diverse sizes, similar to that observed on regular agar plates. The use of a stereomicroscope is advised when performing colony counts. Count all colonies on the filter when there are 2 or fewer colonies per square. For 3 to 10 colonies per square, count 10 squares and obtain an average count per square. For 10 to 20 colonies per square, count five squares and obtain an average count per square. Multiply average count per square by 100 times the reciprocal of the dilution to give colonies per milliliter of water.

All bacteria that produce colonies that are pink to dark red and have a metallic sheen on total coliform filters are considered members of the coliform group. This sheen may cover an entire colony or it may consist of a concentrate in the center of the colony (see Color Plate 28). Typical sheen colonies may be produced occasionally by non-coliform organisms. Atypical colonies (dark red or nucleated colonies without sheen) occasionally may be at least 10% of the colonies when there are more than 50 coliform-type colonies present from a wastewater sample. For coliform counts of more than 5 per 100 ml from drinking water, verify a minimum of 5 colonies. Verification (**optional**) involves the use of lauryl tryptose broth and brilliant-green lactose broth. Transfer a suspected coliform colony to a tube of lauryl tryptose broth containing a fermentation tube (Durham tube). Gas or acidic growth in the tube within 48 hr constitutes a **positive presumptive test** for coliforms. With a sterile loop, transfer a loopful of culture to a tube containing brilliant-green lactose bile broth. Formation of gas in the inverted Durham tube within 48 hr constitutes a **positive confirmed test** for coliforms.

Colonies produced by fecal coliform bacteria are various shades of blue (see Color Plate 29). Nonfecal coliforms are gray to cream colored. Normally, few nonfecal coliform colonies will be observed on the fecal coliform medium because of the selective action of the elevated temperature and the presence of rosolic acid. Record the densities as fecal coliforms per 100 ml of water.

Fecal streptococci will produce colonies that are light pink, or dark red with pink margins (see Color Plate 30). Count colonies using a fluorescent lamp and a magnifying lens. To verify (**optional**), pick selected colonies from the membrane and streak for isolation onto the surface of a brain-heart infusion agar (BHIA) plate and incubate at 35°C for 24 to 48 hr. Transfer a loopful of growth from a colony on BHIA into a brain-heart infusion broth tube and to each of two clean glass slides. Incubate the brain-heart infusion broth at 35°C for 24 hr. Perform a catalase test by adding a few drops of H_2O_2 to the loopful of broth on one of the slides. The appearance of bubbles is a positive test and indicates that the organisms are *not* fecal streptococci. If the catalase test is negative, perform a Gram stain of the smear on the second slide. Fecal streptococci are gram-positive organisms, mostly in pairs or short chains. Transfer a loopful of growth from the brain-heart infusion broth to each of the following media: bile esculin agar (incubate at 35°C for 48 hr); brain-heart infusion broth (incubate at 45°C for 48 hr); brain-heart infusion broth with 6.5% NaCl (incubate at 35°C for 48 hr). Growth of catalase-negative, gram-positive cocci on bile esculin agar and in brain-heart infusion broth at 45°C verifies that the organism is a fecal streptococcus. Growth in the 6.5% broth at 45°C indicates that the organism is an enterococcus.

ALTERNATIVE METHOD FOR DETECTION OF COLIFORMS AND FECAL COLIFORMS IN DRINKING AND SURFACE WATER

Simple one-step methods* have been devised for the determination of the presence or absence of total coliforms and *Escherichia coli* in drinking water. These methods involve sampling 100 ml of water or five 20-ml volumes. The water is added to a dehydrated medium, incubated at 35°C, and inspected for acid reactions after 24 and 48 hr. The Colilert system uses two indicator nutrients: ortho-nitrophenyl-β-D-galactopyranoside (ONPG) and 4-methylumbelliferyl-β-D-glucuronide (MUG). After incubation in the presence of the indicators, if coliforms are present, the enzyme β-galactosidase cleaves ortho-nitrophenyl from the ONPG. This release imparts a **yellow color to the solution**. *E. coli* also possesses the constitutive enzyme glucuronidase, which hydrolyzes MUG. In this hydrolysis, MUG is cleaved into glucuronide, a nutrient, and the indicator, methylumbelliferone, which fluoresces under ultraviolet light.

The Colisure method also detects the presence of the same two enzymes. The Colisure medium contains a chromogenic substrate, chlorophenol red β-D-galactopyranoside (CPRG). Upon hydrolysis by the β-galactosidase enzyme, CPRG releases a chromogenic compound, chlorophenol red, which turns the medium **from yellow to a red or magenta color**. The Colisure medium also contains the fluorogenic substrate MUG. MUG is hydrolyzed by the β-glucuronidase enzyme, releasing 4-methylumbelliferone, which fluoresces when exposed to ultraviolet light. Since β-glucuronidase is specific for *E. coli*, the fluorescence differentiates it from the rest of the coliforms.

Materials (per group)

Colilert Procedure

1. Sterile, transparent, non-fluorescent borosilicate glass bottle (1)
2. Colilert reagent
3. Long-wavelength ultraviolet light (366 nm)

*Colilert®, Environetics, Inc., Branford, Connecticut. Colisure™, Millipore Corporation, Bedford, Massachusetts.

4. UV protective goggles
5. 3% solution of sodium thiosulfate ($Na_2S_2O_3$)
6. Sterile 1.0-ml pipettes (1 canister)

Colisure Procedure

1. Sterile, transparent, non-fluorescent borosilicate glass bottle (1)
2. Colisure P/A Medium
3. Long-wavelength ultraviolet light (366 nm)
4. UV protective goggles
5. 3% solution of sodium thiosulfate ($Na_2S_2O_3$)
6. Sterile 1.0-ml pipettes (1 canister)

Procedure

In both testing methods, 0.1 ml of sodium thiosulfate must be added to any drinking water sample to neutralize any chlorine that may be present.

Colilert Procedure

1. Aseptically add the contents of a tube of the Colilert reagent to a 100-ml water sample in the borosilicate glass bottle.
2. Shake well and incubate at 35°C.
3. Read the color of the bottle within 24 to 48 hr. The presence of a yellow color indicates the presence of coliforms. Shine the long-wave ultraviolet lamp on the bottle. Any fluorescence indicates the presence of *E. coli.*

Colisure Procedure

1. Aseptically add the Colisure medium to the borosilicate bottle containing the 100-ml water sample to be tested. If the manufacturer's premeasured bottle is used, add 100 ml of the water sample directly to the bottle.
2. Shake well and incubate at 35°C. Read the color change reaction in the bottle after 24 hr. Change of color from yellow to red or magenta indicates the presence of at least 1 CFU of coliform bacteria. Shine the long-wave ultraviolet lamp on the bottle. Any fluorescence observed indicates a positive result and the presence of at least 1 CFU of *E. coli.*

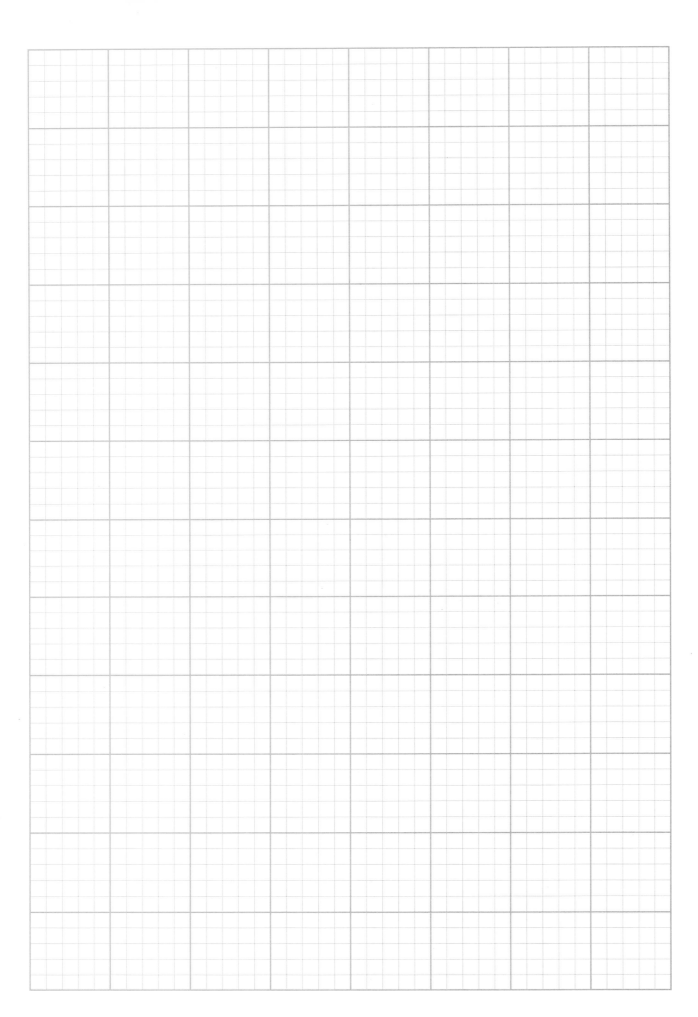

Laboratory Report 47

Name Date Section

MICROBIOLOGICAL QUALITY OF WATER

Water Source	Sample Volume	Microbiological Count			
		Total Heterotrophic	**Total Coliform**	**Fecal Strep.**	**Fecal Coliform**

RESULTS OF ONE-STEP METHODS

	Color Change	Fluorescence (+/−)
Colilert		
Colisure		

Review Questions

1. Were any of the water samples polluted with human waste? How was this determined?

2. What are the limitations in using the membrane-filter technique for the microbiological analysis of water?

3. What are the advantages and disadvantages of the one-step methods to detect coliforms and *E. coli* in water?

4. Why is a 100-ml sample volume used in the one-step tests?

LABORATORY

48 Microbiology of Air

Safety Issues

- Keep alcohol away from any open flame.

OBJECTIVES

1. Become familiar with techniques used in microbial air sampling, including gravity technique, impaction method, and liquid impingement.
2. Compare mean populations of microorganisms at various sites and determine statistical significance.

Air currents spread microorganisms that are carried in aerosols. For example, a single sneeze releases millions of microorganisms carried on droplets in the air.

Even though no standards exist as to what constitutes acceptable numbers of microorganisms in air, considerable interest about the quality of air has been generated for the following reasons:

- Airborne spread of microorganisms can contribute to nosocomial (hospital-acquired) infections.
- Exposure of workers to bacteria-laden dust and dander in many industrial settings may subject them to higher rates of infection, particularly due to the presence of gram-negative organisms.
- The quality of indoor air is of concern in modern, tightly sealed facilities where the filtration of air or replenishment with outside air is not adequate. It has been estimated that 10 million office workers suffer health problems due to indoor air pollution and inadequate building ventilation. This pollution may arise from volatile organics released from office furniture, building materials, and adhesives. Dust and fibers may also contribute to the pollution. Bioaerosols of intact molds (*Aspergillus*) and bacteria (*Thermoactinomyces* or *Legionella*) or their toxic components have also been implicated as sources of indoor pollution and disease. Collectively the increased incidence of such disease symptoms among workers has become known as "sick building syndrome."
- The air quality in "clean rooms" must be free of microorganisms. Areas used in the packaging of pharmaceutical products, in calibration of missile guidance systems, or for surgery must be routinely monitored for the presence of microorganisms.
- *Legionella* released from air conditioning heat reaction systems and hot water systems can be infectious in compromised individuals.

There are basically two ways one can monitor air for the presence of microorganisms. In one method, referred to as impingement, air is bubbled through a liquid to trap any organisms present in a defined volume of air. The second basic method involves impacting air onto the surface of an agar plate. Both of these methods should be used for comparative purposes, since the efficiency for either method is relatively low and probably is compromised by damage done to the organisms during sample acquisition.

The simplest of the impaction methods is the **gravity method**. This method involves the exposure of prepared agar plates for a selected interval. As the name implies, the technique is dependent on microorganisms falling via gravity onto an agar plate. Other impaction devices, such as the SAS sampler (see Color Plate 31) and the Andersen sampler (see Color Plate 32), have been designed to allow for the more efficient sampling of larger volumes of air. One of them, the Andersen sampler, also enables the investigator to size the particles trapped.

Depending upon the sampling device available, perform an analysis of the microbial quality of air in various environments. Try areas that are quiet versus heavily trafficked. Compare the quality of air in the lab environment, food-handling areas, restrooms, and offices. Use your imagination, but remember to first seek permission before doing any sampling.

Develop a simple hypothesis for air sampling. The efficiency of the different samples could be tested or a comparison made at different locations.

GRAVITY TECHNIQUE

Materials (per group)

1. Nutrient agar and Sabouraud dextrose agar plates (40)
2. Stopwatch

Procedure

1. Identify four sampling areas and mark ten plates for each sampling area.
2. Remove lids of Petri plates and time the length of exposure.
3. Remember that plates exposed for extended periods of time (hours) may undergo significant dehydration.
4. Incubate all plates at 30°C for 48 to 72 hr.
5. Compare the number of microorganisms present in each sampled environment by comparing colony counts.
6. Determine the number of different kinds of microorganisms based upon different colonial morphologies.
7. Calculate the mean population for each area. Using a statistical program, compare the significance of mean differences by calculating an analysis of variance (ANOVA).

SURFACE AIR SYSTEM (SAS) IMPACTION METHOD

This battery operated device is extremely portable and easy to use. Rodac plates filled with any agar serve as the impaction surface. Air flow is at the rate of 180 L/min, so large volumes of air can be sampled in short periods.

Materials (per group)

1. SAS Sampler (Spiral Biotech, Bethesda, MD)
2. Rodac plates with tryptic soy agar (TSA), Sabouraud dextrose agar (SDA) and eosin methylene blue agar (EMB) (20 of each)
3. 70% ethanol (100 ml)

Procedure

1. Plug the sampler into the charged battery pack.
2. Remove the protective plastic cover and swab the sampler cover and the inside of the plastic cover with 70% ethanol to sterilize.
3. Replace the plastic cover.
4. Unscrew the lid of the sampler with the plastic cover in place.
5. Insert the Rodac plate into the sampler so that it is held securely in place by the clamps.
6. Remove the lid of the Rodac plate and replace the cover of the sampler with the plastic lid in place.
7. Turn on the battery pack and use the thumbwheel switch to set the sampling time. Each numerical setting corresponds to 20 sec. This means that, when operating at setting 6, the sampler will run for 2 min, collecting 360 L of air. (Exposing plates for 2 min in classrooms, offices, and laboratories yields countable plates. Reduce the sampling interval in areas that are suspected to be more contaminated.)
8. Remove the plastic cover and depress the start button.
9. After sampling has been completed, replace the plastic cover, unscrew the cover of the sampler, and remove the exposed Rodac plate. It is best to replace the lid of the Rodac plate before removing to avoid contamination.
10. Reload with another plate and continue sampling. Sample each area using at least five plates of each medium. In that way a comparison can be made of the total microbial population, coliform count, and the mold count.
11. Incubate the TSA plates at 30°C, the EMB plates at 37°C, and the SDA plates at room temperature.
12. Reserve plates exhibiting mold growth for further identification in Laboratory 56.
13. Count colonies and calculate the number of microorganisms present by using the following equation:

$$\frac{\text{Number of colonies} \times 1000}{60 \times (\text{numerical setting})} = \frac{\text{number of organisms}}{\text{per cubic meter}}$$

LIQUID IMPINGEMENT METHOD

The system described here uses an all-glass impinger (Millipore Corp., Bedford, MA). It allows for the collection of air in a liquid by use of a vacuum device attached to a limiting orifice (see Color Plate 33).

Materials (per group)

1. All-glass impinger
2. Vacuum device
3. Impingement fluid (50 ml tryptic soy broth)

4. Sterile absorbent pads (10)
5. Sterile Petri plates (10)
6. (Total count medium) Tryptic soy or plate-count agar (10)
7. Forceps
8. 70% ethanol (50 ml)
9. Sterile distilled water (200 ml)
10. 1.0-ml sterile pipettes (1 canister)
11. 4.5-ml sterile saline dilution tubes (10)
12. 10-L/min limiting orifice (Millipore Corp.)

Procedure

1. Pipette 50.0 ml of sterile impingement fluid into the impinger which has been rinsed with sterile distilled water.
2. Identify the sampling site and attach the impinger to the vacuum source. A 10-L/min limiting orifice is used to regulate air flow.
3. Turn on the vacuum and collect the sample.
4. Turn off the vacuum and note the sampling time.
5. Disconnect the impinger from the vacuum source. Determine the number of microorganisms present by using either a membrane-filtration technique (Laboratory 47) or a viable plate-counting technique (Laboratory 16).

If using the membrane-filter technique, remove a sample from the impinger, filter through a 0.45-μm membrane filter, and place the filter on an absorbent pad saturated with 2.0 ml of tryptic soy broth. If a viable plate-count technique is used, plate undiluted and 10^{-1} samples of the impingement fluid on the surfaces of tryptic soy agar plates.

Laboratory Report

Name ______ Date ______ Section ______

48

Calculate mean, variance, and ANOVA for each area. Two common techniques used in biology in the analysis of quantitative statistical data are calculations of the mean and median.

The **mean** (arithmetic mean) $(X) = \sum x/N$
$\sum$ is used to indicate addition, and x denotes a sample, and N indicates the total number of samples. So, X is the sum of all the samples collected divided by the total number of samples.

The **median** (md) is the middle measurement in a set of data. There are as many observations greater than the median as there are smaller.

Variance (s^2) is one way to measure the dispersion of values around the arithmetic mean. It is calculated as the average squared deviation of all observations from their mean value. The square root of variance is the **standard deviation (s)**. The formula for calculation of variance is:

$$s^2 = \sum \frac{(\bar{x} - x)^2}{N - 1}$$

where $x - x$ represents the deviation of each individual value from the mean. The sum of those numbers squared divided by the total number minus 1 is the variance.

In this experiment, we seek to determine whether the population variation between air-sampling sites is significantly greater than within a single sampling site. The statistical method that allows us to answer the question is a procedure that is called the **analysis of variance (ANOVA)**. The assumption is that there has been random sampling of the populations at each site and that each population has a normal distribution.

All statistical tests have what is referred to as a **null hypothesis (H_0)**, an **alternative hypothesis (H_1)**, and an **alpha level (a)** associated with them. In our situation, the null hypothesis is that a difference does not exist among the mean population of organisms at each sampling site (N) compared to the population mean (μ). The alternative hypothesis is a statement that a difference does exist among the means at each site compared to the population mean. The alpha level is a probability level set by the investigator.

Using a statistical program such as MINITAB or JMP Statistics for the Apple Macintosh (SAS Institute, Inc.), calculate the means and ANOVA to

Gravity Sampling Technique

	Plate Counts in Areas Sampled			
	1	2	3	4
Mean				
Variance				
ANOVA				

SAS Sampling Technique

	Plate Counts in Areas Sampled			
	1	2	3	4
Mean				
Variance				
ANOVA				

determine whether the means of each sampling area are significantly different at the 95% confidence interval. Means whose confidence levels do not overlap are considered to be significantly different. The information is usually presented graphically and in table form.

Review Questions

1. What advantages does the impaction air-sampling technique have over the liquid impingement method?

2. Some impaction air-sampling devices allow for the sizing of particles sampled. Why would this be important?

3. You are called into a residence where the inhabitant is having a number of health problems that seem to be related to air quality. How would you go about setting up a program to monitor the air quality in the residence and to determine whether the problem may be caused by bacteria or molds?

4. Does an environment that contains 5000 microorganisms/m^3 necessarily pose a greater health risk than an environment that contains 500 microorganisms/m^3? What other factors must be considered before establishing potential health risks?

5. Is there an environment sampled that clearly has a larger population of microorganisms than the others? Is there some characteristic of that environment that helps to explain the larger population?

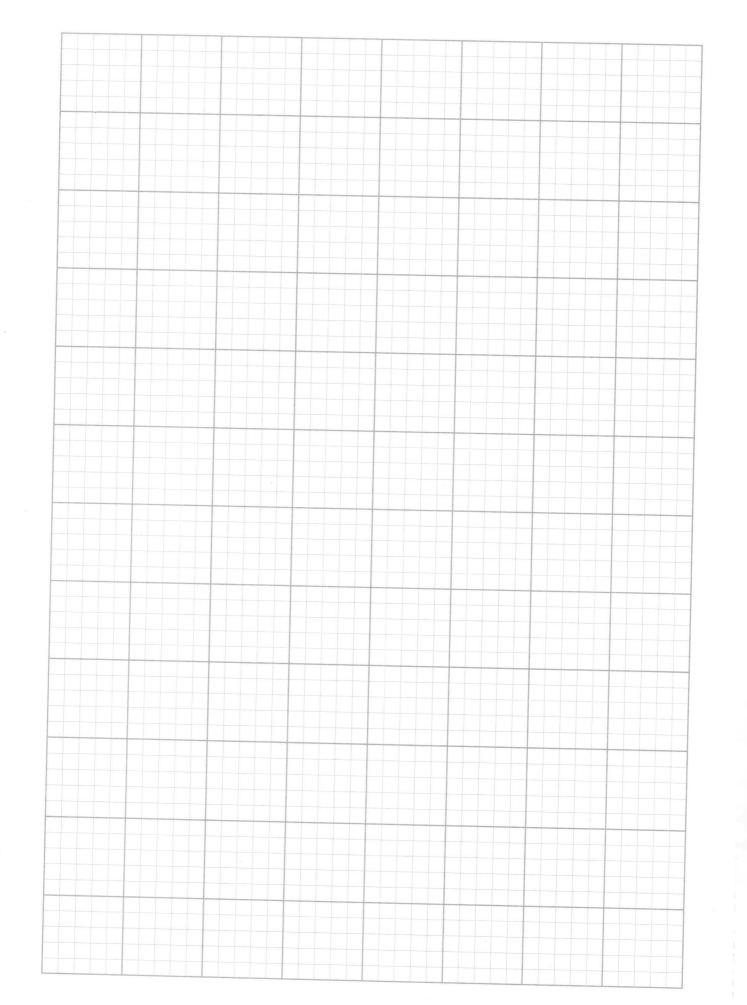

LABORATORY

49 Microbiology of Foods

Safety Issues

- Use eye protection when removing methylene blue reduction tubes from the incubator, as accumulation of gases may cause the rubber stoppers to pop off.
- Use a mechanical pipetting device when transferring solutions. Never attempt to mouth pipet any material.

OBJECTIVES

1. Determine the quality of milk samples by using the methylene blue reduction test.
2. Evaluate the microbial population of various foods by employing the pour-plate technique.

The various types of food that mammals use as a source of nutrition can also be used by bacteria. It is common to experience the growth of microorganisms in foods that are unrefrigerated, or even those kept refrigerated for long periods of time. All fresh foods have a measurable population of microorganisms associated with them. Some arise because they were intentionally added during the preparation of the food, others are introduced as a consequence of handling the food during packaging and storage. In most cases, this bacterial population consists of harmless bacteria that impart the characteristic aroma and flavor to many foods such as yogurt, sauerkraut, or cheeses. However, the occurrence of pathogenic bacteria in food products must be carefully monitored. The presence of pathogenic microorganisms, or their products, can lead to serious cases of food poisoning and perhaps even to death, as recently occurred with the meat contaminated with *Escherichia coli* 0157:H7, a specific subtype (strain) of *E. coli*. In the experiments that follow, the bacterial population in milk, meats, and vegetables is determined.

MICROBIOLOGICAL QUALITY OF MILK

A widely utilized test for the quality of milk is the standard plate count of milk. This procedure is recognized by the American Public Health Association as the official method for assessing the quality of milk. It essentially involves the enumeration of bacteria present by performing viable plate counts. We will use a simpler, non-quantitative type of test to assess quality, known as the methylene blue reduction test. When methylene blue is in a reduced state, it loses its color. This property is utilized to evaluate milk samples. By placing a few drops of methylene blue in a test tube of milk and tightly stoppering it, an environment can be created in which the metabolic, aerobic respiration of bacteria present can be measured. The higher the population of organisms present in the milk, the more rapid will be the depletion of the oxygen. Once all the oxygen has been consumed, the methylene blue dye will act as an artificial electron acceptor and be reduced to its colorless form. The rate at which this color change occurs is an indication of the numbers of bacteria in the milk sample. The time that it takes the methylene blue to be reduced is referred to as the **methylene blue reduction time**. Any milk that has a reduction time of less than 30 min is a poor-quality milk. Any reduction time over 6 hr usually indicates a milk of good quality (see Color Plate 34).

Materials (per 20 students)

1. Sterile test tubes with rubber stoppers (40)
2. Milk samples (2)
3. Sterile methylene blue dye (20 mg/ml) (30 ml)
4. Sterile 1.0- and 10.0-ml pipettes (5 canisters of each)

Procedure

1. Pipette 10 ml of each of the milk samples to be tested into the sterile rubber-stoppered tubes.
2. Add 1.0 ml of the methylene blue dye to each tube of milk.
3. Tightly close the tubes with rubber stoppers.
4. Invert the tubes two or three times to mix adequately.
5. Place the tubes in a water bath set at 35°C.
6. After 5 min in the bath, invert the tubes once more and then reincubate without disturbing.
7. Observe the tubes every 30 min and note any color changes.
8. Record the time it takes for at least 80% of the tube to become white.
9. Compare the quality of the milk samples analyzed.

BACTERIAL POPULATIONS IN FOODS

In this experiment, a number of different types of foods will be selected for analysis. To make the investigation more interesting, use some of your own food samples. The technique employed in this experiment is referred to as the **pour-plate technique**. The pour-plate technique is the method most commonly used in food analysis. Samples of food are macerated and dilutions of the product are pipetted into sterile Petri plates. Molten agar is then poured into the plate, hence the name pour-plate. After incubation, colonies of bacteria appear both on the surface of the agar as well as embedded in the agar. Embedded colonies appear saucer- or disk-shaped.

Materials (per group)

1. Sterile Petri plates (2)
2. Samples of food (ground beef, chicken, fish, or vegetables) (3)
3. 99.0 ml sterile distilled water in dilution bottles (2)
4. Flask of sterile distilled water (500 ml)

5. Bottle containing 180.0 ml of sterile distilled water (1)
6. Molten standard plate-count agar held in 50°C-water bath (50 ml)
7. Food blender
8. Weigh boats

Procedure (See Figure 49.1)

1. With conditions as aseptic as possible, weigh out 20 g of the food you selected for analysis.
2. Place the food along with 180 ml of the sterile distilled water into a blender that has been rinsed with sterile distilled water, and blend for at least 5 min. This creates a 1:10 dilution.

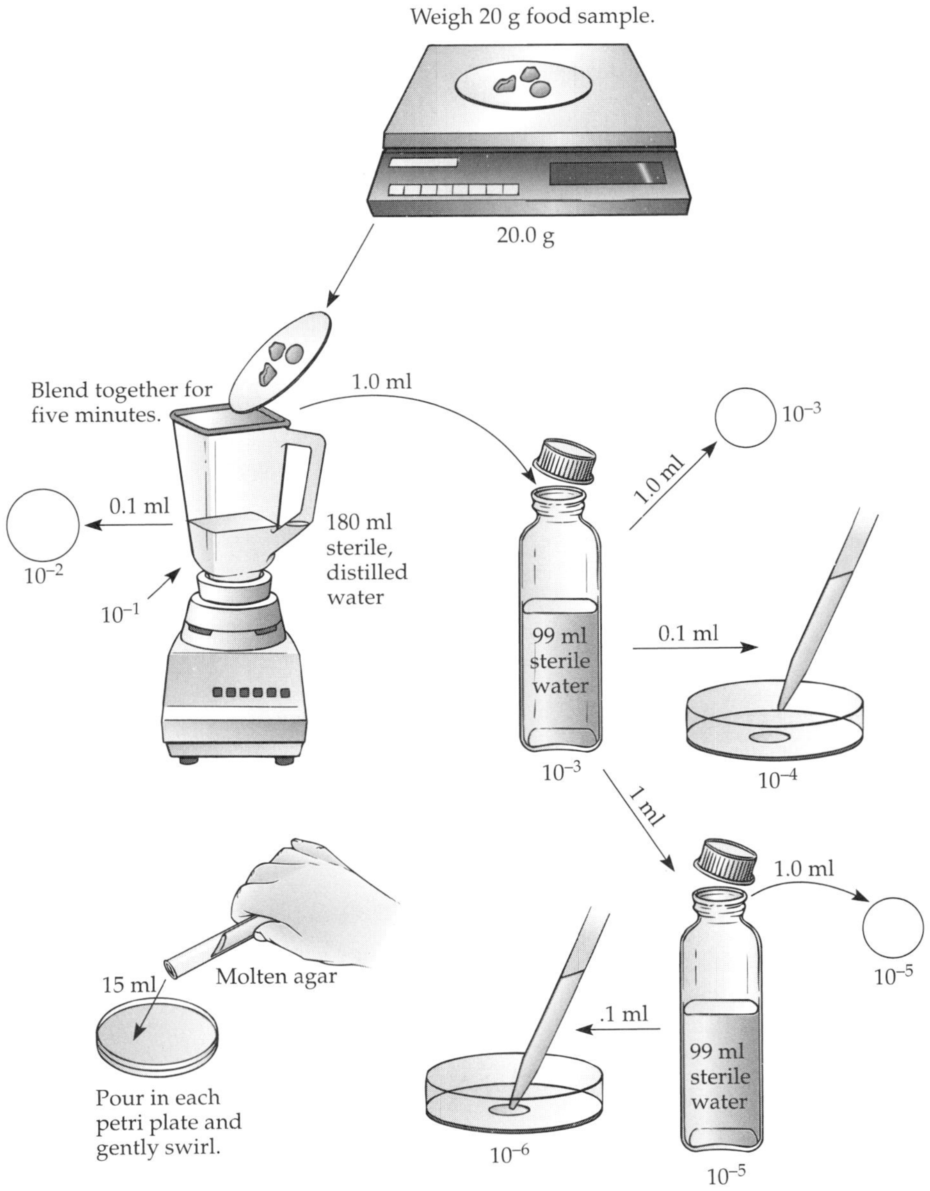

FIGURE 49.1 Food analysis procedure.

3. Transfer 1.0 ml of the 1:10 dilution to a 99-ml water dilution bottle. What dilution have you created? Transfer 0.1 ml to a plate marked 10^{-4}. Transfer 1.0 ml to a plate marked 10^{-3}.
4. Transfer 1 ml from the first sterile dilution bottle to another 99-ml sterile water dilution bottle. What dilution does this represent of the original? Also pipette 0.1 ml from this bottle to a plate marked 10^{-6}. Transfer 1.0 ml to a plate marked 10^{-5}.
5. Pour about 15 ml of the molten agar (45–50°C) into each of the Petri plates and gently swirl in a number of different directions to adequately mix the dilution with the agar. Care must be taken, however, not to splash any of the agar out of the plate. Allow to harden on the bench.
6. Incubate the plates in an inverted position at 35°C for 24 hr.
7. Count the number of colonies. For statistical significance, count the plates that have between 30 and 300 colonies.
8. Calculate the number of microorganisms per gram of food.
9. You may wish to analyze a second food sample. To do so, carefully rinse the food blender with tap water and then rinse with sterile distilled water. Use the same procedure, starting with step 1.

Laboratory Report 49

Name Date Section

Quality of Milk

Time of Incubation (min)	Color of Methylene Blue in:	
	Milk Sample A	**Milk Sample B**
0		
30		
60		
90		
120		
150		
180		
210		

Bacterial Population of Foods

Food	Dilution Plate	Number of Colonies	Microbial Titer

Review Questions

1. Does the methylene blue reduction test for milk reveal whether there are any pathogens present in milk?

2. Why does milk "spoil" even when it is kept refrigerated?

3. What are some pathogenic microorganisms that are associated with foods? Were there any significant differences in the populations of microorganisms present in the various samples of ground beef that were tested? Does a high population of organisms in ground beef necessarily impart a health risk?

LABORATORY

50 Production of Yogurt

Safety Issues

- Wear hot gloves when heating the milk to prevent burning of hands.
- Yogurt should not be eaten in the laboratory. Taste testing should occur outside of the lab.
- Remember to keep alcohol deodorizer away from burner flames.

OBJECTIVES

1. Observe the microorganisms present in yogurt and understand their roles in fermentation.
2. Prepare yogurt using starter cultures of microorganisms.

Yogurt is a fermentative end product formed by the action of microorganisms on lactose (milk sugars). The microorganisms involved in the fermentation are *Lactobacillus bulgaricus* and *Streptococcus thermophilus*. The latter rapidly ferment lactose to lactic acid using the glycolytic pathway. Under acidic conditions, the casein (milk protein) coagulates to form a semi-solid curd. *Lactobacillus bulgaricus* carries out additional fermentation, producing additional lactic acid and other aromatic compounds which impart the distinctive flavor and aroma to yogurt.

Materials (per group)

1. Commercially available unflavored yogurt with live, active bacterial cultures (1 container per lab)
2. Skim or low-fat milk (500 ml)
3. Non-fat dry milk powder (1 box per lab)
4. Hot plate or ring stand (5)
5. Thermometer
6. 400-ml beaker
7. Plastic spoons
8. Paper cups
9. pH paper
10. 45°C incubator
11. Tubes of sterile saline
12. Gram stain reagents (3 sets per lab)
13. Glass stirring rod
14. Aluminum foil
15. Flavorings (preserves, honey, or fruit)

Procedure

⚠ Be careful not to burn hands or fingers on the hot plate. Handle only with protective gloves.

1. Pour 100 ml of skim or low-fat milk into a beaker.
2. Bring the milk to the boiling point using a hot plate or ring stand/burner setup. Take care not to boil the milk.
3. Allow the milk to cool to about 60°C and add 3.5 g of the non-fat dry milk powder. Stir vigorously with the stirring rod to dissolve the powder.
4. Allow the mixture to cool to 45°C.
5. Add 1 to 2 tsp of the commercial yogurt. Mix well.
6. Cover the beaker with aluminum foil and incubate at least 6 hr at 45°C. The mixture should become firm.
7. Refrigerate the yogurt at 4°C. The yogurt may be stored overnight if desired.
8. Measure the pH of the yogurt product using a stirring rod and pH paper.
9. Add a sample of the yogurt to a paper cup.
10. The yogurt product may be tasted outside of the laboratory. If desired, add some fruit preserves or honey before tasting. (Some people do not like the taste of plain yogurt.)
11. Prepare a Gram stain of the yogurt to determine the types of bacteria present. Since there is considerable protein present, the smear prepared should be thin. Some of the yogurt may be placed in a tube of sterile saline and loopfuls transferred to a glass slide to prepare a more dilute preparation.
12. **Optional**: Prepare an agar medium suitable for the propagation and isolation of the organisms found in the yogurt. Remember the types of organisms normally encountered in yogurt. Streak the organisms on the medium, then cultivate and Gram stain organisms from the isolated colonies. How do the morphologies compare with those observed from the direct yogurt stain?

Laboratory Report 50

Name Date Section

1. Describe the consistency of the starting material and the end products.

2. Draw the morphological types of bacteria present in the yogurt as determined by the Gram stain.

3. Are the organisms gram-positive or gram-negative?

4. Compare the taste, consistency, odor, and pH of your end product with that of the original commercial yogurt.

Review Questions

1. Why is the milk heated to boiling prior to the addition of the commercial yogurt inoculum?

2. What effects would microbial contaminants have on the end product?

3. What types of medium would be suitable for the isolation of bacteria found in the yogurt?

4. Would you expect any change in the population of organisms during storage? How could such change be determined?

LABORATORY

51 Production of Sauerkraut

Safety Issues

- Sauerkraut should not be sampled in the laboratory. If you wish to taste it, do so outside of the lab.

OBJECTIVES

1. Learn the process involved in the fermentation of cabbage to produce sauerkraut.
2. Observe the succession of microorganisms that occur during the fermentation process.

The fermentation of cabbage to sauerkraut involves the action of a mixed population of microorganisms. These organisms are present on the leaves of the cabbage plant, so that when cabbage is cut, they are distributed throughout the cabbage tissue. Gram-negative coliforms, *Enterobacter* and *Erwinia*, gram-positive coccal-shaped *Leuconostoc*, and gram-positive *Lactobacillus* are all involved in the process. In the initial phases of fermentation, salt-tolerant bacteria proliferate to produce an anaerobic environment. The production of lactic acid follows, causing a lowering of the pH. After about one week when the acid concentration reaches about 0.8%, the activity of *Leuconostoc* is replaced by the fermentation of available sugars by *Lactobacillus*. Final concentrations of acids reach 1.5 to 2.0%.

Materials (per group)

1. Cabbage (1)
2. Vegetable graters (1)
3. Large bowls (1)
4. Large jars (Mason jars) (1)
5. Non-iodized salt (1 container per lab)
6. Balance
7. Weigh boats

Procedure

1. Remove the outer leaves of the cabbage. Cut the cabbage into three to four pieces and shred into a large bowl. Do not shred the core.
2. Weigh the cabbage in clean, large weigh boats or in a large, clean beaker and return to the large bowl.
3. Add 2.4% NaCl by weight to the cabbage (2.4 g NaCl/100 g of cabbage).

4. Pack two jars with the cabbage. Make sure all of the plant leaves are covered with the brine juice and that there are no air bubbles present.
5. Cover each of the jars with a small watch glass or plate. Add weight to make sure the leaves are compressed and covered with liquid.
6. Incubate at room temperature for two days.
7. Using a sterile pipette, obtain a brine sample from one of the jars and perform a Gram stain on it.
8. Incubate the jars for two to three weeks, periodically performing Gram stains on the brine in the sample jars.
9. After each sample, be sure that pressure is kept on the leaves so that they remain submerged in the brine.
10. After the incubation period, discard the first few layers of leaves from the undisturbed jar and note the aroma, color, and taste of sauerkraut.

Laboratory Report

Name Date Section

51

Record the appearance of the fermentation vessel and the morphological types of microorganisms present.

Time (days)	Appearance	Morphologies and Gram Stains of Organisms
2		
4		
6		
8		
10		
12		
14		

Describe the color, aroma, and taste of the final product.

Review Questions

1. What is the purpose of using salt in the process?

2. Describe the changes in the morphological types of bacteria that appear through the stages of fermentation.

3. Batches of home-prepared sauerkraut can become spoiled. What might be the cause of the spoilage?

4. Why is the maintenance of an anaerobic environment so important?

LABORATORY

52 Microbiological Populations in Soil

Safety Issues

- Always use a pipetting device for pipetting procedures. Never attempt to pipet any material by mouth.

OBJECTIVES

1. Learn techniques for the enumeration of bacteria and fungi in soil.
2. Be able to differentiate and identify the bacteria and fungi isolated from soil.

The first few inches of fertile topsoil contain an abundance of microorganisms. The number and types of organisms present is influenced by a variety of environmental conditions including (a) temperature, (b) pH, (c) moisture, and (d) nutrient availability. The microorganisms present play an important role in nutrient recycling. Heterotrophic bacteria can convert organic carbon and nitrogen into decomposition products that can be recycled for use by autotrophic bacteria and plants. Other heterotrophic organisms play an important role in a process known as bioremediation. In this process, unwanted pollutants are metabolized into harmless end products. Many industrial bioremediation companies sell bacterial powders to be used as the agents for hydrocarbon, detergent, or other chemical degradation. These organisms can be isolated directly from the soil in a polluted environment, or their populations can be enhanced by enrichment-culture techniques. In some cases, microbes have been genetically engineered to degrade the unwanted pollutants.

Soil not only contains a mixed population of bacteria, but also many types of protozoa, fungi, algae, viruses, and a group of filamentous bacteria known as actinomycetes. No one sampling strategy is sufficient to allow for the cultivation of all of these types. In this exercise, the pour-plate technique will be used to enumerate the total population of bacteria, fungi (molds), and actinomycetes.

Tryptic soy agar is used for the enumeration of heterotrophic bacteria in the soil sample. Sabouraud dextrose agar is used for the isolation of fungi; and glycerol yeast extract agar is used for the isolation of actinomycetes. These agars have their limitations, as estimates have been given that microbiologists are able to cultivate only 10 to 30% of all microorganisms present in nature. The remainder, while still being viable, are referred to as non-cultivatable, and must be studied by other techniques.

Materials (per group)

1. Two types of soil
2. Dilution bottles with 99.0 ml of sterile distilled water (15)
3. Molten tryptic soy agar for bacterial isolation (120 ml)

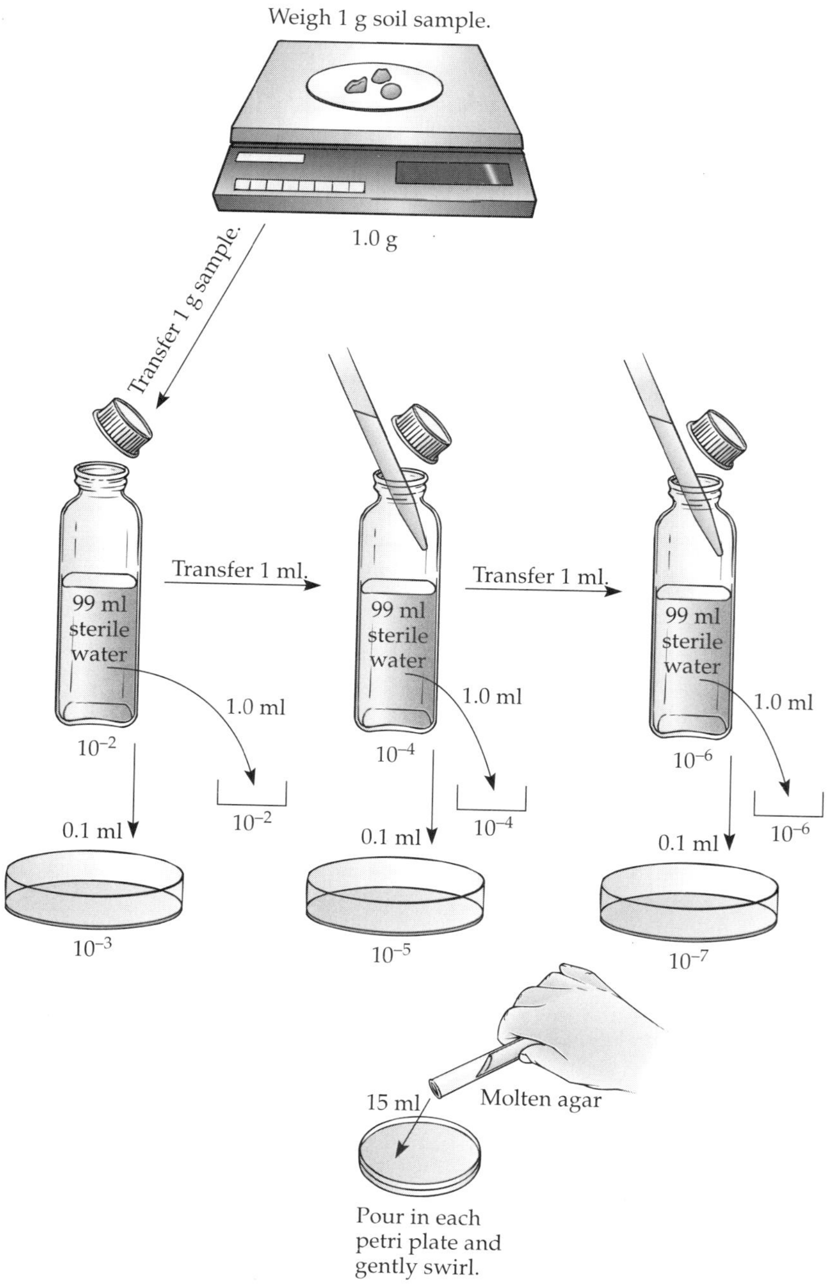

FIGURE 52.1 Dilution scheme for soil sampling.

4. Molten Sabouraud dextrose agar for fungal isolation* (120 ml)
5. Molten glycerol yeast extract agar for actinomycete isolation* (120 ml)
6. Sterile 1.0-ml pipettes (1 canister)
7. Water bath set at 50°C
8. Petri plates (18)

Procedure (See Figure 52.1)

1. Select 1.0 g of a soil sample and place it in 99.0 ml of sterile distilled water. Mix vigorously for at least 2 min. Calculate the dilution obtained.
2. Make serial dilutions of the sample by using a sterile pipette and transferring 1.0 ml from the first dilution bottle to a second bottle. Once again mix vigorously. A 10^{-4} dilution has been prepared.
3. Using a sterile pipette, proceed in a similar manner to prepare a 10^{-6} dilution.
4. Pipette appropriate volumes of each of the dilution samples into labeled sterile Petri plates.
5. Pour 15 to 20 ml of the appropriate molten agars into the Petri plates. (Be careful not to allow the agar to harden after removing from the water bath.) Rotate the plates in several directions while they are lying flat on the lab bench. Be careful not to slosh the agar out of the dish or onto the lid.
6. Allow the agar to harden and incubate at room temperature in an inverted position for seven days.
7. Check the plates periodically and count the plates that have between 30 and 300 colonies. The actinomycetes will usually form opaque rough colonies on the agar. In some cases, they form a ramifying network of filaments similar to the mycelia of filamentous fungi.
8. Determine the number of each type of organism present per gram of soil sample.

*Containing 15 μg/ml of aureomycin or streptomycin to prevent bacterial growth.

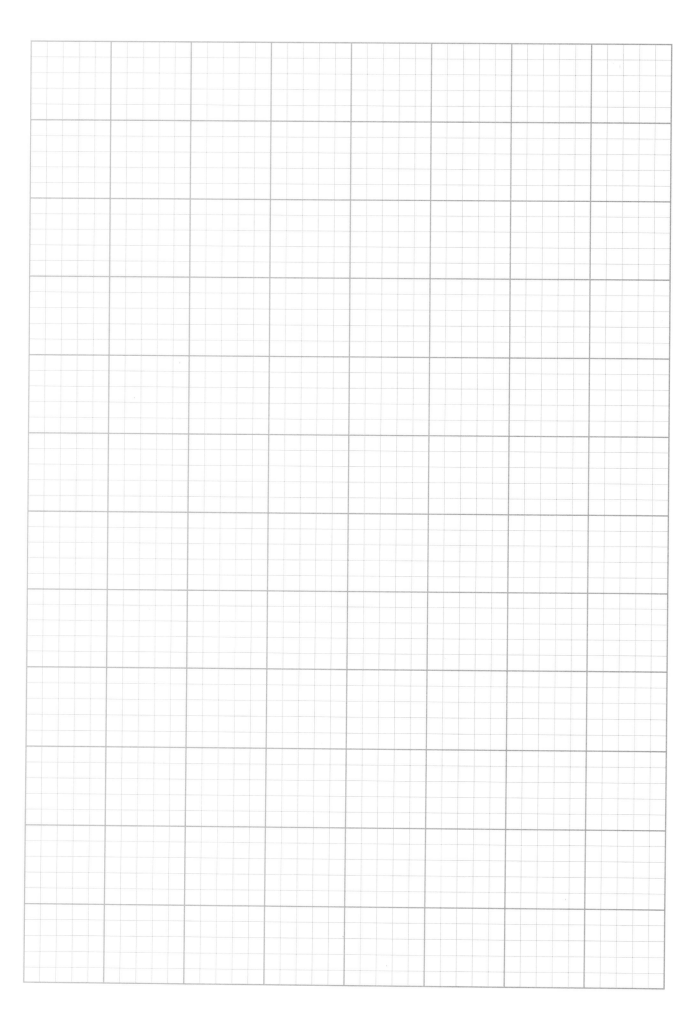

Laboratory Report 52

Name Date Section

Total Microbial Population of Soil

Soil Type	Medium	Microbial Population
	TSA	
	TSA	
	SDA	
	SDA	
	GYE	
	GYE	

Review Questions

1. Why were different media used for the propagation of the soil microorganisms?

2. What is the significance of the ratio of bacteria, fungi, and actinomycetes in the soil samples that were tested? What factors would influence the ratio found in a particular soil sample?

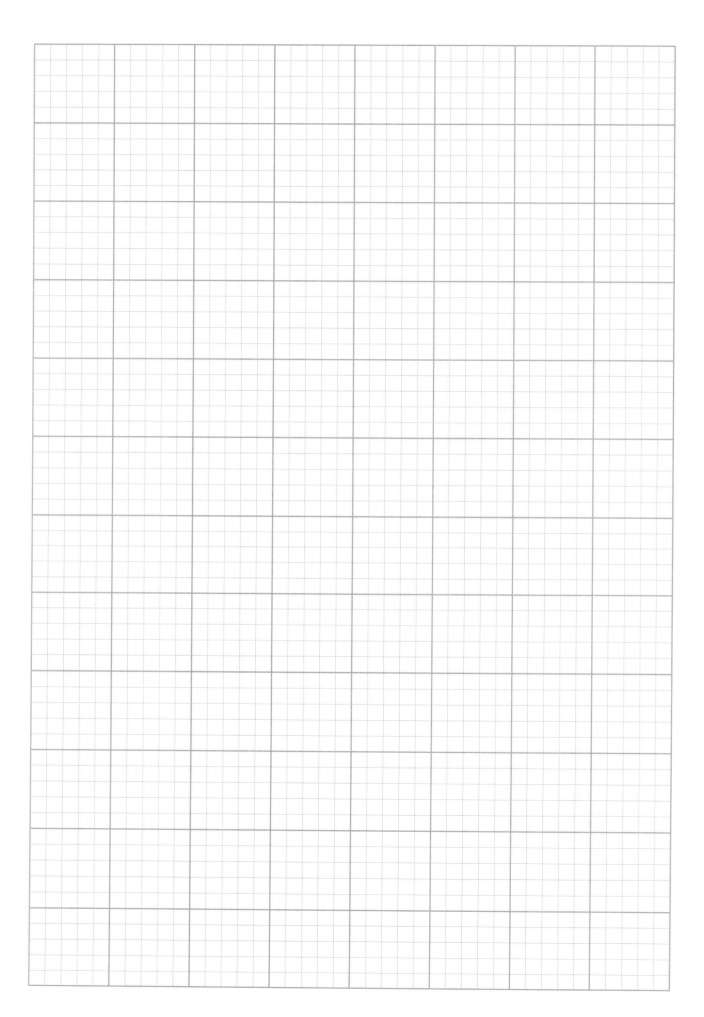

LABORATORY

53 Isolation of an Antibiotic-Producing Microorganism

Safety Issues

- Always use a mechanical pipetting device for pipetting procedures. Never attempt to pipet any material by mouth.

OBJECTIVES

1. Isolate antibiotic-producing organisms from soil.
2. Measure the effectiveness of antibiotics produced by soil microorganisms against gram-positive and gram-negative microorganisms.

Antibiotics are chemicals that are produced by certain microorganisms that inhibit the growth of or kill other microorganisms. Most antibiotics currently in use are produced by soil microorganisms belonging to one of the following genera: *Bacillus*, *Penicillium*, *Streptomyces*, or *Cephalosporium*. The production of antibiotics is one of the most economically important and therapeutically significant industrial fermentations.* Some antibiotics are still produced by microbial fermentation, while others are prepared semi-synthetically or by complete chemical synthesis. The techniques used in the discovery of a new antibiotic-producing microorganism are similar to the techniques first used by Fleming, Florey, and Chain in the original isolation and use of *Penicillium*. Large numbers of microorganisms must be screened in order to select appropriate candidates that produce diffusible products that may be inhibitory for other bacteria. This screening process is usually conducted on agar plates. Colonies that produce inhibitory agents are selected and grown in pure culture for further study. The suspected producer is then tested against a battery of pathogenic microorganisms. Areas of inhibition surrounding the suspected producer are indicative of effectiveness against the test organisms. Alternately, a filtrate from the antibiotic producer can be added to broth cultures of test microorganisms to assess potential effectiveness of any antibiotic in the filtrate. Particular attention is paid to any isolates that represent potential new genera of antibiotic-producing microorganisms.

Promising isolates are then grown in bench-scale fermenters to determine the extent to which the antibiotic is freely diffusible. If not readily diffusible, then extraction and purification techniques need to be designed. Before large scale production can take place, toxicity and therapeutic testing is conducted using animals.

*Hundreds of thousands of tons of antibiotics are produced yearly, with sales recorded in the billions of dollars.

High-yielding mutant strains have been isolated from many primary isolates and are used extensively in industrial fermentations. The application of genetic engineering techniques has also led to the selection of high-yielding strains.

Once the chemical structure of a new antibiotic has been determined, chemical modification techniques can be applied to produce alternative forms of the antibiotic. Immobilized enzymes are also used to catalyze critical steps in the synthesis of new types of penicillin.

In this experiment, soil samples will be screened for the presence of organisms that produce substances inhibitory to the growth of other microorganisms. Test organisms are then used to test for effectiveness and the range of activity of the inhibitory compounds.

Materials (per group)

1. Rich garden soil
2. 99.0-ml water dilution bottle (non-sterile)
3. 1.0-ml pipettes (1 canister per group)
4. Molten tryptic soy agar (60.0 ml)
5. Molten Sabouraud maltose agar (60.0 ml)
6. 9.0-ml tap water tube
7. 5.0-ml tap water tube
8. Tryptic soy agar slants (5)
9. Petri plates (6)
10. Tryptic soy agar plates (5)
11. Slant cultures of
    Escherichia coli
    Micrococcus
    Pseudomonas
    Bacillus

Procedure

1. Weigh out 1.0 g of soil and add it to the bottle containing the 99.0 ml of tap water and shake vigorously.
2. Prepare a 10^{-3} dilution by pipetting 1.0 ml from the dilution bottle to the tube containing 9.0 ml of water. There is no need to utilize aseptic technique in this experiment. Vortex to mix well.
3. Transfer 5 ml from the 10^{-3} dilution tube to a tube containing 5.0 ml of tap water. You should recognize that you have created a 1:2000 dilution of the original sample. Make sure to vortex once again to mix well.
4. Transfer 1.0 ml from each of these dilutions to duplicate plates marked 1:100, 1:1000, and 1:2000.
5. Pour approximately 20 ml of the molten tryptic soy agar into one series of plates, and Sabouraud maltose agar into the other series. Mix by rotating the plates. Allow to solidify.
6. Incubate in an inverted position at room temperature for four to six days, inspecting frequently.
7. Observe for zones of inhibition on any of your plates. If no zones are apparent, check with another member of the class.

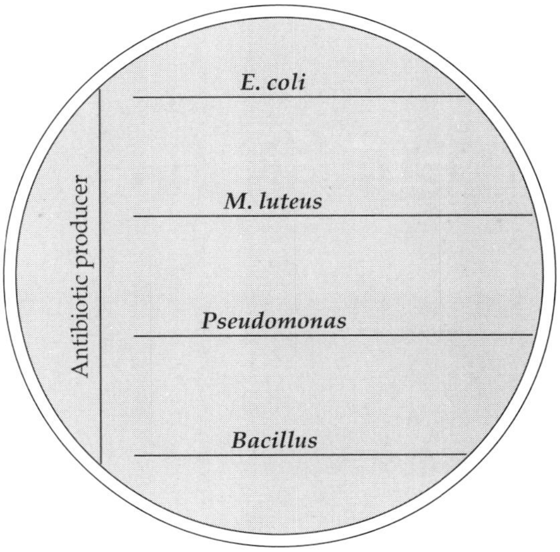

FIGURE 53.1 Streak method for measuring antibiotic effectiveness.

8. Carefully "fish" any colony that is showing a zone of inhibition with an inoculating needle. Transfer to a tryptic soy agar slant and incubate for two to four days at room temperature.
9. Perform a Gram stain on your suspected antibiotic producer, noting the stain and morphology of the isolate.

Determine the spectrum of activity of the antibiotic producer by making a straight-line inoculation onto a tryptic soy agar plate as illustrated in Figure 53.1. At right angles to the line of inoculation of the suspected antibiotic producer, streak the following organisms: (a) *Escherichia coli*, (b) *Micrococcus luteus*, (c) *Pseudomonas*, and (d) *Bacillus*. Be careful not to overlap the streak of the suspected antibiotic-producing organisms. Incubate at room temperature for 24 to 48 hr. Examine the plates and determine the spectrum of effectiveness against the test organisms.

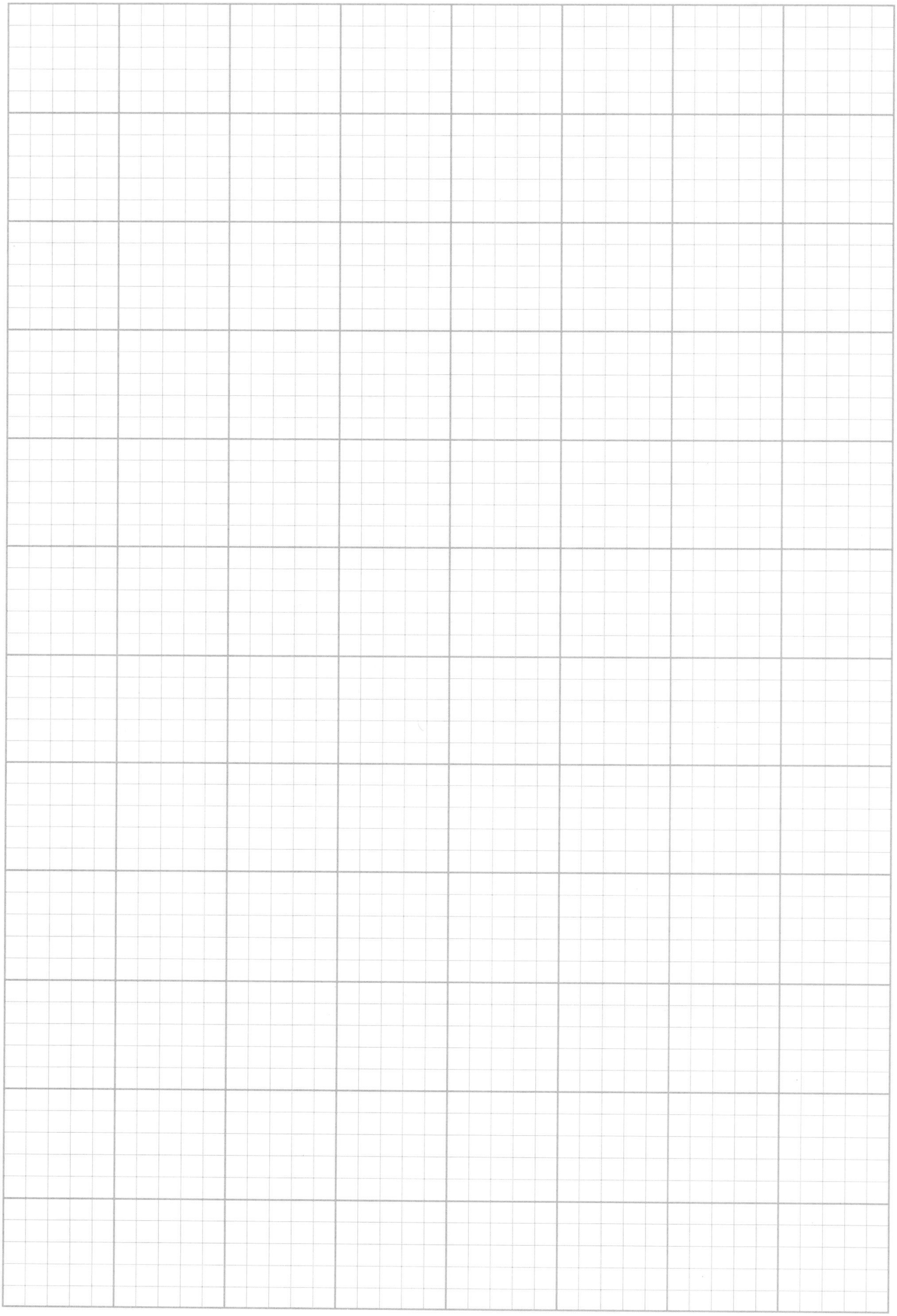

Laboratory Report 53

Name _______________ Date _______________ Section _______________

Antibiotic Producers on Dilutions

Media	1:100	1:1000	1:2000
TSA			
SMA			
Gram Stains of Antibiotic Producers:			
Plate ____		**Plate** ____	
Dilution ____		**Dilution** ____	
Stain ____		**Stain** ____	

Range of Activity of Antibiotic Producer:

(a) *E. coli*
(b) *Micrococcus*
(c) *Pseudomonas*
(d) *Bacillus*

Review Questions

1. Were any of the antibiotic-producing isolates capable of retarding the growth of any of the three test organisms used?

2. How could any of the antibiotic-producing organisms be preserved for further study? How could the range of activity of the antibiotic producer be more precisely determined?

3. Is the search for new antibiotic producers in industrial laboratories done in the same way as in this laboratory? What are some of the other methods currently being used by pharmaceutical companies to develop new antibiotics?

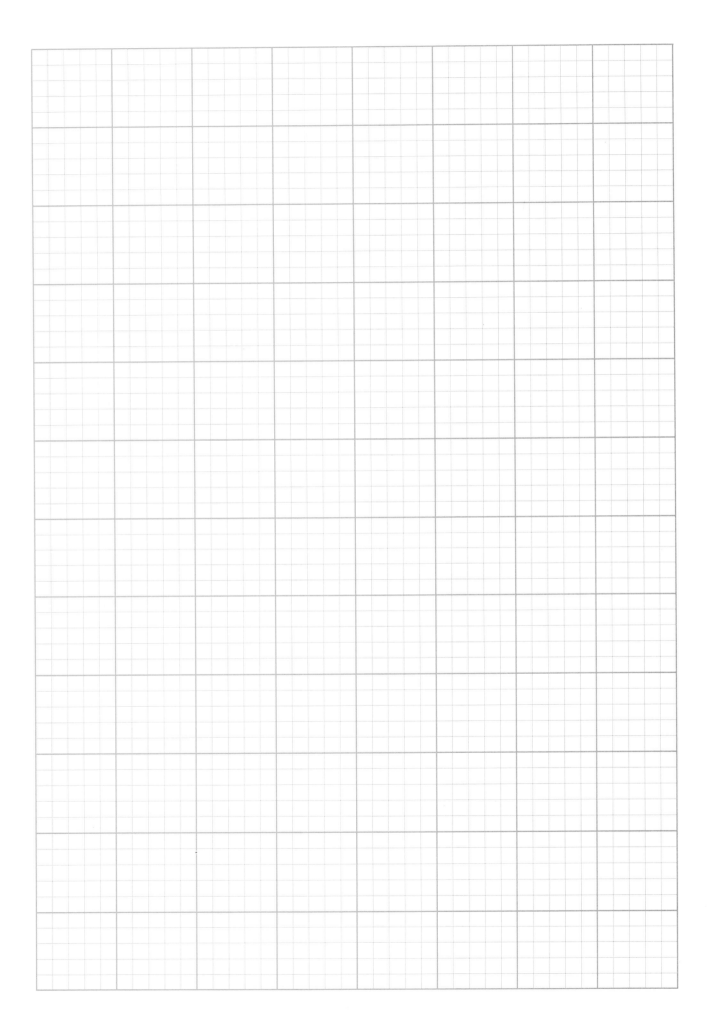

LABORATORY

54 Enrichment-Culture Technique

Safety Issues

- Always use a mechanical pipetting device for pipetting procedures. Never attempt to pipet any material by mouth.

OBJECTIVE

Learn how to use the enrichment-culture technique to isolate a specific organism from soil.

Microorganisms can adapt (acclimatize) to grow under various environmental conditions. One can vary the composition of the medium to select for a particular microorganism. The basic strategy involves a large population of microorganisms in a rich medium containing the particular chemical to be degraded. For example, if selecting for a microorganism that is capable of degrading a detergent, a small amount of detergent is added to the medium. After growth has occurred, successive transfers are made to media containing smaller amounts of the complete nutrient, while holding the concentration of the detergent constant. Further transfers are made until the final step, which is the addition of the organisms to a medium containing only minimal salts and the detergent. At that point, organisms that utilize detergent as a sole source of carbon and energy have been isolated.

The enrichment-culture technique can be employed for the isolation of microorganisms that are able to metabolize pollutants which may be toxic to plants and animals. These organisms can then be used in bioremediation procedures to rid a particular environment of an unwanted, toxic pollutant.

Materials (per group)

1. Minimal salts medium (500 ml)
2. 250-ml Erlenmeyer flasks (5)
3. Soil sample
4. Incubator shaker
5. Nutrient broth
6. Carbon sources to be degraded
7. Minimal salts agar plates
8. Sterile 1.0-ml pipettes (1 canister)

Procedure

1. Set up a series of 250-ml flasks containing 100 ml of minimal salts medium plus varying concentrations of nutrient broth, as illustrated in Figure 54.1.

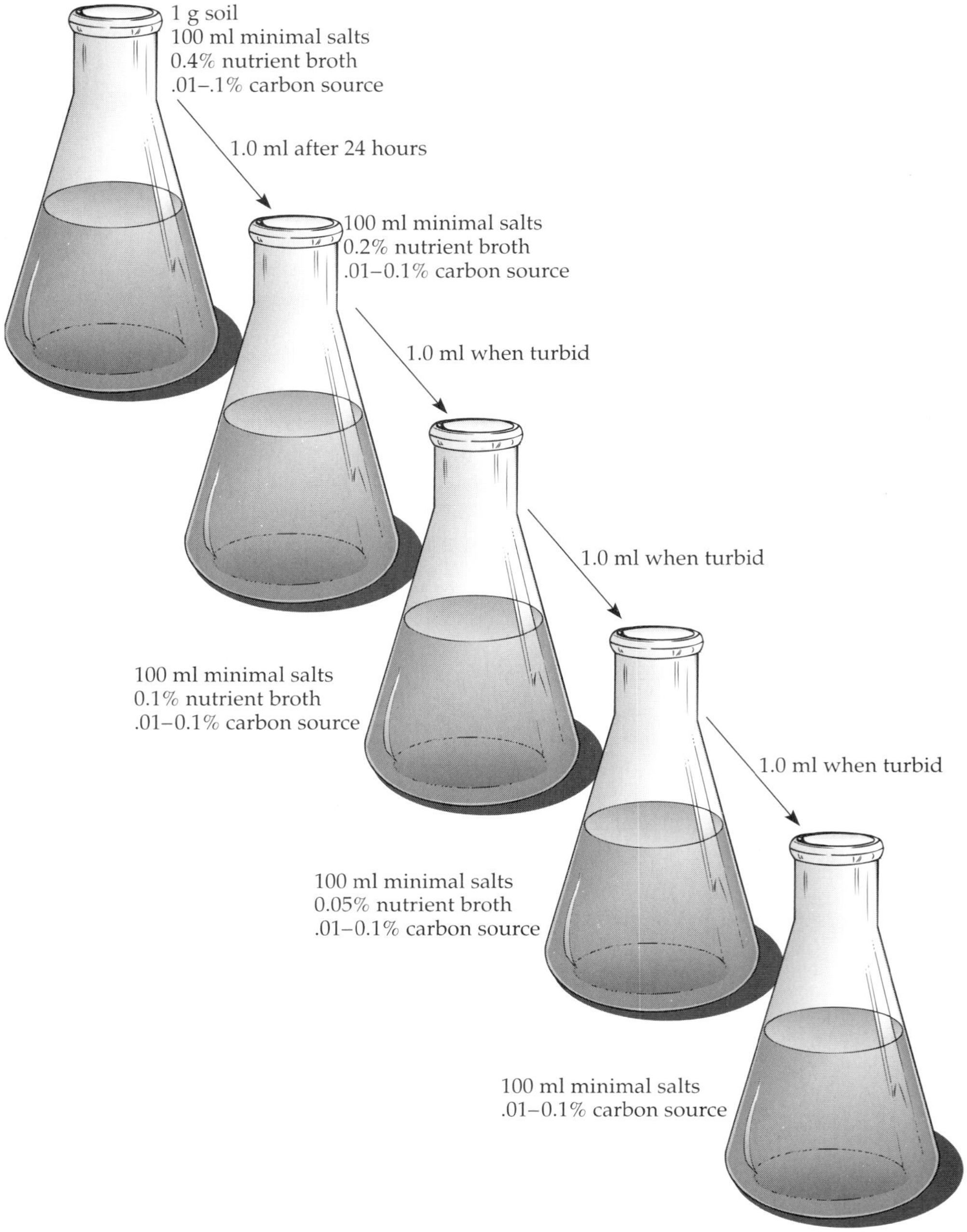

FIGURE 54.1 Enrichment-culture technique protocol.

2. Add the carbon source to be degraded (0.01–0.1% w/v)
3. Add 0.4% w/v nutrient broth (NB)
4. Inoculate with 1 g of soil or other appropriate sample.
5. Incubate with shaking at 30°C for 24 hr.
6. After growth has occurred, transfer 1.0 ml to a second flask containing 0.2% nutrient broth (NB) in minimal salts (MS) plus additional carbon source (C).
7. Examine regularly for growth. Once turbidity is evident, transfer 1.0 ml to the third flask containing 0.1% NB, MS + C medium.
8. After growth has occurred, transfer 1.0 ml once again to a flask containing 0.05% NB + MS + C.

9. The final transfer should be made to a flask containing only MS + C.
10. Incubate until growth is apparent. To verify that the C source is being degraded, streak a loopful of the medium onto the surface of a mineral salts-C-source-agar plate.
11. If growth is apparent on the agar plate, a microorganism that is capable of degrading the selected C source has been isolated.

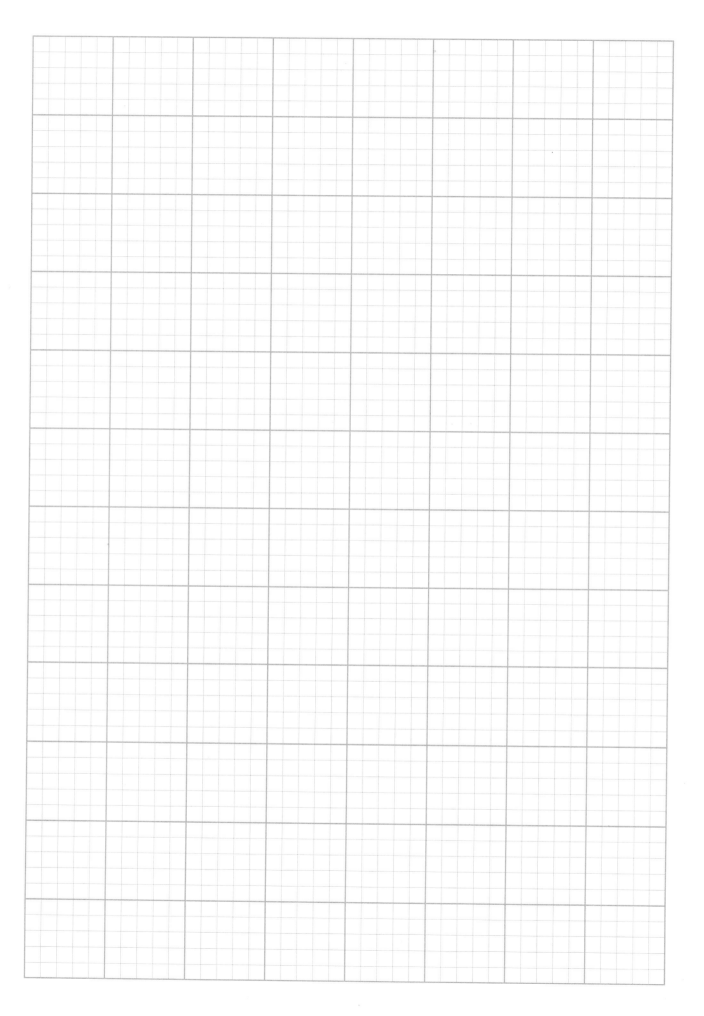

Laboratory Report 54

Name ______ Date ______ Section ______

Carbon Source: ______

Appearance of Growth in Flasks

Time	0.4% NB + C	0.2% NB + C	0.1% NB + C	0.05% NB + C	MS + C

Review Question

If the enrichment-culture technique was successful, what further experiments could be run using the isolate? If not successful, does this mean that organisms capable of degrading the selected carbon source do not exist? How might the experiment be modified to better optimize chances of success?

UNIT VIII

Algae, Fungi, and Protozoa

This unit examines the eukaryotic microorganisms including the algae, molds (filamentous fungi), and protozoa. These organisms are likely to be encountered in habitats where bacteria exist. The algae contain chlorophyll and carry out oxygenic photosynthesis. Unicellular, colonial, and filamentous forms are likely to be found in the environments such as pond water. Algae have been used as research organisms, as sources of food, and as sources of agar. Certain algae may grow in water causing distasteful drinking water and contaminated swimming pool water. They are also responsible for contamination of building products and for algal blooms in ponds.

Molds are filamentous fungi that lack chlorophyll and form asexual spores (conidia) which are resistant to drying and allow for the spread of the organisms. These airborne spores are responsible for many allergies. Molds are important decomposers of organic matter in the soil. They are also used in the production of various cheeses and biochemicals.

The protozoa lack cell walls and chlorophyll, and are motile. They are found in many water habitats, and as parasites in animals. They move by amoeboid movement or by flagella.

All of these eukaryotic microorganisms play a significant role in nature and impact on our daily lives. The next three laboratories present opportunities to study a few of these organisms.

LABORATORY

55 Algae and Cyanobacteria

Safety Issues

- Keep alcohol away from open flame. Do not place hot forceps into alcohol container.
- Be careful of sharp edges on glass slides or coverslips. Dispose of broken glass in appropriately marked containers.

OBJECTIVES

1. Differentiate the common types of algae found in pond water.
2. Isolate and propagate algae from contaminated surfaces such as roofing shingles.

Algae are a diverse group of photosynthetic organisms which are important as primary producers of organic material and as oxygenators of water during periods of daylight. They exist in fresh water and marine environments and can also be found in terrestrial habitats. Algae vary in complexity from microscopic single cells to large seaweeds. Algae have a long history dating back 3.1 to 3.3 billion years to the Precambrian era.

Algae have beneficial and detrimental roles for humankind. Besides being primary producers and evolvers of oxygen, their structural components are important sources of nutrients for other life forms. The vitality of coral reefs is influenced by the photosynthetic activity of algae. Some algae are used as food and for commercial purposes. Diatomaceous earth (remains of diatoms) is used in filtration processes. Agar is extracted from red algae and is used widely as the primary, inert solidifying agent in microbiological culture media. Other compounds isolated from marine algae are used as thickening agents and stabilizers in food and cosmetics.

One of the negative effects of algae is algal blooms, that is, the growth of massive numbers of algae ponds, lakes, and coastal waters. The algae can produce toxins that are harmful to aquatic animals; can impart an unpleasant taste to drinking water; and can cause the formation of thick mats, resulting in formation of anaerobic conditions in water leading to the death of animal life. Algae and cyanobacteria are also responsible for discolorations found on roofing material and building sidings.

Most fresh water algae encountered are grouped into one of the divisions described in the following sections.

Division Chlorophyta

The green algae represents one of the major groups of algae found in aquatic environments. They possess chlorophylls a and b found in association with thylakoids in chloroplasts. There is considerable morphological diversity.

They may be unicellular, colonial, flagellated and non-flagellated, filamentous, and tubular. *Chlamydomonas* is a unicellular flagellated member of this division. Colonial types include: *Volvox, Gonium,* and *Pandorina. Ulothrix, Stigeoclonium,* and *Oedogonium* are examples of filamentous types.

Division Chrysophyta

The golden brown algae possess chlorophylls a and c but lack chlorophyll b. Carotenoid pigments like fucoxanthin are responsible for the golden to yellow-green pigmentation. The primary food storage product is leucosin, a β-1,3 polymer of glucose. Lipid reserves are also found. Most of the common forms are flagellated single cells (chrysococcus) or colonies of flagellated cells (synura). Diatoms form one class and are important components of many aquatic algae communities. They occur as individual cells or in small colonies. The distinguishing feature is the presence of silica walls (usually composed of two valves) surrounded by a mucilaginous material. The two valves fit together like the lid and bottom of a Petri dish. Examples of diatoms include *Diatoma, Navicula,* and *Nitzschia.*

Division Phaeophyta

Of the 260 genera of brown algae, about 6 have been reported as fresh water types. Photosynthetic pigments include chlorophylls a and c, β-carotene, fucoxanthin, and violaxanthin. The food storage reserve is laminarian, a β-1,3-linked glucans with some β-1,6-linkages. Sizes range from microscopic forms to the large seaweed *Macrocystis,* which may grow to more than 50 m in length. Since most are marine specimens, it is improbable that any will be observed in a pool water sample.

Division Euglenophyta

Members of this division possess chlorophylls a and b, lack a cell wall, and have paramylon, a β-1,3 polymer glucose, as their carbohydrate reserve. They have a proteinaceous pellicle under the membrane and an eyespot. All members, except for one type, are unicellular flagellates. Many biologists categorize members of this division as zoo flagellates. Examples of this division include *Euglena, Phacus,* and *Trachelomonas.*

Division Pyrrophyta

Members of this division are dinoflagellates. The minority are marine forms, but many species occur in fresh water. Only about half possess the photosynthetic pigments, chlorophylls a and c, while others are strict heterotrophs. The storage materials include starch and lipids. An interesting anatomical feature are flattened plates of cellulose called theca which lie beneath the cell membrane. Examples or organisms in this division are *Peridinium, Ceratium* and *Glenodinium.*

A description of cyanobacteria is also included here since they are common in most aquatic and terrestrial environments and are isolated with algae. Members of the genera *Gloeocapsa* and *Scytonema* have been implicated in the discoloration of many building materials. The cyanobacteria are true oxygenic photosynthetic bacteria. Both unicellular and filamentous forms exist with considerable morphological diversity apparent. The cell walls of cyanobacteria are similar to gram-negative bacteria containing a peptidoglycan layer. Many of the organisms are surrounded by a viscous sheath which may be pigmented. Some members form heterocysts (thick-walled structures within which nitrogen fixation occurs). Some representative genera are *Gloeocapsa*, *Microcystis*, *Nostoc*, *Anabaena*, and *Oscillatoria*.

EXAMINATION OF POND WATER FOR ALGAE AND CYANOBACTERIA

Samples of pond water are to be analyzed for algae and cyanobacteria by using either a wet mount or a membrane-filtration technique. Care should be taken to use a fresh pond water sample as deterioration of algae may occur in old samples.

Materials (per 20 students)

1. Pond water samples (at least 5)
2. Known unicultures of algae and cyanobacteria (5–10)
3. Microscope slides
4. Glass coverslips
5. Sterile membrane-filtration apparatus (5)
6. Sterile 0.45-μm filters (20)
7. Immersion oil
8. Forceps
9. 70% ethanol

Procedure

A. Slide Technique

1. Obtain a sample from the bottom of the pond water container and place a few drops on a slide. Cover with a glass coverslip.
2. Observe types of algae and cyanobacteria present under low power (4–10×) and high power (40×–97×). Draw the types observed.
3. Repeat procedure with the known cultures.
4. Use drawings of representative types of organisms in Figure 55.1 and observation of known cultures to help identify the algae observed in the pond water sample.

B. Filtration Technique

1. Hook up the sterile membrane-filter flask to either a vacuum pump or an aspirator.
2. Sterilize forceps by dipping in alcohol. Flame off alcohol by briefly passing the forceps through the flame of a Bunsen burner.
3. Using the sterilized forceps, place a sterile 0.45-μm membrane filter on the base of the filtration apparatus.
4. Filter 10 to 100 ml of a pond water sample. For filtration volumes less than 50 ml, add 50 ml of sterile distilled water to the funnel before placing sample in and filtering. If the filter clogs, a smaller volume of water may need to be filtered.

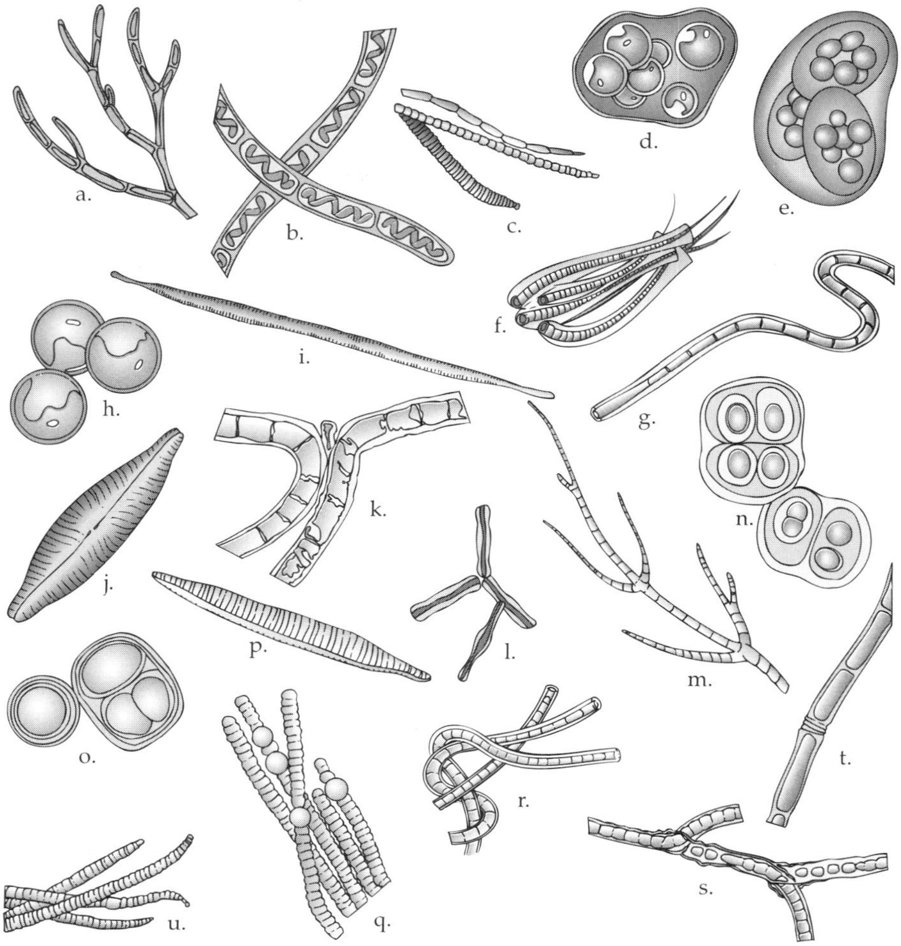

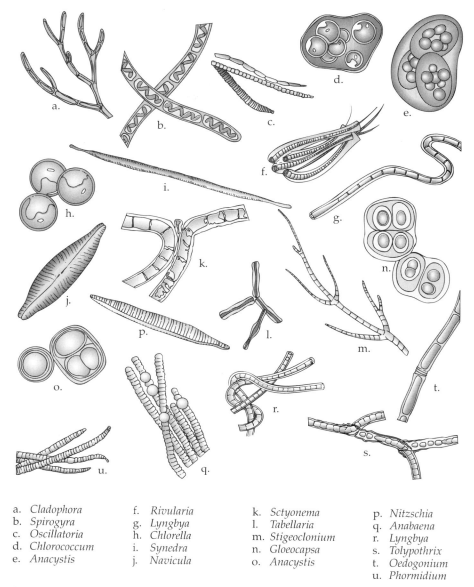

a. *Cladophora*
b. *Spirogyra*
c. *Oscillatoria*
d. *Chlorococcum*
e. *Anacystis*
f. *Rivularia*
g. *Lyngbya*
h. *Chlorella*
i. *Synedra*
j. *Navicula*
k. *Sctyonema*
l. *Tabellaria*
m. *Stigeoclonium*
n. *Gloeocapsa*
o. *Anacystis*
p. *Nitzschia*
q. *Anabaena*
r. *Lyngbya*
s. *Tolypothrix*
t. *Oedogonium*
u. *Phormidium*

FIGURE 55.1 Representative types of algae.

5. After filtering, use alcohol-sterilized forceps to remove the membrane filter from the support, and place in a sterile dish. Allow the filter to completely dry (20-30 min).
6. Add a few drops of immersion oil to the dried filter. As the oil is absorbed the filter should become transparent. If the filter is not completely dry, it will remain opaque and additional drying time may be necessary.
7. The filter can then be microscopically examined directly in the Petri dish, using the oil-immersion objective (100×); or sections of the filter may be cut and placed on a microscope slide for examination. Exercise caution so as to not contaminate any of the low-power objectives with immersion oil.
8. Draw the types of algae observed on the filter.

CULTIVATION OF ALGAE AND CYANOBACTERIA FROM SURFACES

Algae are known to cause contamination of building materials such as siding and roofing shingles. The contaminated areas usually appear as dark discolorations. A suspected area of contamination can be scraped, and material transferred to a growth medium for the propagation of algae or cyanobacteria. Those types that have been implicated are *Gloeocapsa*, *Scytonema*, and *Lyngbya*.

Materials (per 20 students)

1. Contaminated (discolored) building products
2. 125-ml Erlenmeyer flasks containing 50 ml of sterile Alga-Gro medium (Carolina Biological Supply Co.) (20)
3. Forceps or razor blade (20)

Procedure

1. Scrape the contaminated product to remove some of the discolored shingle granules or surface film from siding.
2. Transfer some of the scrapings to flasks of Alga-Gro medium.
3. Incubate at room temperature under a fluorescent grow-light bank.
4. Periodically sample the media for the presence of algae by using the wet-mount slide technique.

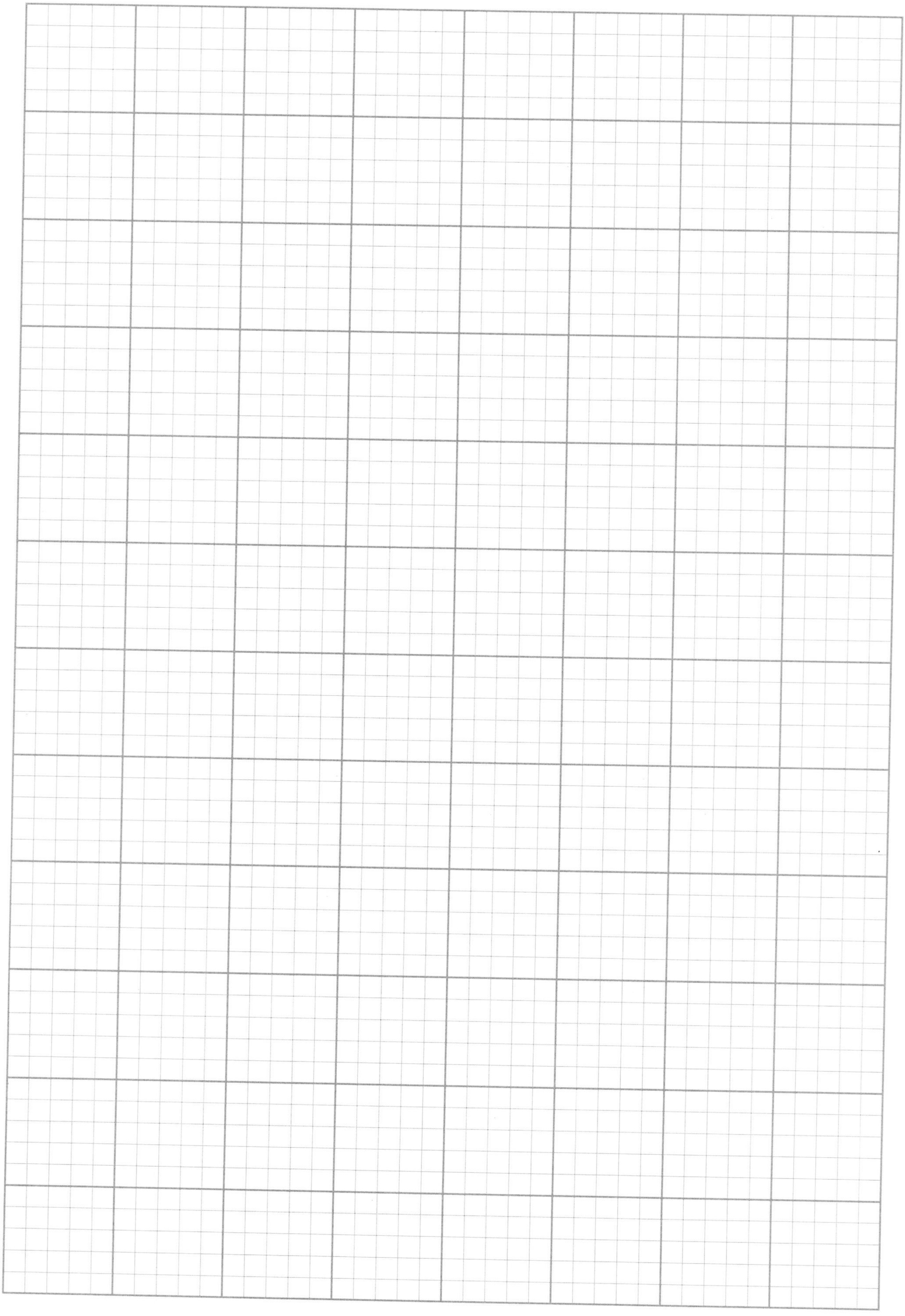

Laboratory Report 55

Name ____________________ Date ____________ Section ____________

Slide Observations. Sketch the types of algae and cyanobacteria observed in pond water samples.

Filter Observations. Sketch the types of algae and cyanobacteria observed.

Describe the appearance of the media in flasks containing the contaminated building material.

Sketch the types of algae cultivated from the contaminated building products.

Review Questions

1. What advantages does the filtration method have over the wet-mount slide technique?

2. What conclusion can be reached about the quality of the pond water sampled?

3. What needs to be done to prove that any cultivated algae are actually the cause of the discoloration of the building material?

4. What controls should be run as part of the experiment on the cultivation of algae and cyanobacteria from surfaces?

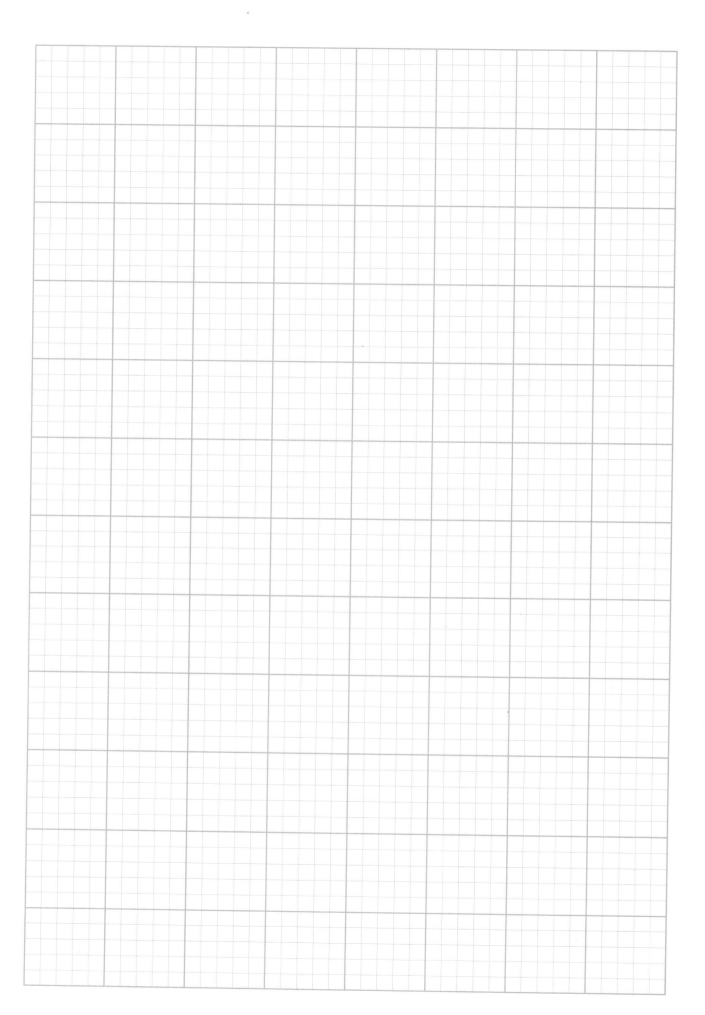

LABORATORY

56 Mycology—An Introduction to the Filamentous Fungi

Safety Issues

- Exercise caution to keep alcohol away from open flame.
- Keep plates closed to avoid laboratory contamination with airborne spores.
- Individuals with known fungal allergies or immunocompromised individuals should not perform this lab.

OBJECTIVES

1. Become familiar with techniques used to propagate filamentous fungi, particularly molds.
2. Learn to identify molds by examination of "known" organisms and to isolate and identify molds from the environment.

At this point in the study of microbiology, it is most probable that you have encountered molds at some point as laboratory contaminants. These organisms form filamentous, cottony, hairlike colonies which can be found growing on many substances including microbiological media, refrigerator gaskets, foods, and building surfaces. Also, a number of fungi are commonly recovered from clinical specimens of skin, respiratory tract, genitourinary tract, cerebrospinal fluid, and blood. Increasing numbers of fungal infections are being reported, especially in patients who are immunocompromised or have long-term intravenous cannulation or chemotherapy. Under microscopic examination, the mass of branching filaments observed are called **mycelia**. The individual filaments that make up the mycelia are referred to as **hyphae**. The process of identifying the various molds involves careful morphological observations. For observation of clinical specimens, stains such as India ink, calcofluor white, or periodic acid-Schiff stain are used for detection of fungi. A 10% solution of potassium hydroxide or sodium hydroxide is used to digest tissue fragments, such as skin scales or membranes. If not digested, they may interfere with detection of fungal mycelia. The hypha that grow on the surface of a nutrient that are specialized for the procurement of nutrients are called the **vegetative mycelium**. Another portion of the mycelial mat differentiates into aerial structures known as the **reproductive mycelium**. This mycelium is involved in the elaboration of reproductive structures which will bear the asexual reproductive **spores** or **conidia**. In some species, the hyphae have cross walls and are called **septate hyphae**. Other hyphae lack cross walls and are known as **non-septate** or **coenocytic hyphae**. These hyphae are easily differentiated from bacteria in that most hyphae are up to ten times as large in diameter as are bacteria. This makes lower-power microscopic observations possible.

The reproductive mycelium and its spores are widely diverse in different fungi, and this serves to classify and identify them. Major taxonomic groupings include members of the subdivision Zygomycotina. These fungi are saprophytic and have non-septate hyphae. A saprophytic fungus grows on decaying organic material. An example belonging to this group is the common bread mold, *Rhizopus*. They form asexual spores (sporangiospores), which are held within a sac called a sporangium. The sporangia are attached to stalks called sporangiophores. Sexual spores (zygospores) can also be formed by the union of nuclear material from the hyphae of two different strains.

The subdivision Ascomycotina comprises organisms that are commonly referred to as the ascomycetes. They have septate mycelium and produce sexual spores (ascospores) in sacs called asci. Members of this group also produce asexual spores (conidiospores), which are formed on specialized hyphae called **conidiophores**. The spores are borne free, and are not contained within a membrane or sac. Examples are *Penicillium* and *Aspergillus*.

The subdivision Basidiomycotina produce sexual spores called **basidiospores**, which are held in club-shaped bodies called basidia. Members include the mushrooms and shelf fungi found on dead tree branches.

Molds are classified principally on the basis of the form of reproductive structures. They also can be characterized by color and shape of the colony, and by the type of hypha. This exercise provides an introduction to the cultivation of molds and the use of techniques to isolate them from the environment.

Since observation of the reproductive structures of the molds is important in identification, techniques have been developed that optimize visualization of these structures. If a regular whole plate of agar were used for sampling, within a short period of time the entire plate would be covered with a vast network of mycelia. It would be difficult to observe reproductive structures under the light microscope. A number of techniques have been used to help better visualize these reproductive structures.

In this exercise, molds isolated from the environment are compared with known mold cultures. Care must be exercised to microscopically identify the presence of mycelia, in order to delineate the types of reproductive mycelia, as well as to identify conidial structures. These observations are critical to establish the presence of mold in a specimen and to verify its presumptive identification.

Materials (per 20 students)

1. Glass slides
2. Concave glass slides (20)
3. Vaseline
4. Glass rods (20)
5. Petri plates (20)
6. Tubes of molten Sabouraud dextrose agar (50 ml)
7. Rodac plates filled with Sabouraud dextrose agar (20)
8. Alcohol
9. Spatulas (10)
10. Known mold samples (5–10)
11. Mold samples isolated previously

REGULAR GLASS SLIDE TECHNIQUE (SEE FIGURE 56.1)

Procedure

1. Clean a glass slide with alcohol.
2. Pipette a few drops of molten Sabouraud agar onto the surface of the slide and allow it to solidify.
3. Scoop off a portion of the circle (half moon) of hardened agar with forceps or an inoculating needle to create a straight edge.
4. Using an inoculating loop, inoculate the straight edge with a sample of a mold previously isolated.
5. Place some Vaseline on three edges of a coverslip and place it over the agar surface, leaving the top, inoculated edge open to the air.
6. Place a piece of filter paper moistened with tap water in the bottom of a Petri dish and position a bent glass rod on top of the paper.
7. Place the inoculated slide onto the surface of the glass rod in the Petri dish. Cover the Petri plate and incubate at room temperature.

Since a very small amount of agar is being used, precautions must be taken to prevent dehydration of the agar. The creation of a saturated chamber accomplishes this. Make sure that the filter paper remains wet over the course of the observational period.

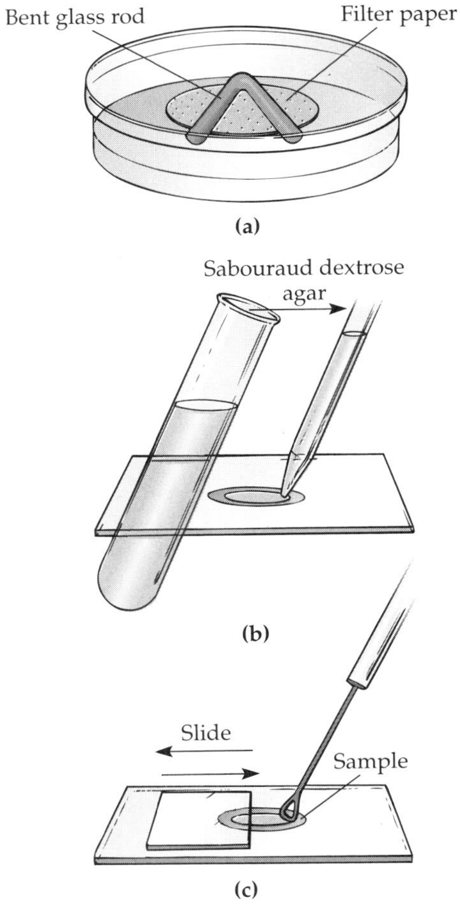

FIGURE 56.1 Regular slide technique.

CONCAVE GLASS SLIDE TECHNIQUE (SEE FIGURE 56.2)

Procedure

1. Place a piece of filter paper wetted with tap water in the bottom of a Petri plate. Put a small bent glass rod onto the surface of the filter.
2. Clean a concave glass slide with ethanol.
3. Apply a thin layer of Vaseline around the edge of the concave slide.
4. Add a few drops of molten Sabouraud dextrose agar to the bottom of the well.
5. Cover with an alcohol-sterilized coverslip.
6. Stand the slide on edge until the agar hardens.
7. Slide the coverslip to make accessible the top surface of the agar. Using an inoculating loop, apply the sample of the isolated mold to the top edge of the agar.
8. Slide the coverslip back into place.
9. Place the slide on the surface of the rod inside the moistened Petri plate and cover.
10. Incubate at room temperature.

Using either of the slide procedures, mycelia can be observed by using a low-power objective on a microscope. Remember, it is important to observe the reproductive structures, and this should be done before abundant growth occurs. This means daily observations should be made of the slide cultures.

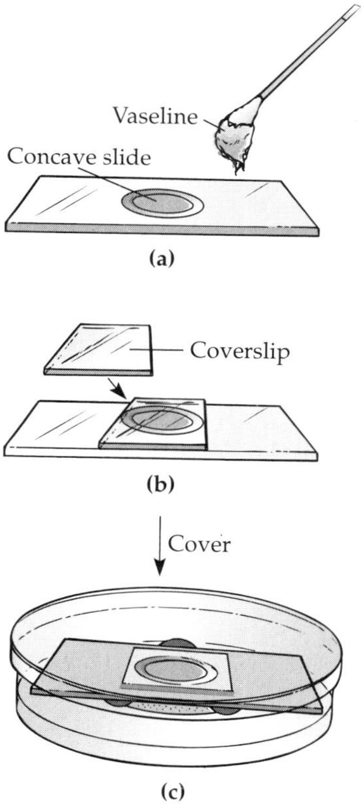

FIGURE 56.2 Concave-glass slide technique.

RODAC PLATE METHOD

Procedure

1. Obtain a Rodac plate containing Sabouraud dextrose agar.
2. Sterilize a flat spatula by dipping into 70% ethanol and flaming in a Bunsen burner. Be careful not to ignite the container of alcohol.
3. Scoop out a half-inch portion of the agar from the center and discard (Figure 56.3).

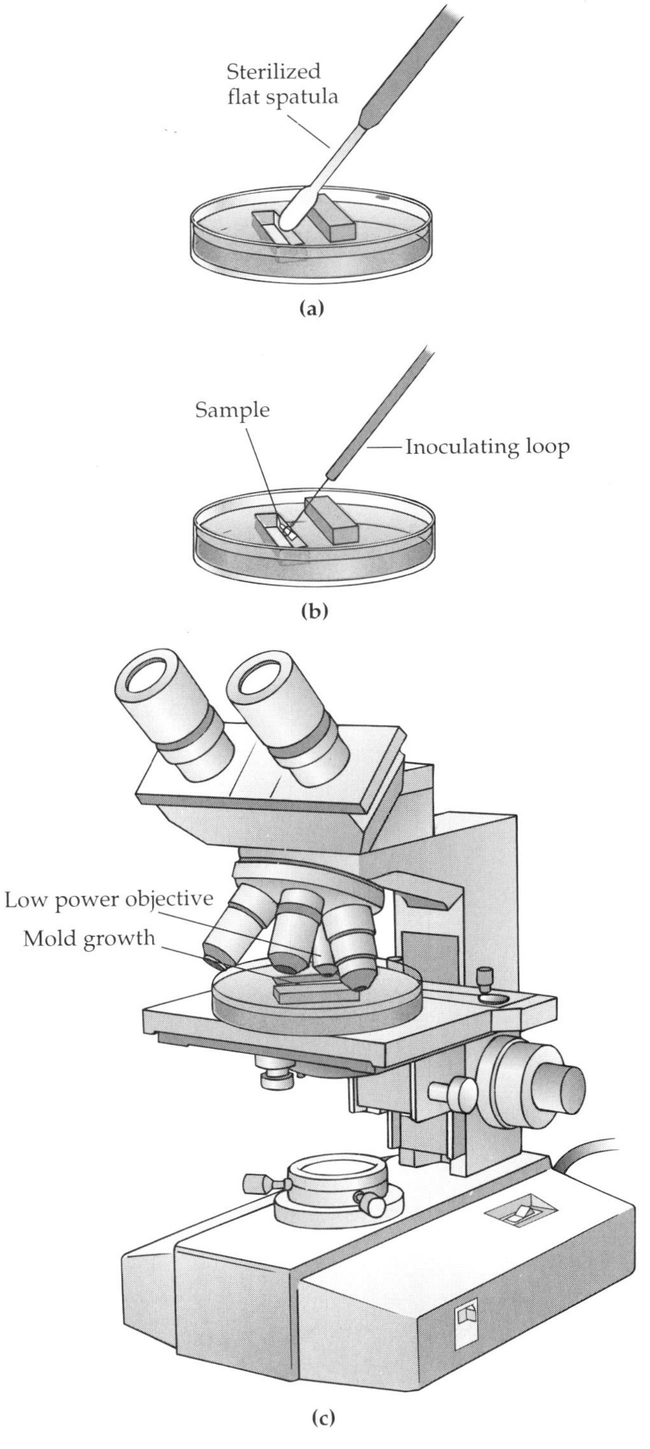

FIGURE 56.3 Rodac plate method.

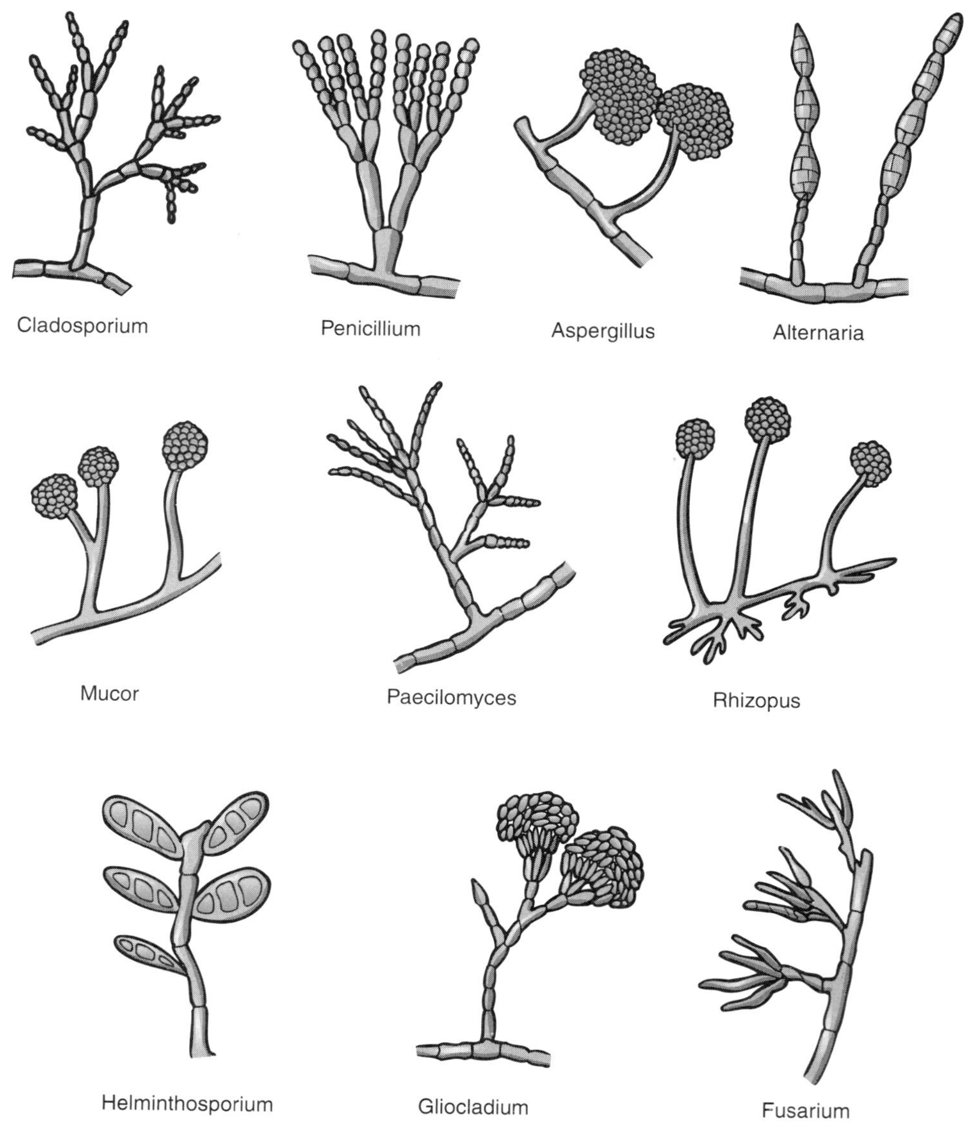

FIGURE 56.4 Representative types of mold.

4. Using an inoculating loop, transfer any sample isolated from the air in a previous lab to the edges of the agar.
5. Incubate at room temperature.
6. Using a low-power objective, observe the mycelia by placing the entire plate on the stage of a microscope. Sufficient light is obtained by positioning the objective near the center of the plate. Make observations of the fabulous overhanging gardens!

There are some sketches of the reproductive structures of some common molds in Figure 56.4. Other manuals of mycology are available in the laboratory for use.

Take time also to examine some "known" slides and plates of mold cultures that have been already prepared. Can you differentiate the reproductive structures and identify the organisms?

Laboratory Report 56

Name Date Section

Make sketches of the following known mold cultures:

Penicillium

Aspergillus

Acremonium

Cladosporium

Alternia

Mucor

Chaetomium

Paecilomyces

Gliocladium

Helminthosporium

Fusarium

Nigrospora

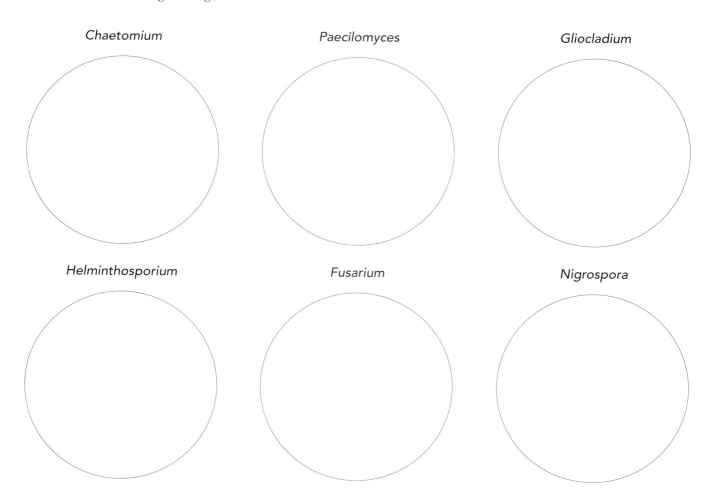

Sketch some of the molds isolated from the environment. Can you identify them?

Review Questions

1. How can one prevent the growth of bacteria on mold slides or plates that need to be incubated for long periods of time?

2. Which of the mold types examined in this experiment have been found in clinical specimens and may be associated with human disease?

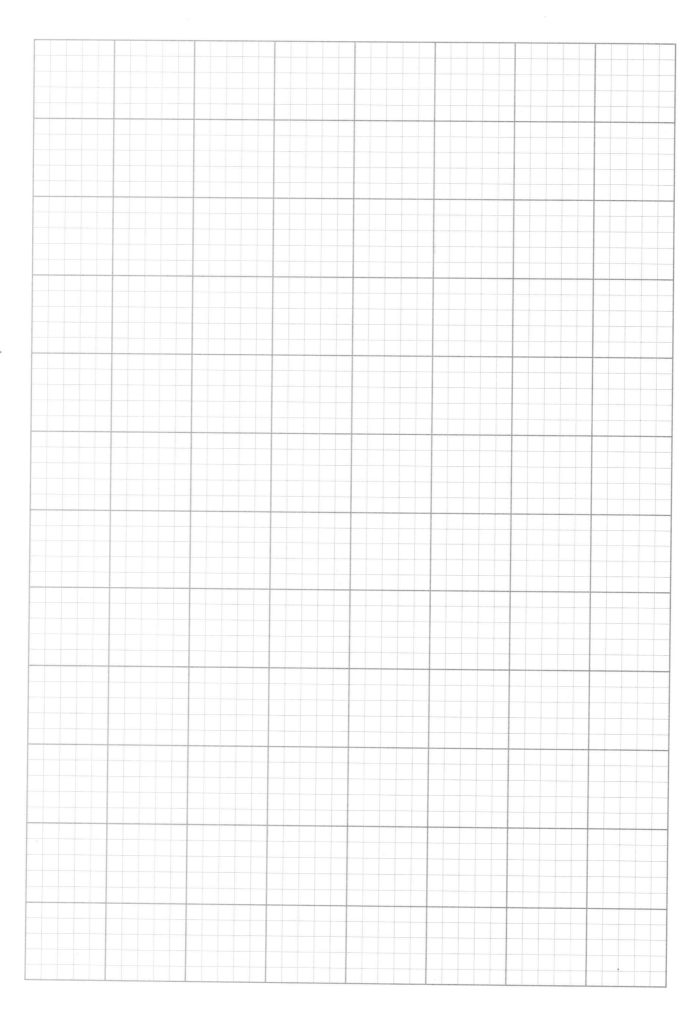

LABORATORY

57 Protozoa

Safety Issues

- Caution should be taken when working with any samples of secondary sewage sludge, as bacteria and viruses pathogenic to humans may be present.
- Dispose of slides in a tray of disinfectant after observations have been made.

OBJECTIVES

1. Become acquainted with the various morphological types of protozoa found in environmental samples.
2. Distinguish protozoa from algae and bacteria found in similar environments.

In most freshwater environments, protozoa are found coexisting with bacteria and algae. Protozoa are single-celled organisms, although some form colonial types consisting of thousands of individuals. Given their ubiquity in nature, they undoubtedly play a key role in the dynamic balance of populations in aquatic communities. The protozoa are placed by most authors in the Kingdom Protista and grouped in the Subkingdom Protozoa. All members possess cell membranes with at least one well-defined nucleus. Cell walls are absent and a variety of organelles can be observed. Since many protozoa obtain food via phagocytosis, food vacuoles are commonly found in the cytoplasm. Organelles of locomotion, and feeding flagella and cilia, are found in most cells. Flagella normally occur in pairs and are longer than the densely packed cilia. The presence of contractile vacuoles help the organisms to control osmotic balance with the environment by eliminating water from the cytoplasm.

Classification of the protozoa is under considerable discussion, but seven major phyla have been identified.

Phylum Sarcomastigophora

This phylum contains two subphyla, Mastigophora and Sarcodina. These types are motile by flagella, pseudopodia, or both. Sexuality is syngamic, if present. (*Trypanosoma*, *Giardia*, *Amoeba*, *Leishmania*)

Phylum Ciliophora

This is the largest phylum, with over 8000 species represented. These are unicellular forms that are motile by cilia. Ciliates reproduce asexually by binary fission, and exhibit several types of sexual reproduction, including conjugation. (*Paramecium*, *Stentor*, *Didinium*, *Tetrahymena*, *Vorticella*)

Phylum Labyrinthomorpha

Members have spindle-shaped or spherical non-amoeboid cells. Some members that move by amoeboid motion glide on a network of mucous tracks. (*Labryinthula*)

Phylum Apicomplexa

Members produce an apical complex and have a spore-forming stage in their life cycles. They are sometimes referred to as sporozoans and all are parasitic. (*Toxoplasma, Plasmodium, Eimeria*)

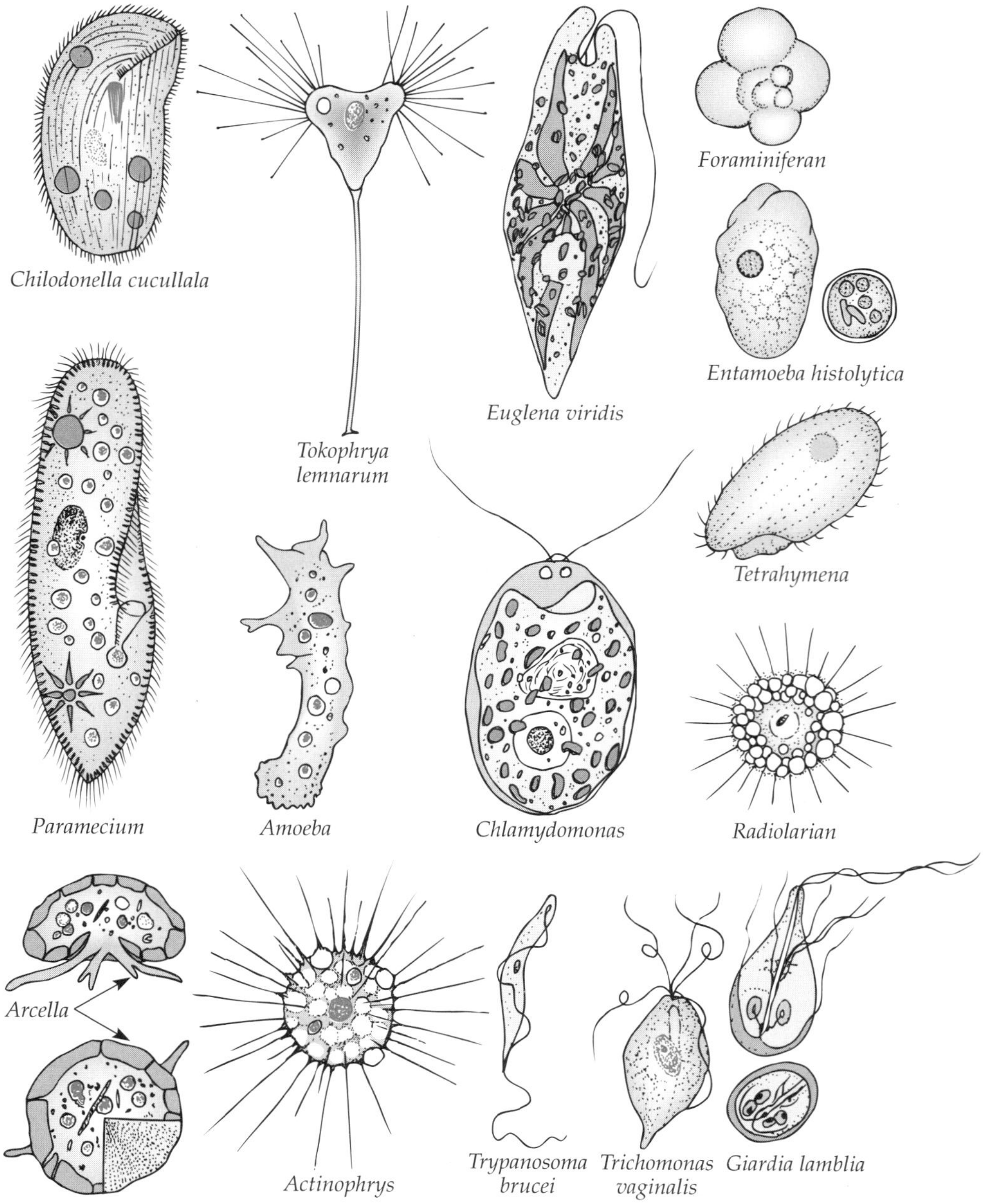

FIGURE 57.1 Representative types of protozoa.

Phylum Microspora

All members are intracellular parasites with unicellular spores. (*Nosema*, *Microsporidium*, *Pleistophora*)

Phylum Ascetospora

All species are parasitic with multicellular spores, lacking polar caps and polar filaments. (*Haplosporidium*, *Paramyxa*)

Phylum Myxospora

All myxozoans are parasitic. Multicellular spores are present with a number of coiled polar filaments. (*Myxidium*, *Myxosoma*, *Kudoa*)

Materials (per 20 students)

1. Samples of pond water or secondary sewage (4–5)
2. Microscope slides and coverslips
3. Slowing agent (a number are commercially available) or 1.5% methyl cellulose (50 ml)
4. Known cultures of protozoa (5–10)

Procedure

1. Prepare a wet mount of each of the various known protozoan cultures by placing a drop on a clean microscope slide. Add one drop of the slowing agent to retard the movement of the protozoa. Cover with a coverslip and observe under the high-dry objective. The use of phase-contrast microscopy is recommended.
2. Draw and describe the morphological features of the known cultures.
3. In a similar manner, prepare wet mounts of the pond water or secondary sewage samples. Observe using phase-contrast microscopy.
4. Draw and describe the morphological features of the protozoan observed.
5. Compare your observations with the drawings available (Figure 57.1), and try to place the protozoa observed into the proper groups.

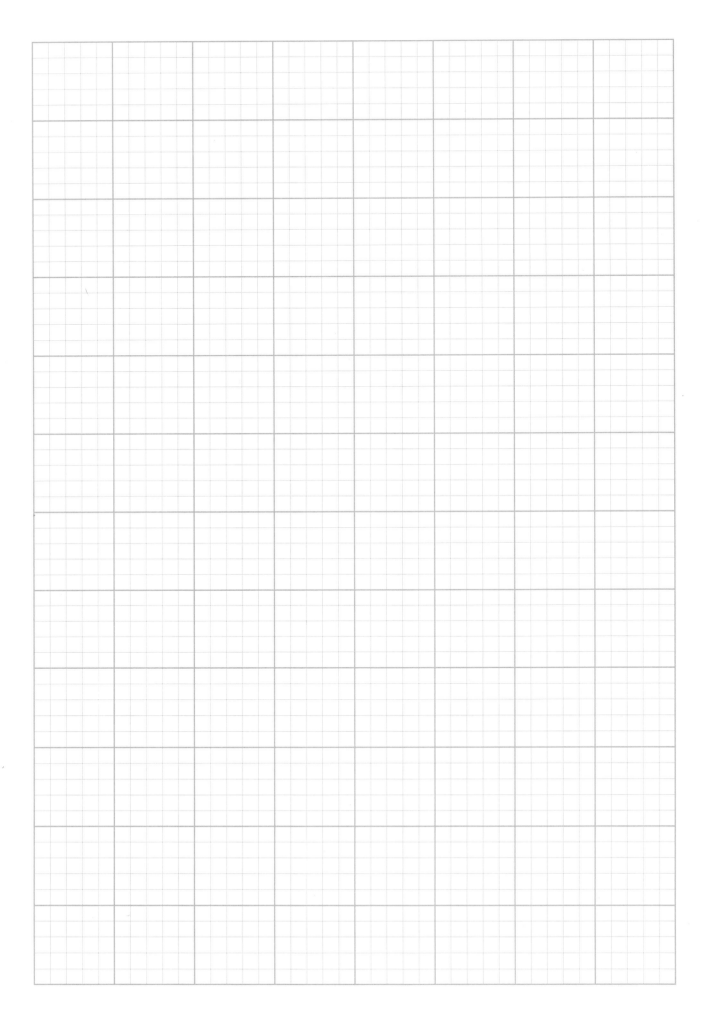

Laboratory Report 57

Name ____________________ Date __________ Section __________

Known Cultures of Protozoa

Environmental Samples

Pond Water

Secondary Sewage Sludge

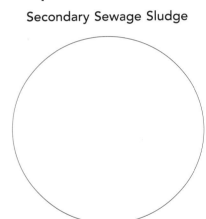

Review Questions

1. What morphological features were most commonly observed? Describe their apparent functions.

2. On the basis of observations made, what seems to be the function of cilia?

3. Explain the occurrence of pigmented substances within the cytoplasm of the protozoa.

UNIT IX

Virology

Viruses are submicroscopic, intracellular parasites that are capable of infecting a wide variety of organisms, including plants, animals, and bacteria. Each virus requires a specific host cell in which to propagate. Plant viruses will not attack animal cells and animal viruses will not attack bacteria. Viruses essentially are composed of a core of nucleic acid surrounded by an outer coat (**capsid**). The nucleic acid core may be either single- or double-stranded DNA or RNA. A complete infectious viral particle is often referred to as a **virion**. Some animal viruses have envelopes surrounding the capsid, and some also possess proteinaceous spikes.

The steps involved in a viral attack of a specific host cell are as follows:

1. The first stage is that of **attachment** to the susceptible host cell. This attachment can only succeed if the best cell bears specific receptor sites.
2. Viruses then must **penetrate** the cell wall or membrane. In the infection of bacteria, the penetration is accomplished by the production of enzymes that allow the virus to penetrate the cell wall and allow nucleic acid to enter. In the infection of animal cells, viruses usually are engulfed by phagocytosis or pinocytosis.
3. If the virus is engulfed, it must be **uncoated** to release the nucleic acid.
4. After the virus is uncoated, the **replication** of the virus genetic information may proceed. Specific enzymes are utilized for the synthesis of new DNA or RNA, and in some cases, for the inhibition of normal host cell synthesis.
5. The virus uses host-cell amino acids, transfer RNA, and ribosomes for the **synthesis of new viral proteins**.
6. **Assembly** or **maturation** of the viral nucleic acids and protein coats occurs next.
7. The **release** of the virus is the last step and varies depending upon the type of virus. There may be lysis (destruction) of the wall or membrane of the host cell; or the viral particles may actually "bud" off a membrane, as is the case in a number of animal cell infections.

LABORATORY

58 The Cultivation and Inactivation of Bacterial Viruses

Safety Issues

- Use pipettes for all pipetting steps. Never attempt to pipet any material by mouth.
- Be careful when vortexing tubes not to spill the contents. If spillage occurs, wipe the outside of the tube and table with a disinfectant and wash hands with antibacterial soap.
- Wear protective UV goggles and gloves to protect eyes and hands while using the ultraviolet lamp.

OBJECTIVES

1. Become familiar with the techniques for the propagation and enumeration of bacterial viruses.
2. Design a procedure to determine the viricidal effect of ultraviolet light on bacterial viruses.

Bacterial viruses (bacteriophages) can be isolated from a variety of environments. Subsequent experiments will involve propagation of bacteriophages from various environments. This experiment focuses on the basic techniques used in determining the number of viral particles in a suspension.

The procedure involves the use of an agar-overlay technique. A base layer of nutrient agar is poured into plates and allowed to solidify. Molten soft agar containing half the concentration of agar is used for the top layer. Dilutions of bacteriophage preparation and the host cell, in this case *Escherichia coli B*, are added to the soft agar prior to pouring onto the base nutrient agar plates. A double layer will be formed, with the top layer, or overlay, containing the host cells and bacteriophage. When the virus confronts a susceptible host cell in the soft agar environment, it will replicate and cause lysis, liberating hundreds of new virus particles. The spreading destruction of bacteria host cells becomes visible as a clear area or **plaque** in the soft agar (see Color Plate 35). If each plaque arises from a single virus particle, the counting of plaques in the soft agar layer provides an indication of the number of virus particles present in the original sample. However, some plaques may arise from more than one virus particle, so rather than reporting the number of viral particles present, we record the number of **plaque-forming units (pfu)** observed.

DETERMINATION OF CONCENTRATION OF BACTERIOPHAGE PARTICLES IN UNKNOWN SUSPENSION

Materials (per group)

1. Nutrient agar plates (10)
2. Tubes of soft agar (10)
3. Sterile 1.0-ml pipettes (1 canister)
4. 45 to 48°C water bath
5. Pipettes
6. 4.5-ml sterile nutrient-broth dilution tubes (10)
7. 18 to 24-hr broth culture of *Escherichia coli B*
8. Solution containing unknown concentration of bacteriophage particles
9. Vortex mixer

Procedure

1. Place ten nutrient-broth dilution tubes in a culture-tube rack and label 10^{-1} to 10^{-10}.
2. With a sterile 1.0-ml pipette, aseptically transfer 0.5 ml of the bacteriophage suspension to the tube labeled 10^{-1}. Discard the pipette into a tray or jar containing disinfectant.
3. Mix well, using a vortex mixer or by hand.
4. Using a fresh pipette, transfer 0.5 ml from the 10^{-1} dilution tube to the tube marked 10^{-2}.
5. Continue in a similar manner until you have prepared all dilution tubes through 10^{-10}.
6. Label nutrient agar plates with the dilutions (10^{-2} to 10^{-11}).
7. Now comes the tricky part. Add phage dilution and 0.5 ml of the *E. coli* broth culture to the molten soft agar and pour it onto the surfaces of the nutrient agar plates before the agar hardens. It is wise to remove the soft agar tubes from the water bath one at a time to prevent premature solidification.
8. Start by adding 0.1 ml of the 10^{-1} phage dilution and 0.5 ml of *E. coli* cell suspension to one molten soft agar tube. Be sure to use a different sterile pipette for each transfer. Mix gently so as to not create air bubbles, and pour onto the surface of a nutrient agar plate labeled 10^{-2} (Figure 58.1). Shake the tube to remove all of the contents. This will form an agar overlay (Figure 58.2). Since the agar will rapidly solidify, gently rotate the nutrient agar plate to distribute the agar before solidification occurs. It is not desirable to have a pile of solidified agar containing bacteria and phage in the middle of the nutrient agar plate. If this happens, discard the plate and prepare the mixture of bacteria and phage dilution using a fresh tube of soft agar.
9. Proceed in a similar fashion until all of the phage dilutions have been plated.
10. Allow the top agar to completely solidify in the plates before inverting and incubating at 37°C. Check these plates after 5 to 8 hr for the presence of plaques. If plates are incubated for a longer period of time, the plaques will spread and overlap, thus making enumeration difficult.
11. Count those plates having between 30 and 300 plaques. The plates may be refrigerated to retard further plaque spread and observed during another regularly scheduled laboratory period.

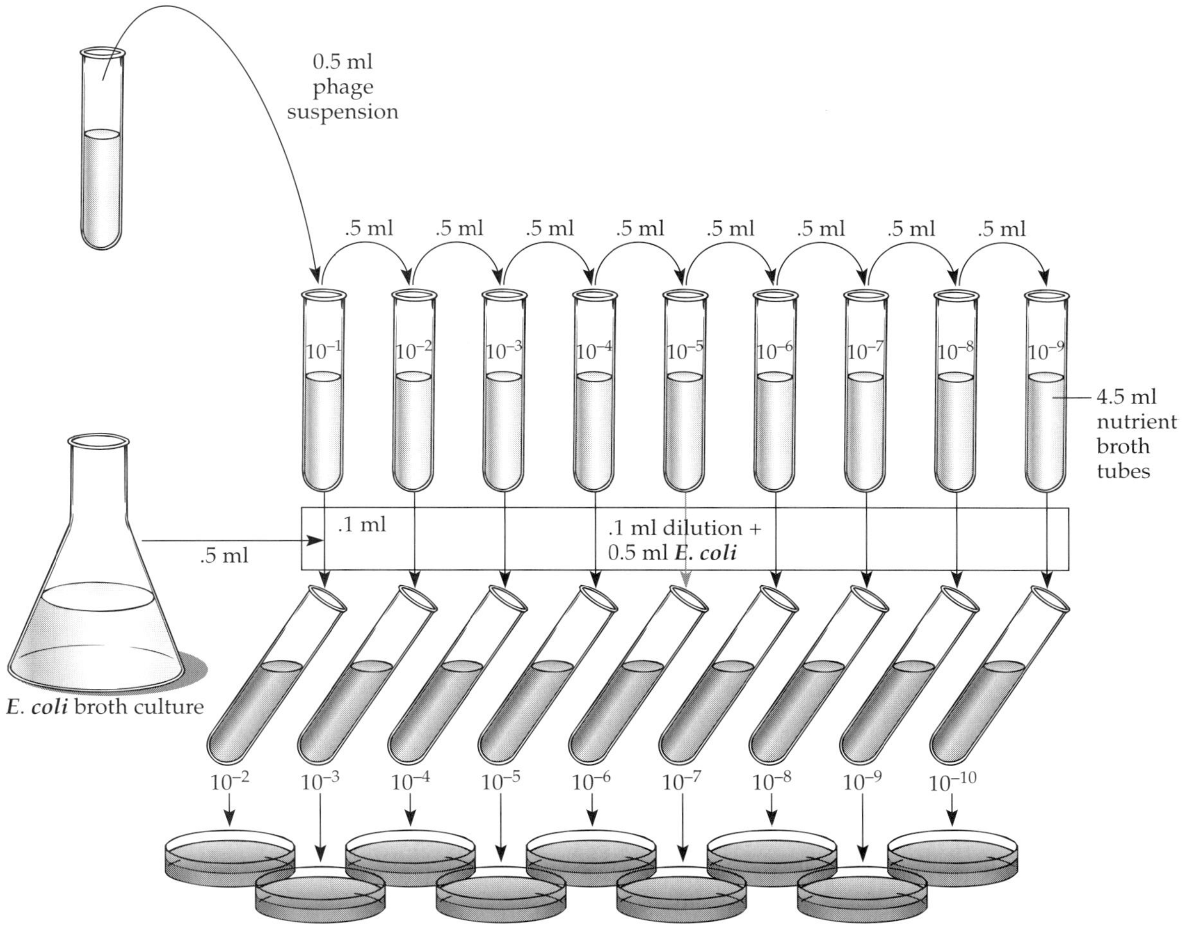

FIGURE 58.1 Dilution scheme for phage titering.

12. Determine the number of plaque-forming units per millimeter (pfu/ml) in the original unknown phage suspension. Make sure to take into account the dilution factor and the amount of phage added to each dilution tube.

$$\text{pfu/ml} = \text{plaques} \times \left(\frac{1}{\text{dilution}}\right)$$

(Plating factor has been taken into account in dilution scheme in Figure 58.1.)

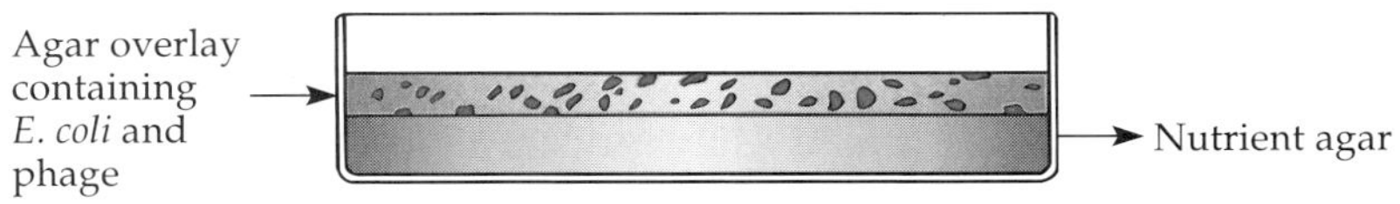

FIGURE 58.2 Agar overlay.

EFFECT OF ULTRAVIOLET LIGHT ON BACTERIOPHAGE

Using information from part of this laboratory and from Laboratory 46, design an experiment to demonstrate the effect of ultraviolet light on bacterial viruses.

Suggested Materials

1. Nutrient agar plates
2. 18 to 24-hour broth culture of *E. coli B*
3. Phage suspension
4. 4.5-ml sterile nutrient-broth dilution tubes
5. Sterile pipettes
6. Tubes of molten soft agar in 48°C water bath

Procedural Suggestions

⚠ Be sure to wear UV-protective eye shields when working with ultraviolet light.

1. Decide on irradiation distance and exposure time.
2. Determine the appropriate controls.

Laboratory Report 58

Name ______ Date ______ Section ______

Determination of Phage Concentration in Unknown Suspension

Phage Dilution	Number of Plaques
10^{-2}	
10^{-3}	
10^{-4}	
10^{-5}	
10^{-6}	
10^{-7}	
10^{-8}	
10^{-9}	
10^{-10}	
10^{-11}	

$$\text{Calculation of phage titer (pfu/ml)} = \text{number of plaque-forming units (pfu) on each plate} \times \frac{1}{\text{dilution}}$$

Effect of Ultraviolet Light on Bacteriophage

Dilution of Irradiated Phage	Number of Plaques

Irradiation Time ______

Irradiation Distance ______

Titer of Bacteriophage after Irradiation ______

What is the percentage of inactivation of the bacteriophage for the irradiation interval used?

Review Questions

1. Why is soft agar used in the top layer?

2. Formulate a hypothesis to explain the appearance of the bacterial colonies in the lowest phage dilution plates. How could your hypothesis be verified?

3. Why can't you just heat a soft agar tube that has prematurely solidified after adding *E. coli* and phage dilution?

4. How would it be possible to verify that the plaques that appear on the plates actually contain bacteriophage?

5. How might your UV irradiation results be used to argue for or against the assertion that viruses are living things?

LABORATORY

59 Isolation of Bacteriophage from Sewage

Safety Issues

- Extreme caution should be taken when working with sewage samples due to the potential presence of pathogenic bacteria and viruses. Wear gloves when handling sewage samples or broth that has been inoculated with sewage.
- Use pipetting devices for performance of all pipetting procedures. Do not pipet any material by mouth.

OBJECTIVE

Apply the techniques for the propagation of bacterial viruses to the isolation of phage from sewage.

Extreme caution must be exercised when handling the raw sewage, even after filtration. Pathogenic organisms are likely to be present.

Bacteriophage can be isolated from a variety of environments. They may be isolated from sewage, soil, foods, insects, or water. In this exercise, an isolation will be attempted from sewage. The first step in the process must be the collection of a sewage sample. The number of phage particles present may be small so an **enrichment procedure** is usually used to enhance the number of phage particles. In order to detect the number of phage, an agar-overlay technique is used.

Materials (per group)

1. Raw sewage (45 ml)
2. Broth culture of *Escherichia coli B* (10 ml)
3. Nutrient agar plates (9)
4. Tubes of soft agar (9)
5. 1-ml and 5-ml sterile pipettes (1 canister of each)
6. 45 to 48°C water bath
7. 4.5-ml sterile nutrient-broth dilution tubes (9)
8. Membrane-filter apparatus (5)
9. 0.22-μm membrane filters (5)
10. Bacteriophage broth (10×) (5.0 ml)
11. Graduated cylinder (1)
12. Sterile 250-ml Erlenmeyer flasks (2)
13. Latex gloves
14. Sterile 50-ml centrifuge tube (1)
15. Centrifuge
16. Pipetting devices

Procedure (See Figure 59.1)

Since the sewage may contain pathogenic organisms, be sure to wear protective gloves when handling.

Day 1. This inoculation may have already been done for you by the lab instructor. If so, then start with step 3.

1. Using a graduated cylinder, carefully add 45 ml of the raw sewage to the sterile 250-ml Erlenmeyer flask. Carefully pipet 5.0 ml of the *E. coli B*

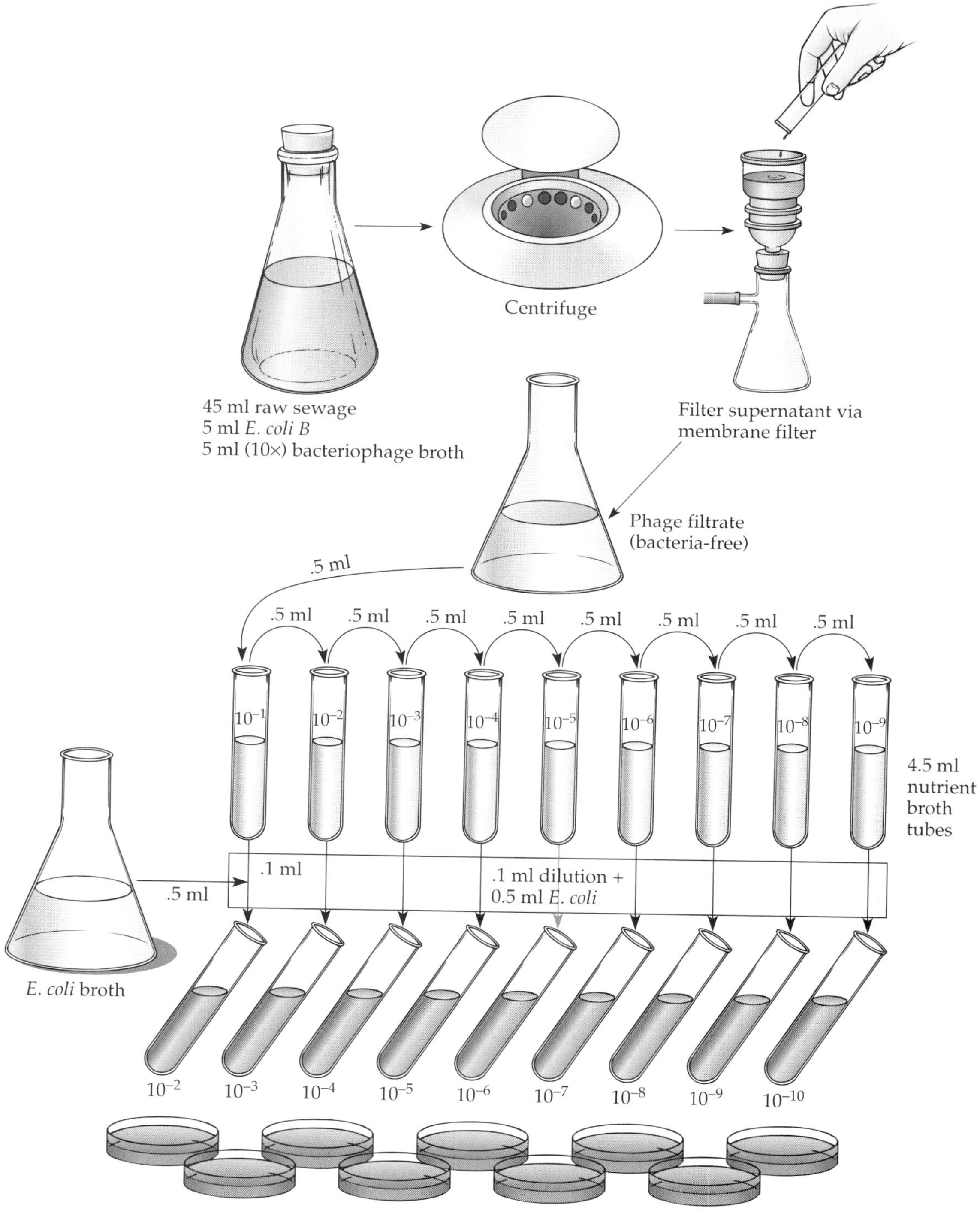

FIGURE 59.1 Phage-isolation procedure.

broth culture and 5.0 ml of the bacteriophage broth to the flask. This creates an enrichment environment for the increase of the number of phage particles.
2. Incubate the flask at 37°C for 24 hr.
3. After incubation, centrifuge the broth at 2500 rpm for 25 min to remove particulates.
4. Carefully decant the supernatant fluid to a fresh sterile flask.
5. Filter the supernatant fluid through a 0.22-μm filter to remove any bacteria present, and collect a filtrate-containing phage. If the supernatant is cloudy, the 0.22-μm filter will rapidly clog. A longer centrifugation may be necessary prior to filtration.
6. Prepare serial ten-fold dilutions through 10^{-9} using the 4.5-ml tubes of nutrient broth, and label "phage dilution."
7. Label nutrient agar plates with the appropriate phage dilutions.
8. Add 0.1 ml of the 10^{-1} phage dilution and 0.5 ml of *E. coli* to one tube of molten soft agar. Be sure to use a separate pipette for each transfer. Mix gently so as to not create bubbles, and pour onto the surface of a nutrient agar plate. Shake the tube to remove all of the contents. Since this agar will rapidly solidify, gently rotate the Petri plate to distribute the agar before solidification occurs.
9. Proceed in a similar fashion until all of the phage dilutions have been plated.
10. Incubate all plates in an inverted position at 37°C. These plates can be checked after 6 to 8 hr for the presence of plaques.
11. Count those plates having between 30 and 300 plaques.
12. Determine the number of plaque-forming units in the original enrichment broth. Report the results as the number of **plaque-forming units (pfu) per milliliter** of enrichment broth. Be sure to make note of the dilution factor and of the amount of phage used in each soft agar tube.

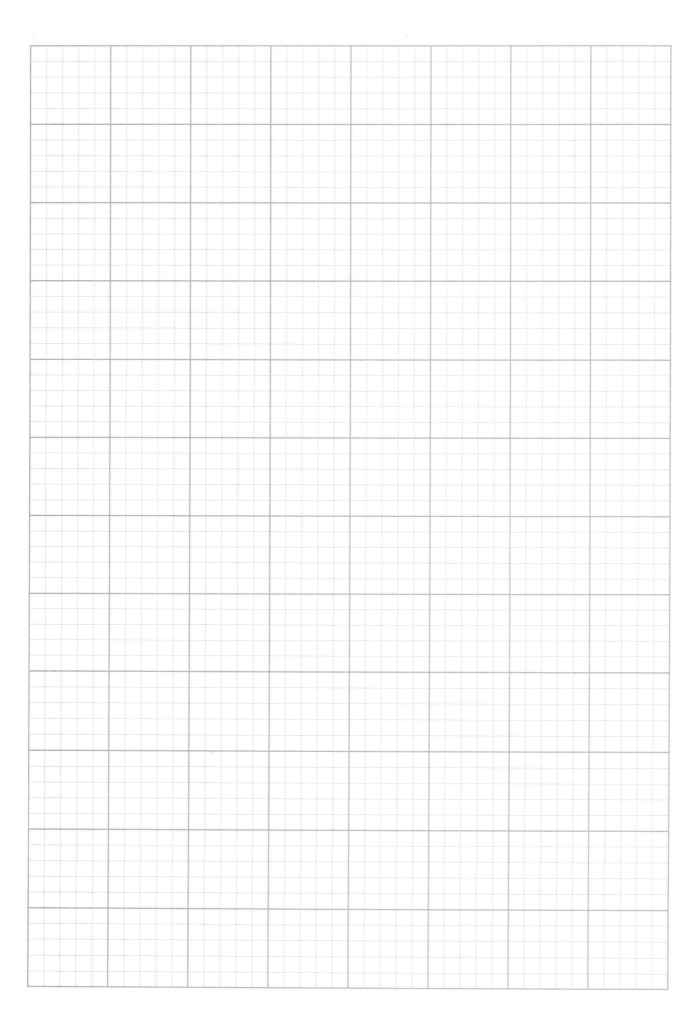

Laboratory Report 59

Name ______________________ Date __________ Section __________

Phage Dilution	Appearance of Plaques	Number of Plaques
10^{-2}		
10^{-3}		
10^{-4}		
10^{-5}		
10^{-6}		
10^{-7}		
10^{-8}		
10^{-9}		
10^{-10}		

Calculate the pfu/ml in the enrichment culture used.

Review Questions

1. What effect does the population of bacteria normally found in sewage have on the initial enrichment culture?

2. How could the presence of phage in a food sample such as ground beef or chicken be detected?

3. What other types of bacteriophage might be expected to be found in a sewage sample? How would detection of these other types be accomplished?

UNIT X

Genetics

The genetic information of bacteria is carried within the **DNA** molecules of the cells. These molecules house **genes** which are responsible for all cell functions. The complete collection of these genes is referred to as the organism's **genotype**. The expression of the genes in visible characteristics such as pigmentation or motility is the organism's **phenotype**. Genes are transcribed to **messenger RNA**, which is respectively translated into a myriad of proteins needed for cell function. The genetic information of cells can be altered by exposure to chemicals or radiation, or may occur spontaneously. These alterations are changes in the genotype of the organism and are referred to as **mutations**. Mutations are inheritable and are responsible for the variability of microorganisms within large populations.

Exchange of genetic information can occur among bacteria. In some cases, the information exchanged is integrated via **genetic** recombination and may allow the recipient organisms to carry out new functions. The primary processes of genetic exchange and recombination include **transformation**, **transduction**, and **conjugation**.

In the laboratories that follow, antibiotic mutants will be isolated, pigmentless mutants will be induced by ultraviolet radiation, and histidine mutants will be used to determine the potential carcinogenicity of chemicals by using the Ames test. Two laboratories involving genetic exchange (transformation and conjugation) are also included. Laboratory 61 demonstrates that variation within populations may not be the result of mutation but rather the response to different environmental conditions. Variation in pigmentation and motility are examined.

LABORATORY

60 Isolation of Antibiotic-Resistant Mutants

Safety Issues

- Use mechanical device for all pipetting steps. Do not pipet any material by mouth.
- Allow heated inoculating loops to cool before touching any growth on an agar plate. Contact with a hot loop may cause splattering of organisms.

OBJECTIVE

Learn how to select for antibiotic-resistant mutants.

Resistance to some antibiotics occurs stepwise, each mutational step conferring a partial increase in resistance. The gradient-plate technique is ideal for the isolation of such mutants since it provides a gradual increase in the concentration of the destructive agent within the confines of a single plate. The level of resistance of a given organism can be easily determined by visible observation of mutant colonies. Mutants of resistant strains exhibiting a higher level of resistance may be isolated by streaking the original colonies along the plate to higher levels of antibiotic. In this experiment, sensitivity and resistance to streptomycin and penicillin using the gradient-plate technique will be demonstrated.

Materials (per group)

1. 3- to 24-hour bacterial broth cultures (*Escherichia coli*, *Staphylococcus aureus*, *Proteus vulgaris*)
2. Filter-sterilized streptomycin solution (25,000 μg/ml; 1.0 ml)
3. Filter-sterilized penicillin solution (50,000 units/ml; 1.0 ml)
4. Sterile Petri plates (3)
5. Flasks with 50 ml tryptic soy agar (for bottom layer) in 50°C water bath (2)
6. Six flasks with 25 ml tryptic soy agar [for top (antibiotic-containing) layers] in 50°C water bath
7. Six tubes containing 3.5 ml soft agar
8. 50°C water bath
9. Tryptic soy agar plates containing 5000 μg/ml streptomycin (3)
10. Tryptic soy agar plates containing 10,000 units/ml penicillin (3)
11. Sterile 1.0-ml pipettes (1 canister)

Procedure

1. Place three empty Petri plates in a slanted position. Pour 25 ml of the tryptic soy agar into the dishes in such a manner that the upper edge of the plate on the incline is barely covered with agar.
2. Place a mark on the Petri plate where the agar is thinnest. This will be the portion of the plate that has the highest antibiotic concentration.
3. After the layer of agar has hardened, return the Petri plates to a flat surface.
4. Using aseptic technique, add 0.5 ml of the streptomycin stock solution to each of the three flasks containing 25 ml of tryptic soy agar (TSA). In a similar fashion, add 0.5 ml of the penicillin solution to each of the other three flasks of agar. Gently mix all flasks, avoiding the formation of bubbles.
5. Pour the streptomycin- and penicillin-containing agars onto the surfaces of the slanted TSA plates. You will have three plates of each (Figure 60.1).
6. Let the agars solidify for at least 30 to 45 min.
7. Inoculate 0.5 ml of the appropriate bacterial culture into the 3.5-ml portions of soft agar and pour onto the surfaces of the gradient plates. Rotate the plates to distribute the molten agar and allow to harden. Incubate at 37°C for 24 to 48 hr.
8. After incubation, observe the plates and record results. Do you notice any antibiotic-resistant organisms on any of the plates? With a sterilized inoculating loop, streak isolated colonies from the portions of the plates containing lower concentrations of antibiotic to regions of the plate containing higher concentrations of antibiotic (Figure 60.2). This is to isolate organisms resistant to higher concentrations of antibiotic. If there is growth

Exercise caution when transferring organisms that are antibiotic-resistant. They may pose a health risk.

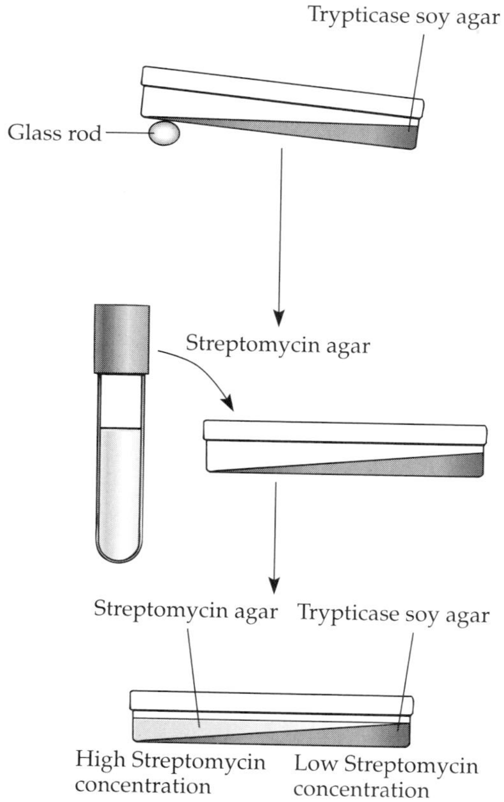

FIGURE 60.1 Gradient-plate technique.

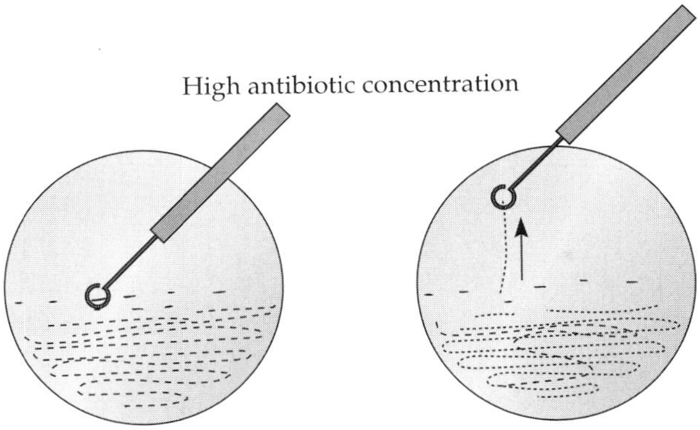

FIGURE 60.2 Streak technique for assessing antibiotic resistance.

in the area of high concentration of antibiotic, streak some of this growth with an inoculating loop onto the corresponding agar plate containing 10× concentration of antibiotic.

9. Incubate all new streaked plates at 37°C for 24 to 48 hr and record results.

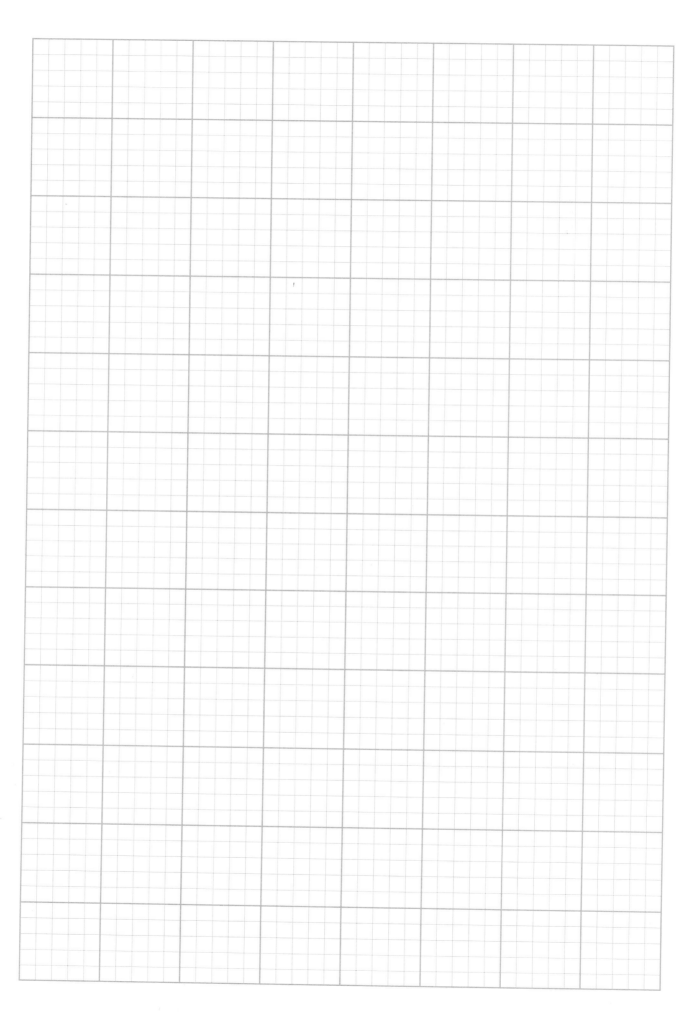

Laboratory Report 60

Name ____________________ Date __________ Section __________

Isolation of Antibiotic-Resistant Mutants

(A) Streptomycin Plates

(Sketch growth on plates)

High Antibiotic Concentration

Low Antibiotic Concentration

Organism Plated:

______________________ ______________________ ______________________

(B) Penicillin Plates

(Sketch growth on plates)

High Antibiotic Concentration

Low Antibiotic Concentration

Organism Plated:

______________________ ______________________ ______________________

(C) Growth on 10× Penicillin Plates:

Organism ________________________________

(D) Growth on 10× Streptomycin Plates:

Organism ________________________________

Review Questions

1. Why has the incidence of antibiotic-resistant bacterial infections increased in recent years?

2. Why are filter-sterilized solutions of streptomycin and penicillin used in the isolation of antibiotic-resistant mutants?

3. Describe another procedure that could be used to isolate antibiotic-resistant mutants.

4. How could it be determined whether the antibiotic-resistant mutants occur in stepwise or single-step fashion?

5. Does the antibiotic cause the mutation, or does it select mutations that have occurred previously? How could you distinguish between the two possibilities?

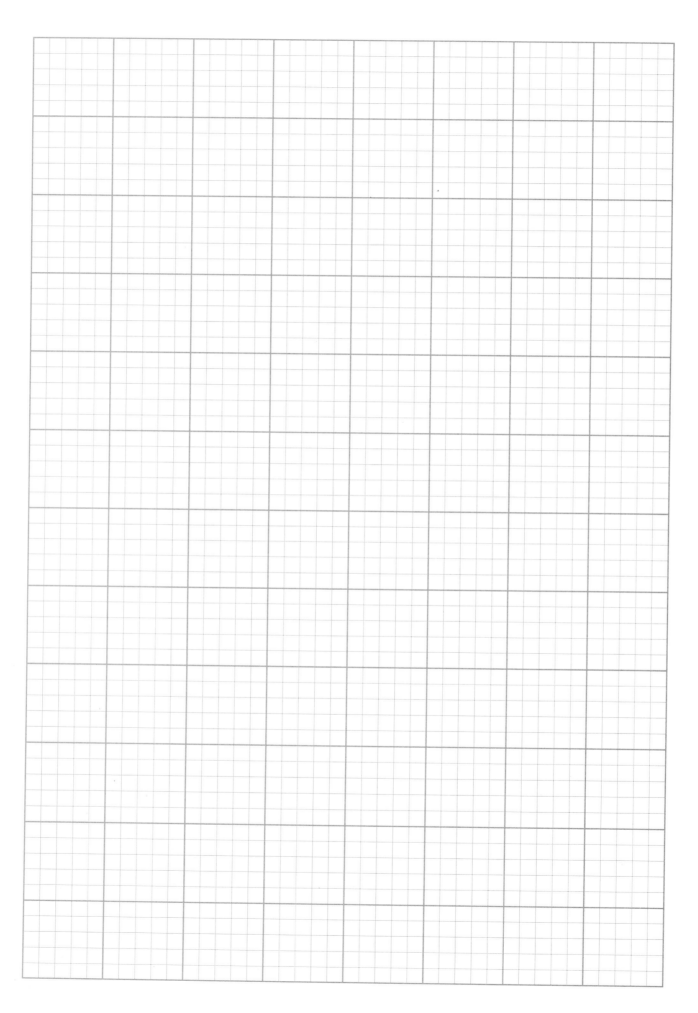

LABORATORY

61 Environmentally Induced Variation

Safety Issues

- Allow heated inoculating loops to cool before transferring any growth on an agar plate.

OBJECTIVE

Observe variations in pigmentation and motility of microorganisms grown under different environmental conditions.

Adaptations, or responses of microorganisms to changes in environment, occur in response to a specific stimulus in the microbial environment. In this exercise, two adaptations of microorganisms, variations in pigmentation and flagellation, are examined.

Materials (per 20 students)

1. Broth cultures of *Serratia marcescens* (5)
2. Tryptic soy agar plates (120)
3. Broth cultures of *Proteus mirabilis* (5)
4. Nutrient agar plates (NaCl-free) (40)
5. Tryptic soy agar (TSA) plates (which contain NaCl) (40)
6. 37°C incubator
7. Inoculating loop
8. Tryptic soy agar + 0.05% phenol

Procedure

A. Variation in Pigmentation

(See Figure 61.1)

1. With a sterilized inoculating loop, streak two TSA plates with the *Serratia* broth culture.
2. Incubate one plate at 37°C and the other at 25°C. Examine each after 48 hr. Record the pigmentation.
3. After 48 hr, transfer a colony from both plates to fresh TSA plates. The plate containing the colony transferred from 37°C should be incubated at 25°C. The other plate, containing the colony transferred from the 25°C plate, should now be incubated at 37°C.
4. Examine the plates after 48 hr. Describe and record the pigmentation.

B. Variation in Motility-Effect of Salt Concentration

1. Inoculate a nutrient agar plate and a tryptic soy agar plate with a loopful of the *Proteus* broth culture. Incubate at 37°C for 24 hr.

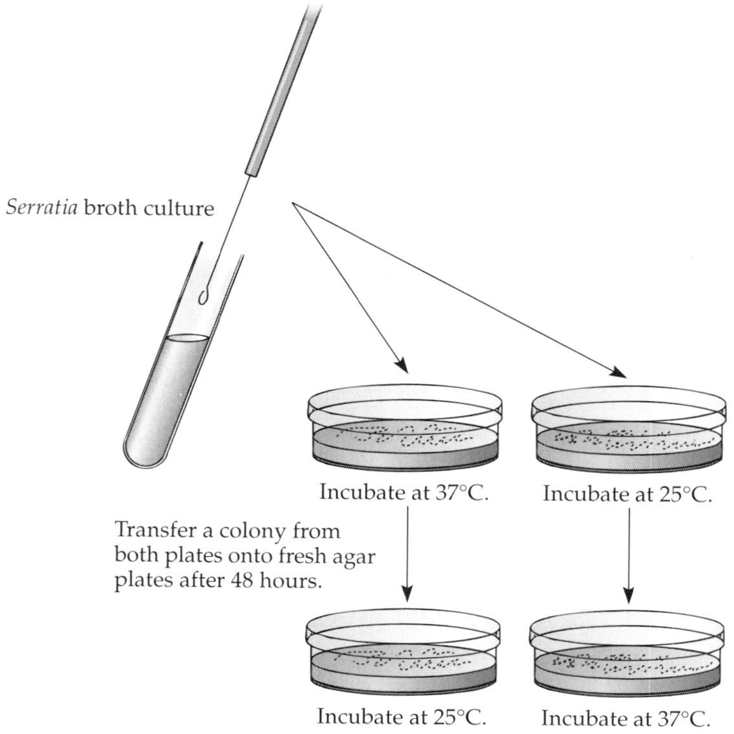

FIGURE 61.1 Scheme for *Serratia* inoculations.

2. After incubation, examine each of the plates for motility.
3. Use a loop to transfer some of the organisms from the nutrient agar plate to a fresh TSA plate. Make a similar transfer from the original TSA plate to a fresh nutrient agar plate. Incubate at 37°C for 24 hr.
4. Record motility observations.

C. Variations in Motility-Effect of Phenol

1. Inoculate a TSA plate and a TSA phenol plate with a loopful of the *Proteus* broth culture, and incubate at 37°C for 24 hr.
2. Record observations, and then inoculate a fresh TSA plate with some of the growth from the original TSA-phenol plate. In a similar manner, transfer some of the organisms from the original TSA plate to a fresh TSA-phenol plate.
3. Incubate for another 48 hr and record observations.

Laboratory Report 61

Name ______ Date ______ Section ______

Environmentally-Induced Variation

(A) Variation in Pigmentation

Pigmentation on Plates Incubated at 25°C

Pigmentation on Plates Incubated at 37°C

Pigmentation After Transfer from 37 to 25°C

Pigmentation After Transfer from 25 to 37°C

(B) Variation in Motility-Effect of Salt Concentration

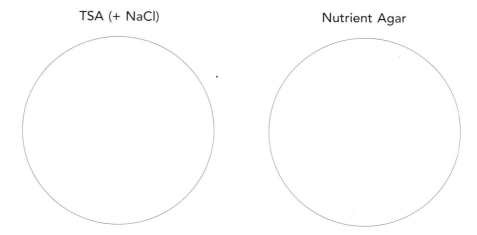

TSA (+ NaCl)

Nutrient Agar

Sketch the extent of *Proteus* growth:

Nutrient Agar

TSA (+ NaCl)

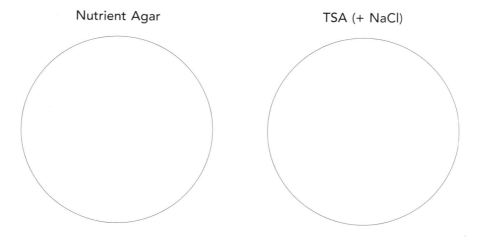

Sketch the extent of *Proteus* growth after transfer:

(C) Variation in Motility-Effect of Phenol Concentration

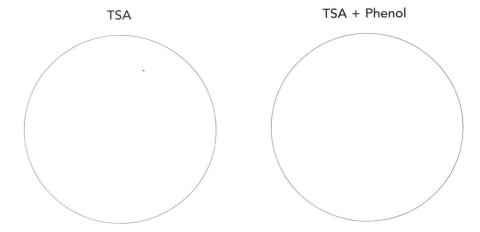

TSA

TSA + Phenol

Sketch *Proteus* growth:

TSA + Phenol

TSA

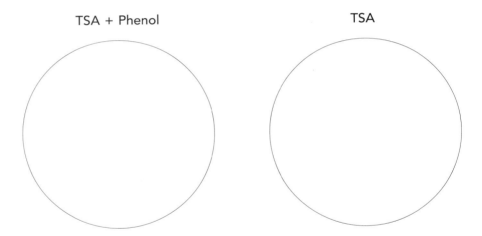

Sketch *Proteus* growth after transfer:

Review Questions

1. How would you determine whether a non-pigmented colony of *Serratia* has arisen by environmentally induced variation or by mutation?

2. What conclusions can be reached about the effect of salt and phenol on bacterial motility? Propose some possible mechanisms of action.

LABORATORY

62 Isolation of Pigmentless Mutants

Safety Issues

- Do not expose ethanol to an open flame as combustion may occur.
- Use a pipetting device for all pipetting steps. Do not pipet any material by mouth.
- Wear UV-protective goggles and gloves while exposing plates to ultraviolet radiation.
- *Serratia marcescens* may be pathogenic in an immunocompromised individual.

OBJECTIVES

1. Understand the role that a mutagenic agent such as ultraviolet light plays in inducing mutations in bacteria.
2. Induce pigmentless mutations in *Serratia marcescens* and determine the ratio of mutants in the surviving population.

In Laboratory 61, observations on the effects of temperature on pigment production in *Serratia marcescens* were made. Lasting inheritable changes (mutations) can be induced by mutagenic agents such as ultraviolet light.

Materials (per group)

1. Saline suspension of an overnight broth culture of *Serratia* grown at room temperature (250 ml)
2. Tryptic soy or nutrient agar plates (16)
3. 4.5-ml sterile saline dilution tubes (16)
4. Sterile 1.0- and 10.0-ml pipettes (1 canister of each)
5. Sterile glass Petri plate (1)
6. Turntable (1)
7. Glass spreaders (2)
8. Germicidal ultraviolet light
9. 95% ethanol (50 ml)

Procedure

1. Each group should receive a saline suspension of the *Serratia* organism.
2. Determine the titer of organisms in the suspension by using the viable plate-count technique (see Laboratory 16). Suggested plating dilutions: 10^{-3} to 10^{-8}.

⚠ Ultraviolet light can cause permanent damage to your eyes. Be sure to wear UV-protective goggles when using the ultraviolet light.

3. Incubate the titer plates at room temperature for 48 hr.
4. Pipet 10.0 ml of the original culture into a sterile Petri dish and position the plate under the ultraviolet light source.
5. Remove the lid from the plate and turn on the light to irradiate the organisms.
6. Groups should use exposure times of 5, 10, 15, 20, and 25 sec. (Each group should select one time interval.)
7. After irradiation, use a sterile 1.0-ml pipette to transfer 0.5 ml of the culture to a sterile saline dilution tube.
8. Prepare appropriate ten-fold serial dilutions and spread plate 0.1 ml of the dilutions onto the agar plates.
9. Incubate covered under a box at room temperature for 48 hr. [Visible light is known to reverse the effects (photoreactivation) of UV-induced mutations.]
10. Uncover plates and count the number of white and red colonies. Count the colonies on the unirradiated culture in order to determine the original titer.
11. Calculate the percentage of surviving organisms and the percentage of mutants in the surviving population.

Laboratory Report 62

Name ______ Date ______ Section ______

Time of Exposure ______								
Unirradiated Culture			Irradiated Culture of Red Colonies			Irradiated Culture of White Colonies		
Dilutions	Plate Counts	Titer	Dilutions	Plate Counts	Titer	Dilutions	Plate Counts	Titer

Review Questions

1. Compare the class results and formulate a hypothesis regarding the relationship of the time of exposure to UV and the induction of pigmentless mutants.

2. What controls should be employed to account for normal pigment variations within the population of cells and any photoreactivation repair?

3. What would be the result if one of the groups forgot to remove the cover of the Petri plate before irradiation of the culture?

LABORATORY

63 Ames Test for Detection of Chemical Carcinogenicity

Safety Issues

- Do not expose ethanol to an open flame as combustion may occur.
- Wear gloves while dipping disks into mutagenic solutions.

OBJECTIVE

Become familiar with the Ames test for the evaluation of potentially mutagenic chemicals.

We have become increasingly aware of the number of chemicals to which we are regularly exposed. Pesticide residues are found in our food and water supplies, industrial wastes contribute noxious chemicals, and cigarette smoke contains suspected harmful chemicals. Many of the chemicals released into the environment are suspected carcinogens; that is, they may induce malignancies or cancers in humans. Many of these carcinogenic chemicals are also known to be mutagenic, that is, cause changes in the nucleotide bases of the DNA molecule.

At the University of California at Berkeley, Bruce Ames and his colleagues have developed a simple microbial procedure to screen chemicals for suspected mutagenicity. The test has become known as the Ames test and utilizes a histidine negative (his$^-$) auxotrophic strain of *Salmonella typhimurium* that will not grow on a medium lacking histidine. The test involves the measurement of the rate of back mutations (reversions) of *Salmonella* (his$^-$) from the auxotrophic to the prototrophic (his$^+$) state. The test organism is also deficient in DNA repair enzymes, which minimizes the potential repair of any mutations that may occur. By using this test, chemicals that are identified as being mutagenic can then be screened using other procedures for carcinogenicity. Not all mutagenic chemicals are necessarily carcinogenic.

It was found that many chemicals are not directly carcinogenic or mutagenic, but rather undergo chemical changes in the human body where they are converted to their active carcinogenic states. These changes take place primarily in the liver. The Ames test incorporates the use of a rat liver enzyme extract that provides a source of activating enzymes to mimic the fate of the chemical in the human body. A liver extract will not be used in this experiment, since agents known to be already mutagenic will be used.

Materials (per group)

1. Plates of glucose-minimal salts agar (GMSA) (4)
2. 45°C water bath containing tubes of soft (top) agar (3.0 ml) with biotin and histidine (4)

3. 1.0-ml sterile pipettes (1 canister)
4. Sterile filter-paper disks (10)
5. Alcohol and forceps
6. Broth culture of *Salmonella typhimurium* strain TA 1538 (ATCC #29631) (5.0 ml)
7. Mutagenic agents:
   Ethidium bromide
   4-nitro-o-phenylenediamine
   2-nitrofluorene
   Hair dye
   Cigarette smoke trappings
   Other agents of your choice

Procedure (See Figure 63.1)

1. Label three glucose-minimal salt agar plates with the names of chemicals you elect to test. Label a fourth plate as a control.
2. Remove a soft agar tube from the 45°C water bath using aseptic technique. Add 0.1 ml of the *Salmonella* broth culture to the agar. Gently mix the tube so as not to create bubbles, and then pour the contents onto the surface of the glucose-minimal salts agar plate. Gently rotate the Petri

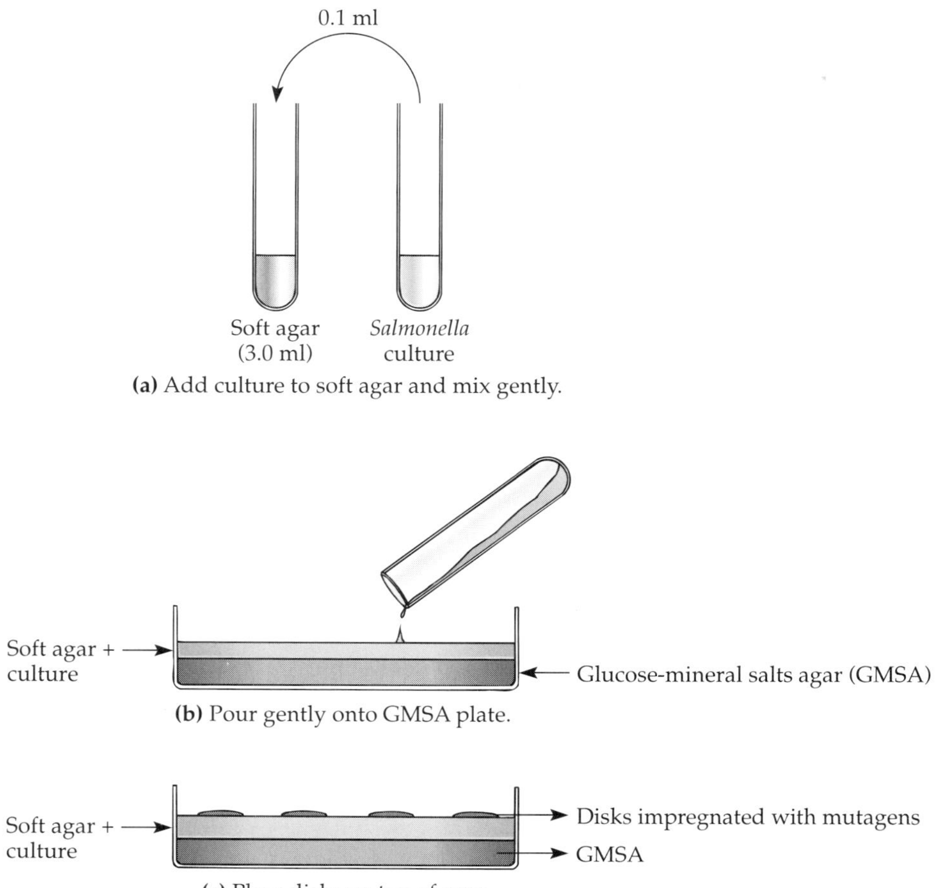

(a) Add culture to soft agar and mix gently.
(b) Pour gently onto GMSA plate.
(c) Place disks on top of agar.

FIGURE 63.1 Technique for Ames test.

dish to distribute the agar evenly over the surface before it hardens. Care must to be taken to work quickly in order to avoid premature hardening and clumping of the agar. If this occurs, acquire a fresh tube of soft (top) agar and start over.

⚠ **Be careful not to ignite the ethanol. Keep the ethanol container away from any open flame.**

⚠ **Since the agents employed are mutagenic, wear gloves when handling.**

3. Repeat the procedure with three other tubes of soft agar and three glucose-minimal salt agar plates.
4. Dip the forceps in alcohol and flame to sterilize. Then use the forceps to dip a filter-paper disk into sterile distilled water. Remove the excess water by pressing the disk onto the side of the tube. Place the disk on the middle of the control agar plate.
5. In a similar manner, use the forceps to apply the suspected mutagenic chemicals to the disks and place on the appropriate labeled plates.
6. Be sure to *gently* press all disks down with the forceps so that they adhere to the agar surface.
7. If the chemical is crystalline, place a few crystals directly on the agar surface.
8. Incubate all plates for 48 hr at 37°C.
9. Examine all plates and determine the degree of mutagenicity of the chemicals by observing the number of revertant colonies in proximity to the filter-paper disks. The revertant colonies appear as a halo surrounding the disk.

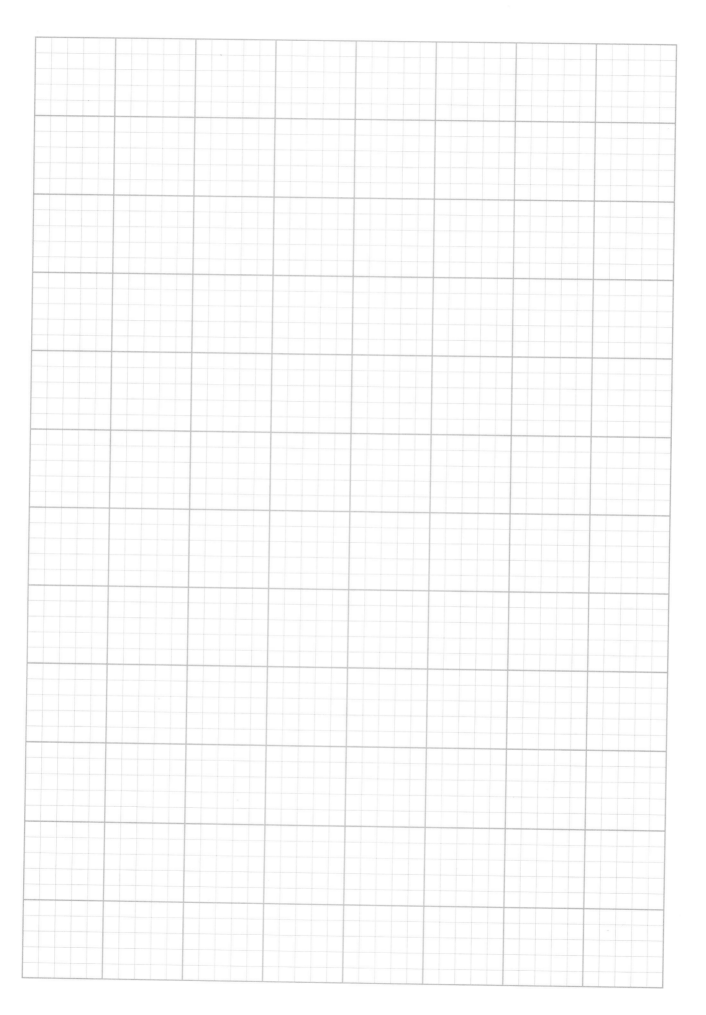

Laboratory Report 63

Name ____________ Date ____________ Section ____________

Ames Test

Sketch the growth pattern surrounding the disks on the four test plates:

Control Plate
Chemical
Chemical
Chemical

Which chemical shows the highest level of mutagenicity?

Review Questions

1. Indicate some of the mutagenic effects chemicals may have on DNA.

2. What can account for any reversions that occur on the control plate in the Ames test?

3. Why is histidine incorporated into the top agar in the Ames test?

LABORATORY

64 Bacterial Transformation

Safety Issues

- Wear gloves when performing this experiment.
- Use a mechanical pipetting device for all pipetting steps.
- Pipet the phenol and chloroform solutions while working under a safety hood. Phenol is extremely toxic and chloroform is very volatile. Inhalation of these solutions can cause respiratory distress and even death. Extreme caution should be taken when working with these solutions.

OBJECTIVES

1. Learn the techniques used in the isolation of DNA from donor bacteria and in the transfer of genetic information to recipient cells.
2. Learn the nature of an auxotrophic mutant.
3. Be able to define competency.

Transformation may be defined as the transfer of genetic information to recipient cells by means of DNA prepared from a donor population. Transformation has been described and investigated in a variety of gram-positive and gram-negative organisms.

In this experiment, a mutant strain of *Bacillus subtilis* which requires either tryptophan or indole will be used as the recipient. The donor cells are the wild type of the same organism, and are independent of tryptophan or indole.

PREPARATION OF TRANSFORMING DNA

Materials (per group)

1. Overnight broth culture of wild type *Bacillus subtilis* (ATCC #23059) grown in 200 ml tryptic soy broth (15 g/L) at 30°C with shaking (150 rpm)
2. EDTA (50 mM) /NaCl (0.1 M) (15.0 ml)
3. Freshly prepared lysozyme (20 mg/ml) (0.2 ml)
4. Sodium lauryl sulfate (saturated solution in 45% ethanol) (0.3 ml)
5. Phenol (5.0 ml)
6. Phenol:chloroform solution (5.0 ml)
7. 3 M sodium acetate (pH 5.2) (0.5 ml)
8. 95% alcohol (ice cold) (15.0 ml)
9. Sterile saline (NaCl) (5.0 ml)
10. Pasteur pipettes (2)

11. Oak Ridge centrifuge tubes (2)
12. Screw cap tube (1)
13. Sterile TE buffer (5.0 ml)

Procedure

1. Centrifuge the ATCC #23059 culture at 7500× g for 10 min to pellet the cells. Carefully pour off the supernatant.
2. Resuspend the pellet in 10.0 ml of EDTA/NaCl and recentrifuge to pellet cells.
3. Resuspend the cells in 5.0 ml of the EDTA/NaCl solution.
4. Add 0.2 ml of lysozyme solution and incubate at 37°C with occasional shaking until the suspension becomes gel-like (about 10–15 min).
5. Add 0.3 ml of sodium lauryl sulfate solution and reincubate for 5 min with shaking.
6. Add 5.0 ml of phenol with a pipetting device and shake vigorously for 1 min.
7. Centrifuge at 10,000× g for 10 min at 25°C.
8. Carefully remove the upper phase with the mouth end of a 10.0-ml pipette and transfer to another centrifuge tube.
9. Add 5.0 ml of phenol:chloroform solution using a pipetting device and shake for 1 min.
10. Centrifuge at 10,000× g for 10 min at 25°C.
11. Remove the top layer with a 10.0-ml pipette as in step 7 and transfer to a screw-cap tube.
12. Add 0.5 ml of 3 M sodium acetate (pH 5.2).
13. Slowly add the ice-cold 95% ethanol. Cap the tube and gently shake.
14. Sterilize a Pasteur pipette by placing in the flame of a Bunsen burner, and form a small loop at the end of the pipette.
15. Wind the DNA around the loop of the pipette.
16. Stand the pipette upright in a test tube rack for a few minutes to allow the ethanol to evaporate.
17. Dissolve the DNA in 2.0 ml of TE buffer.
18. The DNA should be kept chilled in an ice bath or refrigerator.

⚠ **The phenol and phenol:chloroform solutions should be transferred in a hood only by using some type of pipetting device. Do not inhale or allow to come into contact with bare skin.**

DETERMINATION OF DNA CONCENTRATION

Dilute your preparation 1:20 in saline (0.85%), and read the absorption at 260 nm in an ultraviolet spectrophotometer. Calculate the amount of DNA in the undiluted suspension by using the following equation:

Concentration of DNA mg/ml = Absorbance @ 260 nm/280 nm
(for cuvettes with a 1-cm light path)

TRANSFORMATION OF AUXOTROPHS

The DNA prepared above is used to transform tryptophan requiring mutants of *Bacillus subtilis* to tryptophan independence. Small amounts of casein hydrolysate provide growth stimulation and aid in the induction of competence (the ability to take up DNA).

Materials (per group)

For The Class:

B. subtilis (ATCC #23857) growing in *Medium A* (g/L) overnight at 37°C with shaking (150 rpm).

K_2HPO_4	10.5
KH_2PO_4	4.5
$(NH_4)_2SO_4$	1.0
Na-citrate · 2 H_2O	0.5
$MgSO_4$ · 7 H_2O	0.1
Sterilized separately	
Glucose	5.0
L-tryptophan	0.05
Acid-hydrolyzed casein	0.2

1. Medium B (10.0 ml) (Same as medium A but substitute 5.0 μg/ml tryptophan for 50 μg/ml and 0.1 g/L casein hydrolysate for 0.2 g/L.
2. 16 sterile tubes (13 × 100 mm) (16)
3. Sterile DNase (20 μg/ml)
4. Sterile minimal agar plates (same as A but minus tryptophan and casein hydrolysate) (26)
5. Tryptic soy agar plates (2)
6. Sterile saline dilution tubes (4.5 ml) (6)

Procedure

1. The lab assistants have added 0.2 ml of the *Bacillus subtilis* auxotrophic culture to tubes containing 2.0 ml of medium A and incubated them for 5 hr at 37°C.
2. Prepare 5 μg/ml solution of DNA extracted from the parental *Bacillus* culture.
3. Set up the following tubes as shown:

Tube	DNA (5 μg/ml)	DNase (20 μg/ml)	Medium B
1	—	—	0.9 ml
2	0.1 ml	—	0.8
3	0.1	—	0.8
4	0.1	—	0.8
5	0.1	—	0.8
6	0.1	0.1 ml	0.7
7	0.1	0.2	0.7

4. Add 0.1 ml of the *auxotroph* to the first 6 tubes. Mix well and *immediately* spread 0.1 ml from tube 2 onto 2 minimal agar plates. Place all tubes except tube 2 in a shaker at 37°C.
5. Spread 0.1 ml from tube 3 onto each of the two plates at 30 min. Repeat with a 10^{-1} dilution.
6. Repeat this plating procedure for tube 4 at 60 min and tubes 1, 5, 6, and 7 at 90 min.
7. Assay tube 1 by spreading 0.1 ml aliquots of a 10^{-5} dilution onto duplicate TSA plates. Incubate all minimal agar plates at 37°C and TSA plates at room temperature.
8. Count the colonies after 48 hr and record results.

Laboratory Report 64

Name Date Section

Tube	Incubation Time	Number of Colonies
1	90 min	
2	0 min	
3	30 min	
4	60 min	
5	90 min	
6	90 min (with DNase)	
7	90 min (with DNase)	

Calculate the number of transformants observed per number of recipient cells:

What was the percentage of transformation observed?

Review Questions

1. What is the function of DNase in tubes 6 and 7?

2. How are competent cells formed in this experiment?

3. How would you explain any overgrowth on the minimal media plates?

4. What is the function of the phenol and chloroform used in the DNA extraction procedure?

LABORATORY

65 Conjugation and Recombination in *Escherichia Coli*

Safety Issues

- Use mechanical device for all pipetting steps. Never pipet any material by mouth.
- Allow sterilized inoculating loop to cool before transferring organisms to avoid splattering of cells.

OBJECTIVES

1. Learn the mating types that are used in recombinations involving conjugations.
2. Observe a genetic exchange and recombination by crossing an Hfr strain and an F^- strain.

Three types of bacterial cells are involved in conjugation, $\mathbf{F^+}$, $\mathbf{F^-}$, and **Hfr**. F^+ cells possess a "fertility" **(F) factor** which is absent in F^-. This factor is located on a **plasmid**, a small piece of double-stranded, circular DNA that is located independent of the bacterial "chromosome." Genes on this plasmid code for surface structures called sex pili which function in conjugation. These pili serve as a conjugation bridge which enables the F^+ cell to attach to an F^- cell. When the cells are joined, an exchange of genetic information may occur. In this type of mating only the plasmid is transferred. It is not clear exactly how the plasmid is transferred. F^+ and F^- matings result in all the cells becoming F^+ (Figure 65.1a). Since plasmids also can carry genes that code for antibiotic resistance in bacteria, the transfer of plasmids during this type of conjugation is an important mechanism in the spread of antibiotic resistance in bacterial populations. In some cells, the F^+ factor may become integrated with the genome of the bacterial cell; such a cell is termed Hfr (high frequency of recombination). Hfr cells may also mate with F^- cells. During conjugation, the genome with the integrated plasmid replicates and the donor genetic information is transferred by the genome entering the recipient cell first, followed by the integrated plasmid (Figure 65.1b). The entire process may take up to hours and the mating may be interrupted so that only a portion of the donor genome may actually wind up being transferred. In a complete transfer, the integrated plasmid enters the recipient cell last. Since the exchange is linear, such a process can be used to map the relative locations of specific genes on the bacterial genome.

Following conjugation, the recipient cell possesses both its own genome and all or part of the genome from the Hfr donor. The donor fragment pairs in the appropriate regions with the genome of the recipient. Recombination may occur with the integration of the Hfr genes into the recipient genome.

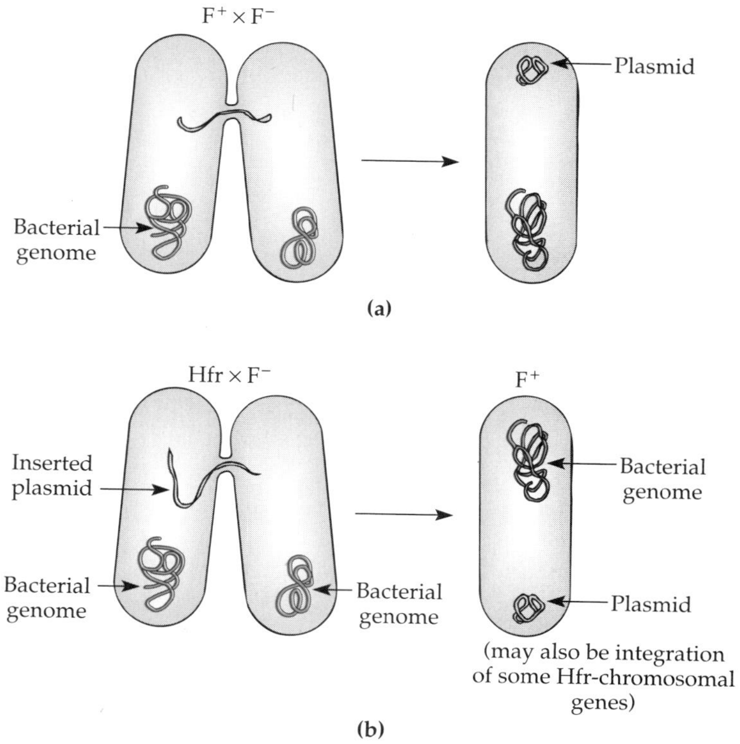

FIGURE 65.1 (a) F^+ and F^- mating; (b) Hfr and F^- mating.

Thus, the recipient cell may permanently acquire some of the characteristics carried into it from the Hfr cell. In matings between Hfr and F^- strains, recombinants appear with very high frequency (this is the source of the Hfr designation).

In this experiment, the *Escherichia coli* (ATCC #25257) derived from *E. coli* K12 is an Hfr strain that is prototrophic, requiring only thiamine for growth, and is sensitive to streptomycin. The *E. coli* (ATCC #25257) is an F^- strain that is auxotrophic and requires several amino acids, including leucine, proline, and tryptophan, as growth factors. The F^- strain also requires thiamine and adenine, is resistant to streptomycin, and is unable to ferment lactose. A minimal medium with a known chemical composition is used. The Hfr strain can grow on the minimal medium with no added growth factors; the F^- strain requires the addition of pro^-, leu^-, try^-, and ade^-.

Platings will be performed at intervals after mixing of the strains under conditions allowing for the recovery of only the recipient bacteria.

Materials (per group)

For The Class:

Log phase cultures of Hfr and F^- grown in tryptic soy broth (TSB) at 37°C overnight, diluted 1:20 in TSB, and aerated at 37°C for 2 hr prior to use.
10 sterile tubes
20 TSA plates containing 200 μg/ml streptomycin

Per Group:

20 sterile tubes
150 ml sterile saline
25 minimal agar plates containing 200 μg/ml streptomycin
25 minimal agar plates + pro+try+ade+strep
25 minimal agar plates + pro+leu+ade+strep
25 minimal agar plates + leu+try+ade+strep
25 minimal agar plates + leu+try+pro+strep
2 EMB agar plates
Minimal agar medium:

K_2HPO_4	7.0 g/L	
KH_2PO_4	3.0	
Na citrate	1.0	+ 1 μg/ml thiamine
$(NH_4)_2SO_4$	1.0	+ 20 μg/ml amino acids and adenine
$MgSO_4 \cdot 7\ H_2O$	0.1	
Glucose	2.0	
Bacto-Agar	15.0	

Second Day, for each pair:
2 EMB-Lactose plates

The cultures have the following genetic markers:
Hfr (ATCC #25257) thi^-, lac^+, pro^+, leu^+, try^+, ade^+, str^s
F^- (ATCC# 25250) thi^-, lac^-, pro^-, leu^-, try^-, ade^-, str^r

Procedure

(Day 1)

1. Place 2.0 ml of the F^- suspension into a sterile tube.
2. Add 2.0 ml of the Hfr strain, mix, and immediately remove 0.1 ml to 9.9 ml sterile saline. This is a 1:100 dilution of the zero-time culture.
3. Place the undiluted culture in the water bath at 37°C. Remove additional samples at 15, 30, 60, and 90 min.
4. Immediately after removing each sample, continue diluting with sterile saline to 10^{-4}. Place 0.1 ml from the 10^{-2} and 10^{-4} dilutions at each time interval onto duplicate plates of the following medium:
   (a) minimal agar + strep
   (b) minimal agar + pro+leu+try+ade+strep
   (c) minimal agar + pro+leu+try+strep
   (d) minimal agar + pro+leu+ade+strep
   (e) minimal agar + pro+try+ade+strep
   (f) minimal agar + leu+try+ade+strep
5. Streak one loopful of a 10^{-5} dilution of each parent culture onto a divided EMB plate.
6. Determine the number of F^- cells at the time intervals by spreading 0.1 ml of a 10^{-6} dilution onto duplicate TSA+streptomycin plates. Incubate all plates at 37°C overnight.

(Day 2). Count the colonies on the best plates of each medium and determine the proportion of recombinants (compared to F^- cells) for each amino acid marker at each time interval.

The F^- strain will grow on minimal medium 5 (all needed growth factors except leucine) if it has received the gene enabling it to synthesize leucine. Analyze each of the other plates in a similar fashion. Divide the EMB plates into sectors and streak a number of colonies from each of the six types of agar onto each of the sectors.

(Day 3). On the third day, examine the EMB plates and determine the relative position of the lactose marker. Can you place the four markers examined in a linear order?

Laboratory Report 65

Name Date Section

Colony Plate Counts

Time (min)	Media 1 10^{-2} 10^{-4}	2 10^{-2} 10^{-4}	3 10^{-2} 10^{-4}	4 10^{-2} 10^{-4}	5 10^{-2} 10^{-4}	F^- colonies
0						
15						
30						
60						
90						

Indicate the position of the markers leu, pro, try, and ade:

Review Questions

1. Explain the fact that recipient cells involved in Hfr $\times$ F^- crosses remain F^-.

2. Why don't F^- cells that have Hfr cells attached grow on the minimal medium?

UNIT XI

Medical Microbiology and Immunology

Medical microbiology is a specialized subdiscipline of microbiology that is sometimes referred to as **clinical microbiology**. It deals specifically with determining the type of microorganism associated with a specific infection.

There are a number of other factors to consider in examining the basis of **infection**. A layperson's understanding of the term "infection" usually includes the participation of some type of microorganism in the establishment of a disease.

The number and type of microorganism is important in the establishment of infection. Most organisms causing infection are **pathogenic** (disease-causing). An assessment of their degree of pathogenicity is referred to as **virulence**. Microorganisms come armed with a variety of virulence factors when they confront a host organism. Some produce diffusible toxins (**exotoxins**), while others carry their toxins as part of their cell wall structure (**endotoxins**). Some produce enzymes and other cell- and tissue-destroying substances that allow them to spread in a susceptible host. Also of importance in the establishment of infection is the manner by which the virulent microorganisms gain access to the host. The term **portal of entry** describes the route by which such access is gained. The portals (respiratory, genitourinary, circulatory, skin) are varied, and infection will not be established if the portal is not conducive to the proliferation or spread of microorganisms. The spores of *Clostridium tetani* can cause tetanus if injected via a puncture wound into deep, anaerobic tissue, but are of little consequence if accidentally contacted with skin.

The human deals with the presence of potentially virulent organisms by using a variety of defense strategies. Collectively these strategies are referred to as "host-defense mechanisms." They may be non-specific or specific. Non-specific mechanisms include chemical and physical barriers to infection (stomach acid, nasal hairs). Phagocytosis, inflammation, fever, and species and racial factors are also types of non-specific resistance. Specific host-resistance mechanisms involve the host's immune system, which is responsible for the elaboration of immune cells and the synthesis of antibodies in response to the invading pathogens.

Shortly after birth, humans acquire a "normal" residential population of microorganisms that they will carry all of their lives. This population is called the **normal flora** and varies depending on the region of the body examined. Occasionally a member of the normal flora can actually cause infection. It is considered to be an **opportunistic pathogen** when behaving in this manner. Ordinarily, it doesn't cause disease, but if the host-resistance mechanisms are weakened or destroyed, the organism gains the upper hand. It should be remembered that the relationship between microbe and host may take many forms and is quite dynamic, being influenced by specific and non-specific factors.

It is impossible to explore all aspects of clinical microbiology in a first laboratory course in general microbiology. In the exercises and experiments that follow, you will become familiar with the isolation and cultivation of some of the normal flora and a few specific pathogens. Fundamental immunological principles are also introduced in this unit.

LABORATORY

66 A Simulated Epidemic

Safety Issues

- Be sure to dispose of all gloves and cotton balls in the appropriately labeled containers.

OBJECTIVE

Understand the role of direct contact in the transmission of infectious agents by monitoring the spread of *Saccharomyces* during hand shaking.

Organisms that have the ability to cause disease (pathogens) must confront the appropriate environment in a host. Pathogens can spread in a variety of ways: by direct contact, by droplets or aerosols, or by vectors such as insects. A common method of pathogen spread in a hospital environment is by direct contact. Most often, disease-causing agents are spread via hands which have not been properly cleaned. Physicians, nurses, lab workers and other health care personnel involved in direct contact with patients must maintain good hygienic technique to prevent such infections.

In this exercise, the transmission of a biological agent via hand shaking will be monitored. The agent used is *Saccharomyces cerevisiae*, common baker's yeast, which has been selected for safety reasons since the chance of infectivity is low. Balls of absorbent cotton have been placed in sterile beakers. One ball has been moistened with a suspension of *S. cerevisiae*. The other balls have been wetted with sterile saline. It will be the assignment of the class to determine which cotton ball is infectious, that is, which contains the yeast.

Materials (per 20 students)

1. Cotton balls in sterile beakers, one of which has been contaminated with yeast (5)
2. Disposable latex gloves (20)
3. Sabouraud dextrose agar (SDA) or potato dextrose agar (PDA) (40 plates)

Procedure

1. Every class member should have an identification number.
2. Each student should place a glove on one hand. This hand will be used to handle cotton balls or to shake hands with other class members.
3. Either three or five students (class of 10 or 20) should remove a piece of cotton from one of the beakers. Make sure to record the number of the sample. Each student should handle the cotton for a minute and then place it back in the beaker.

4. Shake hands with another class member, being sure to keep a record of the number of the person whose hand you shook. This is the first round of hand shaking.
5. Each student should then shake hands with another class member. After the second hand shake, each class member should inoculate one PDA or SDA plate with the fingers of the gloved hand. Plates should be labeled with student number and marked as second-round hand shake. Incubate the plates at 30°C for 72 hr.
6. Shake hands again with another person. Be sure to record the number of the person. Inoculate the plates and label as third-round hand shake. Incubate at 30°C for 72 hr.
7. After incubation, check all plates for the presence of *Saccharomyces*.
8. On a chart provided in the lab, indicate whether plates are positive or negative after the second- and third-round hand shake.
9. From the analysis of the class data, determine who was the original "infectious" person.
10. Trace the pattern of the simulated epidemic.

Laboratory Report 66

Name Date Section

PLATES SHOWING *SACCHAROMYCES* GROWTH

Student Number	Second Shake	Third Shake

Path of Infection:

Review Questions

1. How is it possible for an individual who shook hands with an "infected" class member not to show infection?

2. What are some of the agents most likely to be spread via the hand contact route?

3. Design an experiment to show how handwashing could decrease the infectious potential of an individual.

4. Why are gloves required?

LABORATORY

67 Throat Cultures

Safety Issues

- Wear gloves when performing a throat swab of another person.
- Dispose of tongue depressors and swabs in containers of disinfectant.
- Exercise caution when handling isolated cultures, as some are likely to be pathogenic.

OBJECTIVES

1. Learn isolation and cultivation techniques for organisms from the upper respiratory tract.
2. Differentiate the types of streptococci likely to be found in normal and diseased throats.

The respiratory system is generally divided into the upper respiratory system (the oropharynx and nasopharynx) and the lower respiratory system (larynx, trachea, bronchi, and alveolar air sacs of the lung). This exercise deals with the isolation of microorganisms from the tonsillar area and the posterior pharynx of the human throat. The normal flora of the area consists of staphylococci, streptococci, diphtheroid bacteria and gram-negative diplococci.

Acute pharyngitis is the most common infection in the upper respiratory tract. The cause of infection may be bacterial or viral. Acute streptococcal pharyngitis is characterized by a fire-red throat with patches of pus and difficulty in swallowing. Streptococci can be identified by examining hemolytic reactions on blood agar, by serological tests for the detection of cell wall or capsular antigens, and by biochemical testing. Hemolysis is best observed by examining colonies grown under anaerobic conditions, or alternatively, by examining subsurface colonies formed by stabbing the agar plate. Four types of hemolysis may be produced by streptococci on sheep red blood cell agar plates.

(a) α-hemolysis:	Partial lysis of erythrocytes surrounding colonies, causing a gray-green or brownish discoloration.
(b) β-hemolysis:	Complete hemolysis of red blood cells surrounding colonies, causing a clearing of the medium.
(c) γ-hemolysis:	No hemolysis, and consequently no color change in the medium.
(d) β prime or wide zone β:	A small zone of intact red blood cells immediately adjacent to the colonies, with zones of complete hemolysis surrounding the zone on intact red blood cells.

Group A β-hemolytic streptococci (*Streptococcus pyogenes*) are the main pathogens associated with acute pharyngitis (strep throat). The antibiotic bacitracin specifically inhibits these organisms and are used in the presumptive identification of group A β-hemolytic streptococci.

Materials (per 20 students)

1. Tryptic soy agar (TSA) plates with 5% sheep blood (60)
2. Tongue depressors (20)
3. Sterile throat swabs (20)
4. Taxo A bacitracin disks, 0.04 units (BD Microbiology Systems, Cockeysville, MD) (20)
5. Culture of *Streptococcus pyogenes*

Procedure

1. The proper method for obtaining a throat swab is shown in Figure 67.1. The person's head should be tilted back and the tongue gently depressed with a tongue depressor to visualize the tonsils and the posterior pharynx.
2. The tonsillar areas and the posterior pharynx should be firmly rubbed with the swab. Care should be exercised not to touch the walls of the buccal cavity or the tongue.
3. Swab a blood agar plate by rolling the swab over 25% of the plates.
4. Sterilize an inoculating loop and further streak out the organisms deposited on the plate. Make two to three stabs into the agar in the inoculated and uninoculated areas of the plate.

Since β-hemolytic and other throat organisms are potentially pathogenic, extreme care should be exercised when transferring the organisms.

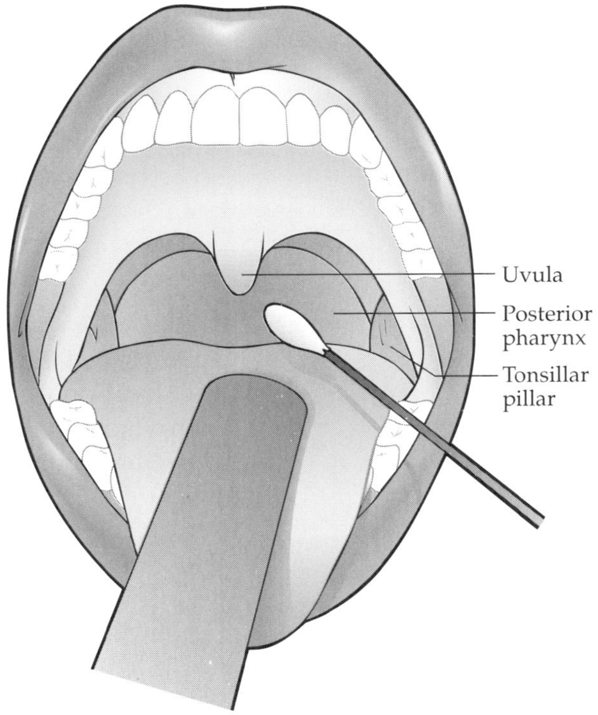

FIGURE 67.1 Acquisition of throat sample.

5. Streak a positive control plate in a similar manner with the *Streptococcus pyogenes* culture.
6. Incubate the plates at 37°C for no longer than 24 hr.
7. After incubation, examine the plates for growth and hemolysis. Observation using a low-power objective is beneficial in interpreting the type of hemolysis.
8. With a sterile incubating loop, remove organisms showing β-hemolysis and streak on a fresh blood agar plate.
9. With sterile forceps, apply a bacitracin (Taxo A) disk to the surface of the plate over the first or second streak.
10. Incubate the plate at 37°C for 24 hr.
11. Observe for zones of inhibition surrounding the disk. This would provide evidence of the presence of group A β-hemolytic streptococci.

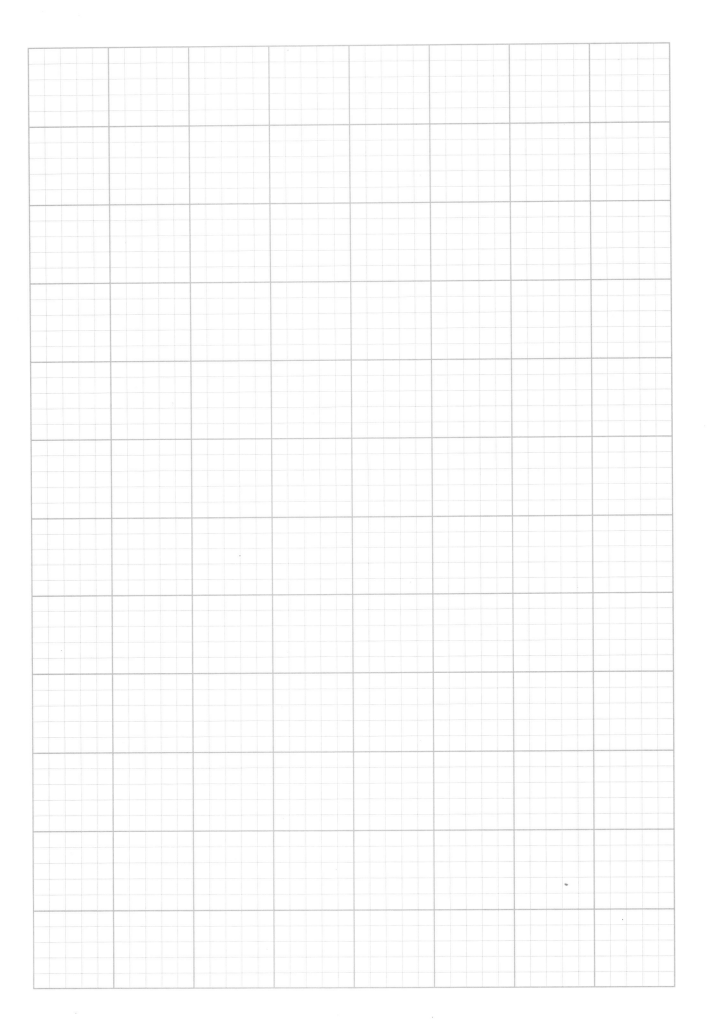

Laboratory Report 67

Name Date Section

Appearance of Blood Agar Plates

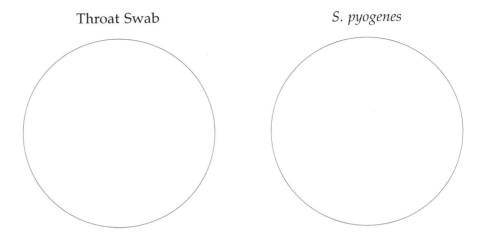

Throat Swab *S. pyogenes*

Review Questions

1. What is the purpose of stabbing the blood agar plates?

2. Describe the different types of hemolysis surrounding the colonies on the surface of the blood plates compared to the types in the stabbed areas.

3. Upon secondary streaking, were there any organisms sensitive to bacitracin? What is the significance?

4. How can β-hemolytic streptococci be differentiated from β-hemolytic staphylococci?

5. What relationship might exist between hemolytic ability and virulence?

LABORATORY 68 Microbiology of Urine

Safety Issues

- Wear gloves during this laboratory and work only with your own urine.
- To prevent splattering of urine, allow inoculating loop to cool before streaking drop of urine.

OBJECTIVES

1. Learn techniques to determine the number of microorganisms present in urine.
2. Use biochemical methods to identify the organisms present in urine.

The urinary tract is divided into the upper tract (kidneys, renal pelves, and ureters) and the lower tract (urinary bladder and urethra) (Figure. 68.1). Upper tract infections result from the spread of organisms into the glomeruli or cortex of the kidneys (**pyelonephritis**), and usually are the result of the ascending spread of microorganisms. Very seldom does infection result from the spread of organisms from the blood. Colonization may also occur dur-

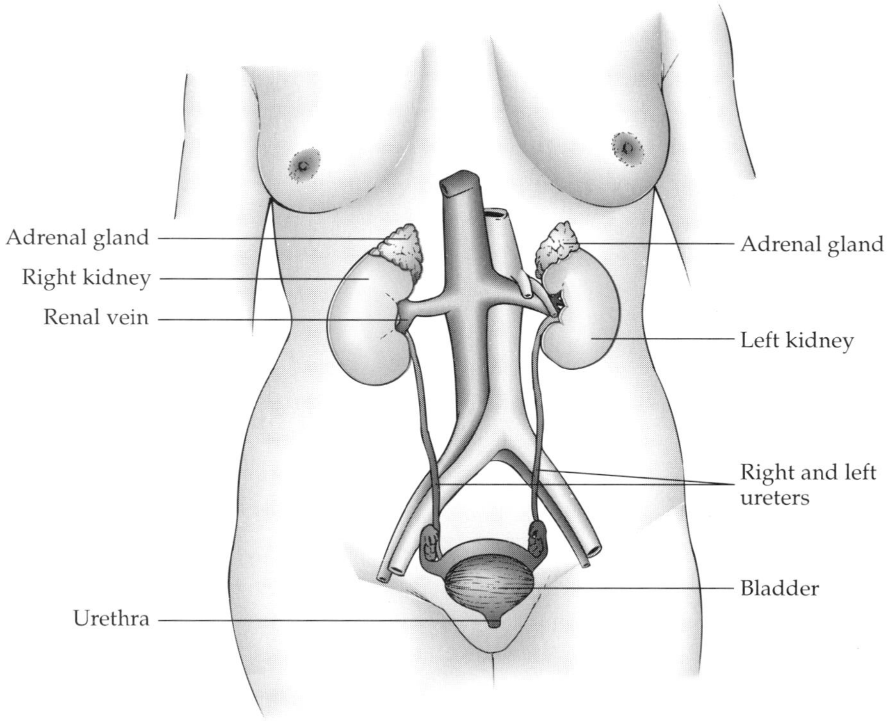

FIGURE 68.1 Urinary tract.

ing the use of a catheter or examination using a cystoscope. Bladder retention due to posture or neurological deficiencies may also be contributory factors to infection. These infections can be quite serious, and a microbiological examination of urine provides important evidence as to the existence and extent of infection. Lower urinary tract infections involve the bladder (**cystitis**). The prevalence of infections varies with the age and gender of the patient. In males, the infection may spread to the prostate gland, while in females the spread may be to the urethra (**acute urethral syndrome**). Pain upon urination is a general indicator of potential lower urinary tract infection. Infections are more common among adult females than among males. There are several reasons for this: (a) In females there is a shorter distance between the anus and the urethral opening; (b) Females have a shorter urethra; (c) Sexual activity can cause an irritation of the urethra with an increased likelihood of fecal contamination; (d) Physical distortion during pregnancy causes retention of urine in the bladder, leading to an increased infection rate.

The normal flora of the urethral opening include coagulase-negative staphylococci, α-hemolytic streptococci, diphtheroid bacilli, enterococci, *Bacillus* sp., saprophytic yeasts, and coliform bacilli. The flora of infected urine can include one or more of the following: coliform bacilli, coagulase-positive staphylococci, *Candida albicans, Mycobacterium tuberculosis*, β-hemolytic streptococci, *Pseudomonas aeruginosa, Proteus sp.*, and *Salmonella* and *Shigella* species.

Urine is normally sterile, and periodic flushing prevents bacterial buildup. The low pH of the urine is also inhibitory to some organisms. There are problems associated with microbiological analysis of urine, the most important being the actual manner of collection of the sample. Contamination of the collected urine is likely to occur unless the samples are correctly obtained and speedily processed. To prevent contamination by organisms surrounding the urethral opening, a **midstream sample** of voided urine is utilized. This is sometimes referred to as a **clean catch** sample. The acquisition of sample is more difficult in females and instruction is necessary to obtain an uncontaminated sample. Aseptic collection requires cleaning of external urogenital surfaces. In males, this is accomplished by sponging the penis with a disinfectant soap and the collection of a midstream sample into a sterile container. In females, the labia should be cleansed with disinfectant soap, and should be held apart while allowing the first few drops of urine to pass into the toilet bowl. Collect the remainder of the midstream sample in a sterile container. If not processed immediately, the urine samples *must be refrigerated*. The urine is likely to be contaminated with a few organisms during collection. Unless the sample is refrigerated, these organisms will multiply and yield misleading results. If the bacterial population exceeds 100,000 organisms per ml, a urinary tract infection is likely.

A number of different types of analyses are available for the detection of bacteria in urine. A quantitative approach uses a calibrated inoculating loop and calls for the enumeration of organisms on blood, MacConkey, or EMB plates.

Materials (per 20 students)

1. Sterile collection vessels (20)
2. Disinfectant soap
3. Sterile wash pads (20)
4. Sterile latex gloves (20)
5. Calibrated inoculating loop (0.001-ml) (20)
6. MacConkey agar or EMB plates (20)
7. Gram stain reagents (5 sets)
8. Hydrogen peroxide (60 ml)
9. Coagulase serum (10 ml)
10. Lactose fermentation broth tubes (40)
11. Tryptone broth tubes (40)
12. Dry Slide Oxidase (Difco Laboratories) (20)

QUANTITATIVE DETERMINATION OF BACTERIA IN URINE

Procedure

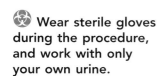

Wear sterile gloves during the procedure, and work with only your own urine.

1. With a sterile 0.001-ml inoculating loop, transfer a loop of urine to the edge of a blood agar and a MacConkey agar plate.
2. Spread the drop with a sterile inoculating loop, streaking for isolation (Figure 68.2).

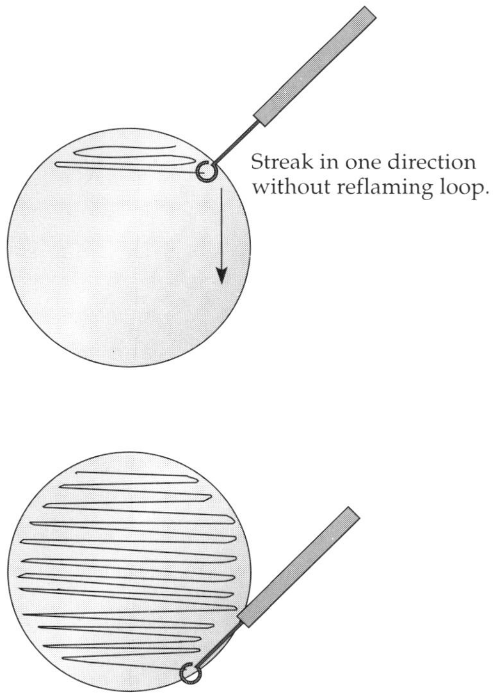

FIGURE 68.2 Loop-inoculation technique for quantitation of organisms in urine.

3. Incubate the plates at 35°C for 24 hr.
4. Count the colonies on the plates. Each colony represents 1000 bacteria per milliliter of urine.
5. Perform Gram stains on organisms from the isolated colonies. Identify the organisms present by conducting the tests and using the identification scheme in Figure 68.3.*
6. If time permits, a portion of the urine can be stored at room temperature for an interval and then sampled for the population of microorganisms present. This will demonstrate the necessity of refrigerating a urine sample if there is a delay in running the analysis.

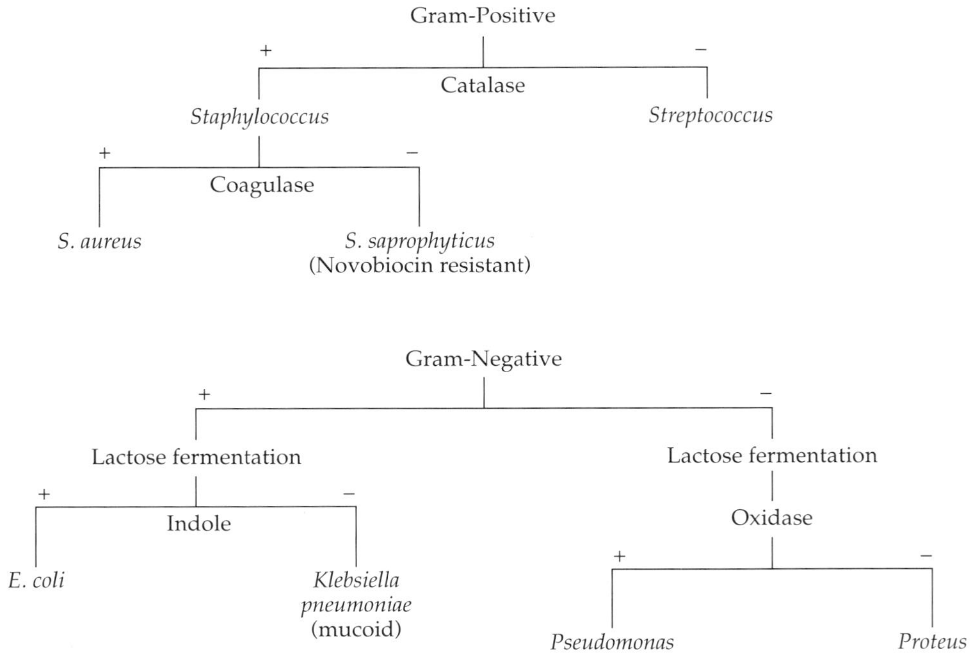

FIGURE 68.3 Identification of organisms cultivated from urine.

*Bacterial ID Scheme provided by Pamela P. Tabery, Northampton Community College, Bethlehem, PA.

Laboratory Report 68

Name Date Section

URINE CULTURE RESULTS

Specimen	Blood Agar	EMB or MacConkey	Gram Stains

Review Questions

1. Describe the differences in the type of growth apparent on plate media used in this experiment.

2. If the quantitative urine count of microorganisms present in urine is greater than 10^5, what technique can be used to determine whether the organisms present are significant human pathogens?

3. How could the presence of gram-positive rods or cocci be detected in a urine sample?

4. Why might it be important for a physician to know that a patient is taking antibiotics prior to ordering a urinalysis?

LABORATORY

69 Dental Caries

Safety Issues

- Do not pipet saliva by mouth. Use a mechanical pipetting device.
- Place beaker containing saliva in disposal area for decontamination.

OBJECTIVES

1. Describe the types of bacterial flora present in the mouth.
2. Determine the susceptibility to tooth decay by using Snyder test agar to culture organisms.

The surface of teeth rapidly become colonized with bacteria shortly after the teeth appear. These organisms are mostly gram-positive and are well adapted to growth on a solid substrate such as tooth surfaces. A biofilm consisting of microorganisms and their extracellular polymers may form on the surface of the teeth. This film is often referred to as **dental plaque**. Production of lactic acid in the plaque can lead to decalcification of tooth enamel. This degradation is called tooth decay (dental caries) and is influenced by the type of diet and other less well described individual traits. The two organisms that have been implicated in dental caries are *Streptococcus mutans* and, to a lesser extent, *S. sobrinus*. The attachment of *S. mutans* is related to the production of a very adhesive polysaccharide dextran, which is produced in the presence of the common disaccharide, sucrose. The common occurrence of sucrose in the human diet helps to explain its association with tooth decay.

Snyder test agar can be used as a reliable predictor of susceptibility to tooth decay. The basis of the test resides in examining the rate of pH reduction in a 2% glucose medium by microorganisms found in saliva. It is known that the decalcification of enamel begins at pH 5.5 and progresses rapidly as the pH decreases due to the anaerobic production of lactic acid. Monitoring the rate of the reaction is facilitated by examining a change in the pH indicator bromocresol green from green at a pH 4.8 to yellow at pH 4.4.

Materials (per 20 students)

1. Paraffin or pieces of Parafilm (20)
2. Sterile 40- to 50-ml beakers (20)
3. 5.0-ml tube of molten Snyder test agar in a 50°C water bath (20)
4. Sterile 1.0-ml pipettes (5 canisters)

Procedure

1. Take a small piece or cube of paraffin and soften it in your mouth for a minute. Chew it for 2 to 3 min, making sure to move it over the surface

of the teeth while chewing. DO NOT SWALLOW YOUR SALIVA WHILE CHEWING.
2. Collect the saliva that was produced into the sterile beaker.
3. Shake the saliva to disperse the bacteria.
4. With a sterile 1.0-ml pipette, transfer 0.2-ml of the saliva to a tube of Snyder test agar.
5. Mix the tube by rolling it between the palms of your hands. Be careful not to introduce excessive bubbles.
6. Incubate the tube at 37°C.
7. Examine the tube on a regular basis for three days to monitor the change in the bromocresol green indicator.
8. Record your results in the table and compare with the caries susceptibility chart (Table 69.1).

TABLE 69.1 SNYDER CARIES SUSCEPTIBILITY CHART

Degree of Caries Susceptibility	Time for Medium to Change to Yellow		
	24 hr	**48 hr**	**72 hr**
High	Positive		
Moderate	Negative	Positive	
Low	Negative	Negative	Positive
Minimal	Negative	Negative	Negative

Laboratory Report 69

Name _______________ Date _______________ Section _______________

Results of Color Change

Hours	Color of Medium	
24		Caries Susceptibility = __________
48		
72		

Review Questions

1. What could you do to reduce your susceptibility to dental caries?

2. A student has performed a Snyder's caries susceptibility test on two different days and has found a considerable difference in susceptibility. On day one, the susceptibility was high; while the next day the susceptibility was minimal. What other factors must be taken into account when interpreting this test?

3. Would drinking large amounts of Diet Coke make you more susceptible to tooth decay?

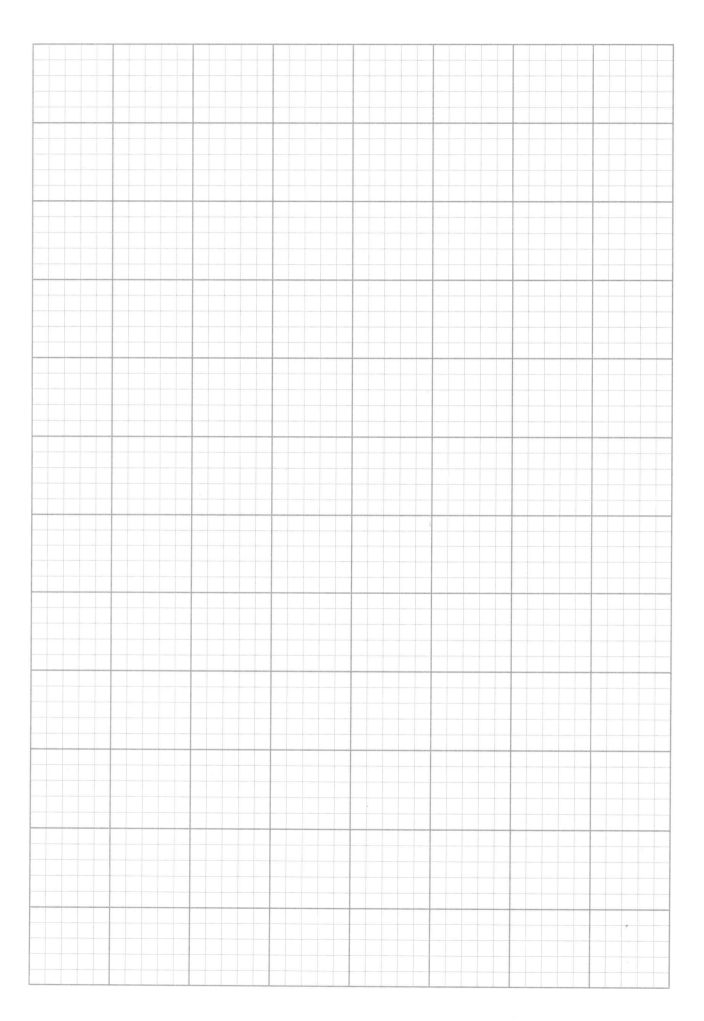

LABORATORY

70 Microbiology of the Skin

Safety Issues

- Dispose of all used swabs in containers of disinfectant.
- Exercise caution when handling isolates, as some may be potentially pathogenic.

OBJECTIVES

1. Identify the normal flora of the skin.
2. Differentiate the gram-positive cocci found on the skin.

The surface of the skin is not generally supportive for the growth of bacteria due to variability in moisture content and the presence of antimicrobial chemicals. The highest population of normal flora on the skin is found in association with sweat glands. Primary locations include the palms of the hands, soles of feet, underarms, genital areas, and hair follicles. The flora can be either **stable** or **transient**. Since the skin is exposed to many different external environments, there exists opportunity for microorganisms to be borne on the surface of the skin. Most of the transient organisms die off quickly, leaving only the stable resident flora. The normal flora consists primarily of gram-positive bacteria, including many types of staphylococci and corynebacteria. *Propionibacterium acnes* is a gram-positive rod that contributes to the formation of acne. *E. coli* is found as a transient fecal contaminant. *Acinetobacter* is one of the few gram-negative organisms found in association with the skin. Members of the genus *Staphylococcus*, while being part of the normal flora, can cause opportunistic infections under appropriate environmental conditions. *S. aureus* may cause a variety of infections, such as infections of hair follicles, impetigo, boils, carbuncles, and cellulitis. It is also frequently isolated from post-surgical wound infections. Strains of *S. aureus* produce a whole host of virulence factors that enable them not only to effectively colonize locally but also enable them to spread systemically. Some pathogenic staphylococci produce the enzyme coagulase, whose detection allows for presumptive identification. Most strains of *S. aureus* are coagulase-positive and historically have been viewed as the primary pathogenic staphylococci. However, *S. epidermidis* and *S. saprophyticus* have also been observed in human disease. *S. saprophyticus* causes 20% of the symptomatic urinary tract infections in sexually active young women (Sherris, *Medical Microbiology*, 3rd ed., 1994). Coagulase-negative staphylococci have recently become recognized as important agents of human disease. They have been found in immunosuppressed patients, in association with indwelling devices (e.g., joint prostheses, heart valves, spinal fluid shunts, and catheters), and they have also been implicated in nosocomial (hospital-acquired) infections. Most of the clinically significant isolates of coagulase-negative staphylococci are *S. epidermidis*. Others have been implicated in wound and urinary tract infections. The emergence of vancomycin-resistant strains has caused considerable interest in the medical microbiological community.

In this experiment, areas of skin will be sampled with specific emphasis on identifying the staphylococci present. Mannitol salt agar (MSA) permits the selection of staphylococci due to the high salt concentration of the medium. Since *S. aureus* ferments mannitol, it can be distinguished from *S. epidermidis* on MSA. Sodium tellurite agar (STA) inhibits the growth of most organisms, but staphylococci and corynebacteria will grow on this medium and form black colonies.

Materials (per 20 students)

1. Mannitol salt agar (MSA) plates (20)
2. Sodium tellurite agar (STA) plates (20)
3. Coagulase plasma (40 ml)
4. Sterile tubes (20)
5. Blood agar plates (30)
6. Cultures of *Staphylococcus aureus* and *S. epidermidis* (5)
7. Sterile swabs (20)

Procedure

1. Wet a swab with some of the sterile saline. Swab at least two different skin areas and inoculate plates of MSA, STA, and blood agar by rolling the swab over 25% of the plate.
2. Streak the remainder of the plate as illustrated in Figure 70.1.

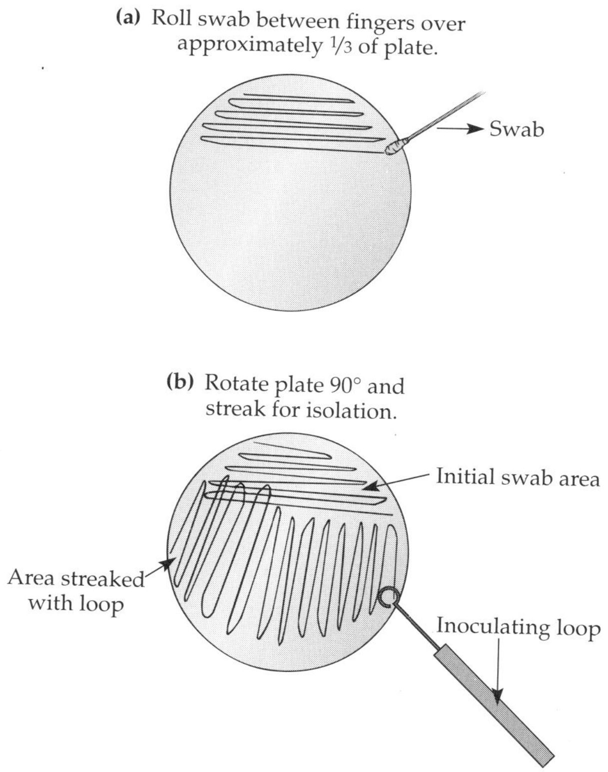

FIGURE 70.1 Streaking of skin sample plates.

3. Streak the cultures of *S. aureus* and *S. epidermidis* in a similar manner on the plates of MSA, STA, and blood agar.
4. Incubate the plates at 37°C for 24 to 36 hr.
5. Observe all the plates and record the results.
6. After 24 hr, transfer any suspected *S. aureus* colonies from the mannitol salt agar to a blood agar plate. Incubate for no longer than 24 hr.
7. Perform Gram stains on the isolated colonies.
8. Place 2.0 ml of coagulase plasma in a sterile tube and inoculate with two to three colonies from the MSA plate.
9. Repeat the procedure with a known *S. aureus* culture.
10. Incubate the tubes in a water bath at 37°C.
11. Observe for solidification. An overnight incubation may be required for detection of a positive coagulase reaction of wild strains.

Laboratory Report 70

Name ____________________ Date __________ Section __________

Growth on Mannitol Salt Agar

Skin	***S. aureus***	***S. epidermidis***

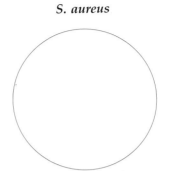

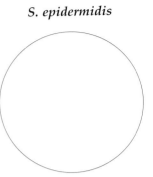

Growth on Sodium Tellurite Agar

Skin	***S. aureus***	***S. epidermidis***

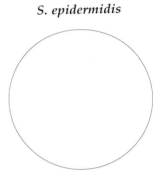

Growth on Blood Agar

Skin	***S. aureus***	***S. epidermidis***

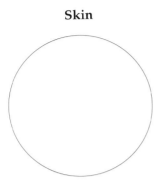

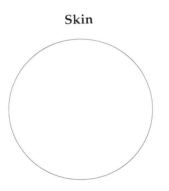

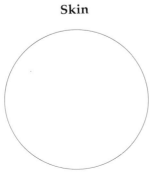

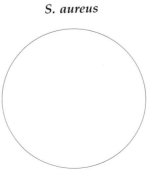

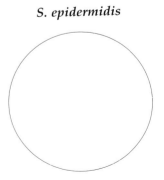

Coagulase Test Results

Organism	Coagulase Reaction
Unknown	
Staphylococcus aureus	

Review Questions

1. What other test could be run to confirm the presence of *S. aureus*?

2. How can the staphylococci be differentiated from the corynebacteria?

3. What percentage of the colonies growing on the MSA plate do you suspect are *S. aureus*?

4. Which area of the skin sampled has the largest population of microorganisms? of staphylococci?

LABORATORY

71 Agglutination: Determination of Human Blood Type

Safety Issues

- It is absolute necessity that you work with your own blood in this experiment. You must protect yourself against blood-borne pathogens such as hepatitis and HIV.
- All lancets, slides, toothpicks, and cotton should be placed in containers with disinfectant.
- Report any droplets of blood that have been spilled on the lab bench or floor to your laboratory instructor.

OBJECTIVE

Understand how human blood is typed and why typing is important.

Agglutination or clumping reactions take place between particulate antigens and specific antibodies. In this type of reaction, a three-dimensional matrix is formed which can either be observed macroscopically or microscopically. In this exercise you are asked to determine your own blood type. **You should work only with your own blood and must be careful to follow all disinfection procedures.**

Human red blood cells (erythrocytes) have on their surfaces a variety of antigens. The two groups usually tested for are the ABO type and the Rh type. Since the agglutination reaction in this case involves red blood cells, the reaction is referred to as a **hemagglutination** reaction. Specifically, the presence or absence of particular glycolipid antigens on the surface of erythrocytes is detected. If antigen A is present, the blood type is A; if antigen B is present, the person has type B; and if both antigens are lacking, type O is the result. Hemagglutination will occur if antiserum containing anti-A antiserum is mixed with type A erythrocytes. A similar reaction will occur if anti-B antiserum is mixed with type B cells. Individuals with type O blood lack antigens A and B and will not react with either antiserum.

Also contained on the surfaces of red blood cells are Rh antigens, a complex of many antigens. Those individuals possessing the factor are called Rh-positive, and those that lack the antigen are referred to as Rh-negative. The antigen is detected by using anti-D antisera.

Materials (per 20 students)

1. Clean glass slides or blood typing slides (20)
2. Anti-A antiserum
3. Anti-B antiserum

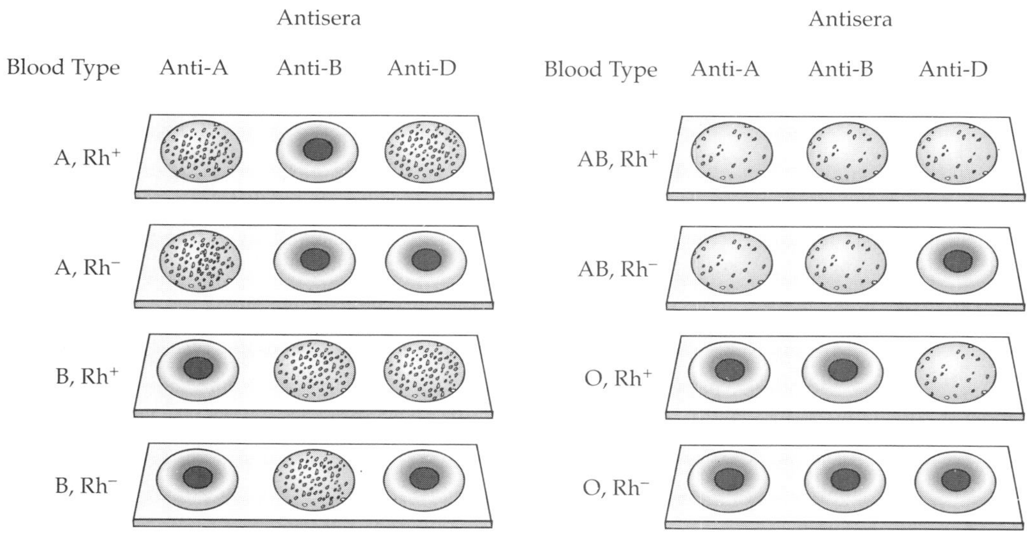

FIGURE 71.1 Human blood typing slide.

4. Anti-D antiserum
5. Sterile cotton balls
6. 70% ethanol
7. Toothpicks or wooden applicator sticks
8. Sterile lancets (20)
9. Bandages
10. Wax pencils
11. Microscopes

Procedure

It is an absolute necessity that you work only with your own blood in this experiment. You must protect yourself against blood-borne pathogens such as hepatitis and HIV.

Place all used lancets, slides, and cotton balls into the disinfectant provided. Wash any spilled blood on the lab bench with the disinfectant.

1. With the wax marking pen, draw three circles on a clean glass slide. If using a blood typing slide, circles are already present.
2. Disinfect the tip of one of your fingers with the cotton ball soaked in 70% ethanol.
3. Obtain a blood sample by pricking your finger with the sterile lancet. Place a drop of your blood in each of the circles on the slide. Stop the bleeding by holding a cotton ball to your finger and then cover with a bandage.
4. Add 1 to 2 drops of anti-A antisera to one circle, 1 to 2 drops of anti-B antiserum to the second circle, and 1 to 2 drops of anti-D antiserum to the third circle (Figure 71.1). Use a **separate** toothpick or applicator to mix each suspension.
5. Observe for hemagglutination. The reaction will proceed more slowly for the Rh factor, since the reaction is weaker than the ABO reaction. You may want to use the low power on a microscope to observe the reaction. Heating the slide slightly on a warming tray or touching the slide to the back of your hand will accelerate the reaction. Can you determine your blood type?

Laboratory Report

Name Date Section

71

Agglutination—Human Blood Typing

	Antisera		
Reactions with Your Blood:	Anti-A	Anti-B	Anti-D
Macroscopic			
Microscopic			

Review Questions

1. You are blood type A, Rh negative. What would be the consequences of you receiving a transfusion of B, positive blood?

2. Why is it important for pregnant females who are Rh negative to know the blood type of their mate? If the partner is positive, what are the possible genotypes in the offspring? What is the genotype of the mother? What is the blood type of a universal donor?

3. You have donated blood to the Red Cross previously and know you have A^+ blood type. However, when you performed the procedure in the lab, it showed you were AB^+. How can this discrepancy be explained?

LABORATORY

72 Agglutination—Serological Typing

Safety Issues

- After making observations, dispose of depression slides and toothpicks into containers of disinfectant solution.
- Report any spills to your laboratory instructor.

OBJECTIVE

Learn the uses of agglutination technique for the identification of bacteria.

In this exercise, specific typing sera are used to identify the particular strain or species of bacteria present by its antigens. If particular antigens are present, the antibodies in the antiserum will bind with the antigens, causing a clumping of the cells (Figure 72.1). This technique is very useful in the rapid identification of many pathogens such as *Salmonella*. The technique is much more sensitive and precise than biochemical methods of identification. In this exercise, *Salmonella* will be identified among the unknown organisms.

Materials (per 20 students)

1. Bacterial unknowns (5)
2. *Salmonella* O antigen, group B
3. *Salmonella* O antiserum, poly A

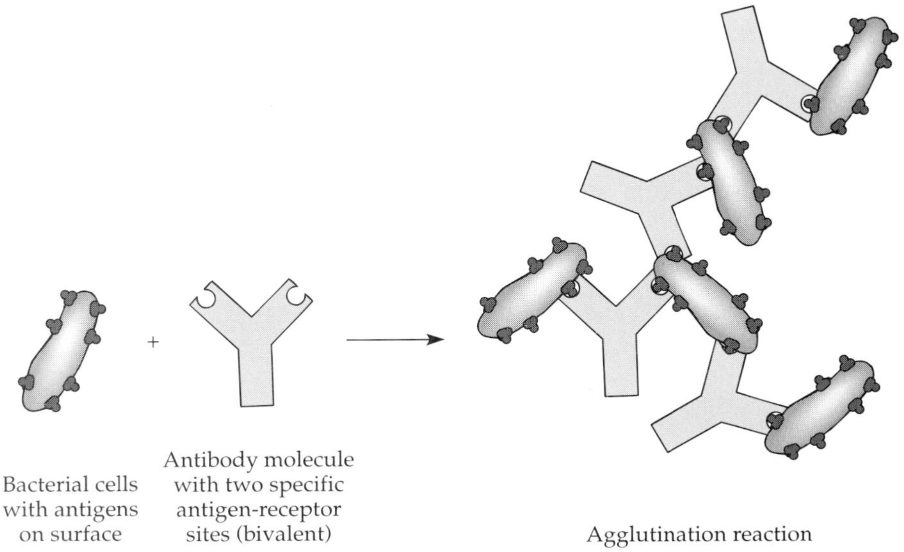

FIGURE 72.1 Bacterial agglutination.

4. Spot plates or depression slides (20)
5. Saline (25 ml)
6. 13 × 100 mm tubes (20)

Procedure

1. Label three depressions (A, B, and C) on the slide or spot plate. Depression A will be the positive control, and C will be the negative control.
2. Prepare a suspension of the unknown organism by taking one or two loopfuls of the organism and suspending it in 1.0 ml of phenolized saline. Make sure to mix well to prevent clumping. Prepare a fairly turbid suspension.
3. To depression A, add one drop of *Salmonella* O antigen. To depressions B and C, add one drop of the unknown suspension.
4. To depressions A and B, add one drop of *Salmonella* O polyvalent antiserum. To depression C, add one drop of the phenolized saline. Be sure you do not touch the dropper to the slide.
5. Mix each of the depressions with a separate, clean toothpick.

Make sure to dispose of all toothpicks in the beaker of disinfectant provided.

Laboratory Report 72

Name Date Section

Serotyping with Agglutination

Depression	Agglutination Reaction (+ or −)
A-*Salmonella* antigen	
B-Unknown	
C-Unknown	

A microscope may be used to distinguish the various reactions. Which of the unknown organisms is *Salmonella*?

Review Questions

1. How could you verify within 24 hr that an outbreak of gastroenteritis in a local campground was caused by *Salmonella*? Describe media and techniques you would use and how the positive results would appear.

2. To what actual structure on the bacterial cell does the anti-O antibody present in the sera attack?

3. What other microorganisms might cause an infection with similar symptoms?

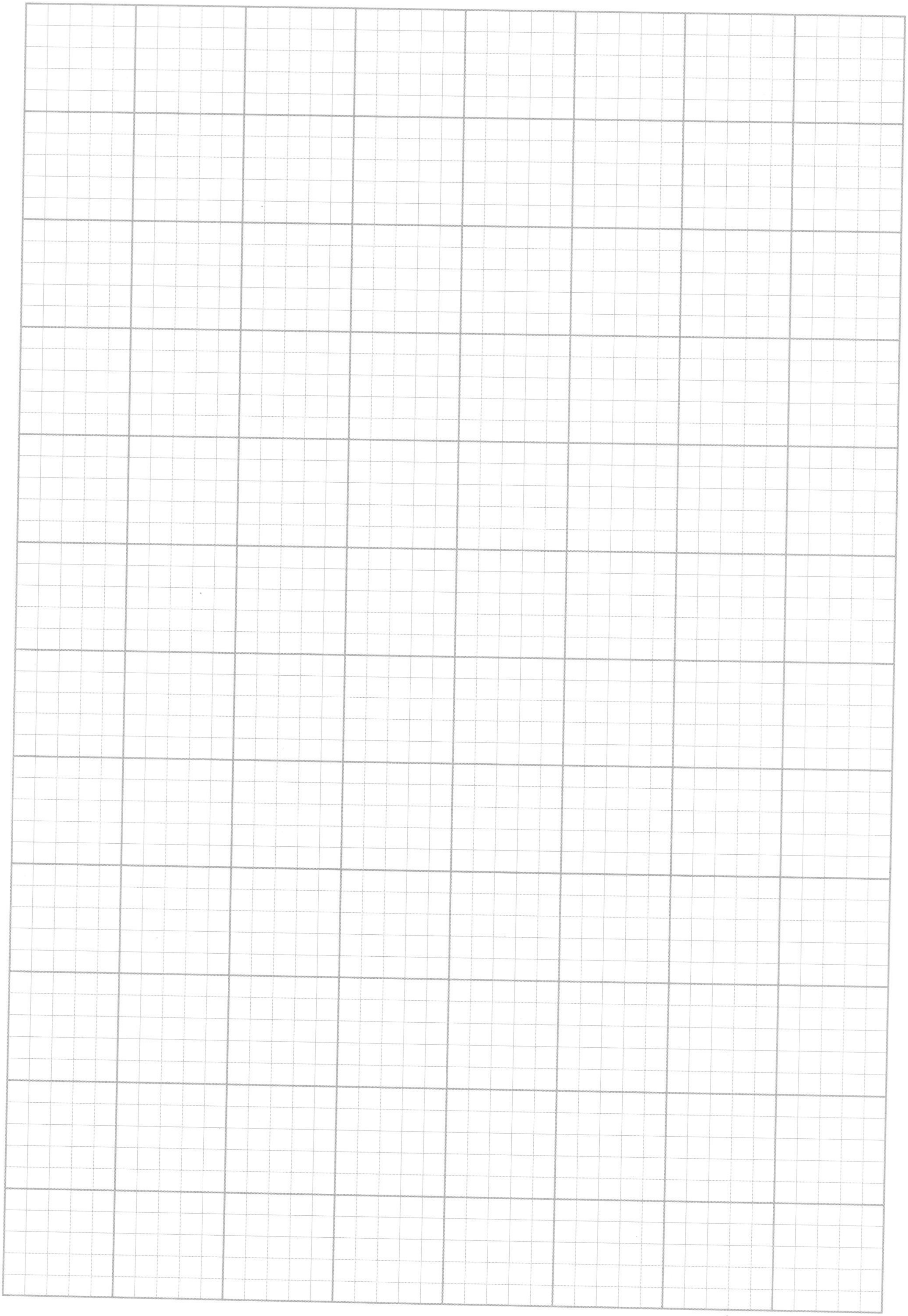

LABORATORY

73 Precipitation-Immunodiffusion

Safety Issues

- Dispose of pipette tips, Pasteur pipettes, and droppers into disinfectant solution that has been provided.

OBJECTIVE

Understand how the technique of immunodiffusion can be used to detect specific antibodies in a patient's serum.

The reaction of soluble antigens and specific antibodies against these antigens forms a visible precipitation. The precipitate is actually a matrix of intertwined antigen and antibody molecules in the form of a lattice (see Figure 73.1). Precipitation reactions differ from agglutination reactions in that the antigen involved has a smaller molecular weight, is soluble, and consists of non-particulate material. In this exercise, a determination of the presence of antibodies will be made against one of the fungal or bacterial agents known to cause a syndrome known as hypersensitivity pneumonitis (HP). The reaction will be carried out in agar, either in a small Petri plate or on the surface of a glass slide. This is a double diffusion method, because the antigens

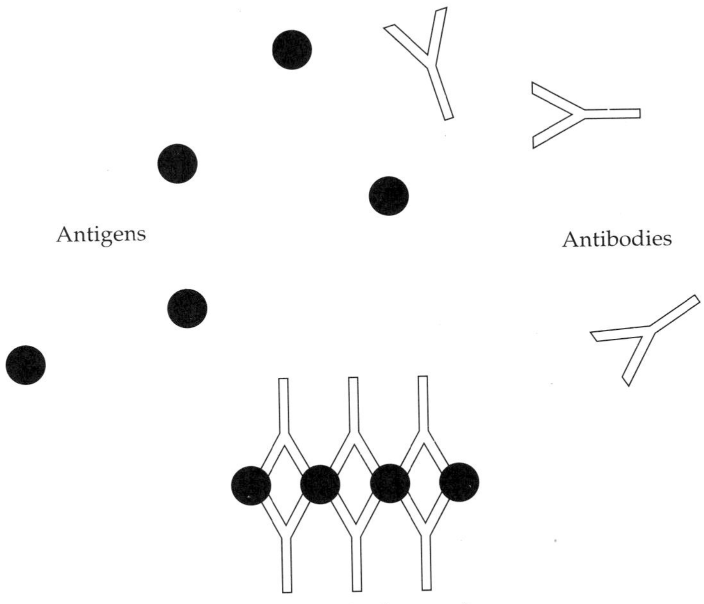

FIGURE 73.1 Lattice formation in immunoprecipitation.

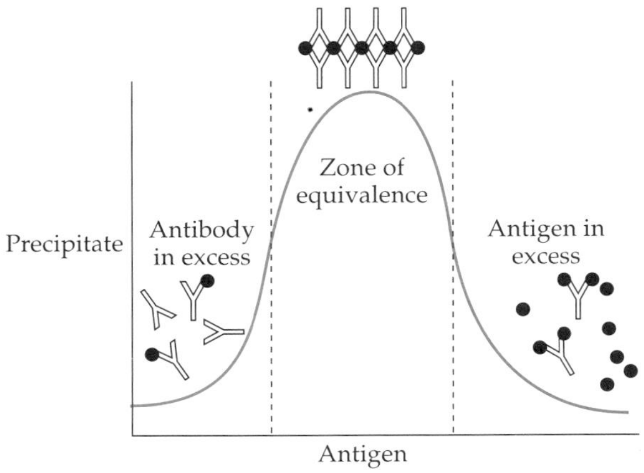

FIGURE 73.2 Zones for equivalence.

and antibodies are placed in wells cut in the agar and allowed to diffuse into the agar. The technique also has been called the Ouchterlony procedure. Antigens and antibodies will diffuse and react with one another and precipitate when they are both at optimal concentrations, at what is referred to as the zone of equivalence (see Figure 73.2). The immunodiffusion technique has wide applications in the clinical, forensic, or food laboratories when it is necessary to detect the presence of soluble antigens or the patient's antibody response to the presence of the antigen. The detection of the precipitation reaction can be heightened by employing a staining technique to better visualize the precipitation band.

Materials (per group)

1. Clean glass slides or small Petri plates (1)
2. Cork borers or Pasteur pipettes
3. Medicine droppers
4. Pasteur pipettes or micropipettes
5. Flask of Noble agar (1.0%) in 0.85% sodium chloride, containing 0.01% sodium azide as a preservative (30 ml)
6. Unknown serum samples
7. Known antigen preparations:**
   Thermoactinomyces vulgaris #1
   Aspergillus fumigatus
8. Known antisera**
   Antiserum to *T. vulgaris*
   Antiserum to *A. fumigatus*
9. Greer templates and kit
10. 70% ethanol (50 ml)
11. Whatman #41 filter paper
12. 0.15 M NaCl (50 ml)
13. Hair dryer or drying oven
14. Coomassie blue (50 ml)
15. Methanol/acetic acid solution (450:100)(bring to 1 L with distilled water for entire class)

**Available from Greer Laboratories, Lenoir, North Carolina.

Procedure

1. If using prepared templates, skip ahead to step 4.
2. To prepare agar plates for immunodiffusion, melt the Noble agar and pour plates. If using slides, wipe the slides with 70% ethanol to clean. Then apply a layer or two of tape around the edge of the slide. Using a pipette, add molten Noble agar to the slide, being careful not to allow the agar to go outside of the tape.
3. After the agar has solidified, prepare the wells. If using agar plates, punch holes in the agar with a cork borer or the blunt end of the Pasteur pipette, using the diagram in Figure 73.3 as a template. Remove the agar plug by using a Pasteur pipette attached to a vacuum device. If using slides, use a medicine dropper to remove the plugs. Gently push the tip of the dropper through the agar while pressing the bulb. Release the bulb and the agar should be sucked off the slide.
4. The center well will be used as the antigen well. The peripheral wells will contain serum samples. Fill the antigen well with the *T. vulgaris* antigen 1 using a thin capillary pipette, micropipette, or a tuberculin syringe with a 25-gauge needle. Hold the pipetting device at a 45° angle. Touch the side of the well with the lower tip of the pipette and then deliver the material. Fill the well until no meniscus can be seen. **Do not overfill.**
5. Label the slide or plate so that you will be able to identify the contents of the sera wells.
6. Fill the sera wells with the following:
   (a) antiserum to *T. vulgaris* (positive control)
   (b) antiserum to *T. vulgaris* (positive control)
   (c) Patient serum sample 1
   (d) Patient serum sample 2
   If using the Greer prepared templates, you will be able to run two additional patient samples:
   (e) Patient sample 3
   (f) Patient sample 4

Extreme caution should be exercised when handling serum samples as blood-borne pathogens may be present. Wear latex gloves when handling the serum samples.

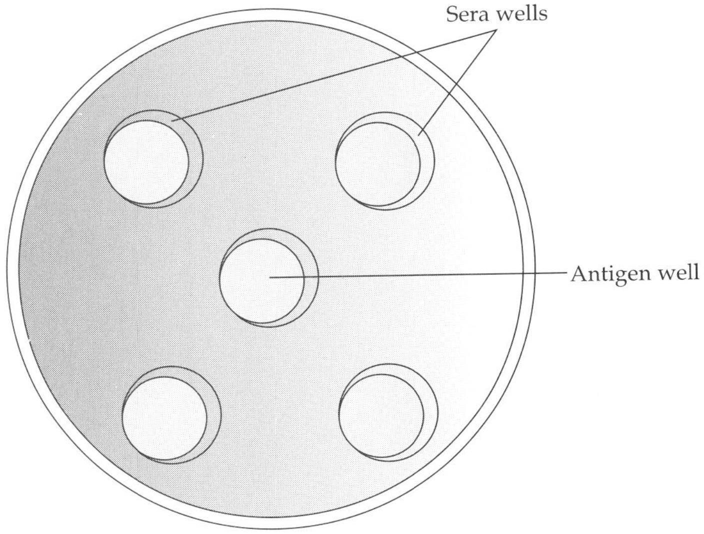

FIGURE 73.3 Template for preparation of immunodiffusion plates.

(Other positive or negative control sera may be used if patient sera is not available.)

7. If using the Greer kit, close the box containing the agar template and carefully insert into plastic bags. Reseal the bags and incubate at room temperature for 24 to 72 hr. Keep plates level during incubation. If using agar plates, replace the lids and incubate at room temperature. If slides are being used, carefully transfer the slide to a 15 × 100 mm Petri dish that contains a piece of Whatman #3 filter paper that has been wetted with sterile distilled water. Carefully place the slide on the filter paper and replace the Petri dish cover. Incubate at room temperature for 24 to 72 hr.
8. Observe daily for the appearance of precipitin bands. Read the plate or slide using a viewer with a dark field background, or hold at an angle to a fluorescent light. Placing the plates in a refrigerator for a few hours prior to viewing sometimes helps to make the bands visible.
9. If you have used the Greer kit or slide technique, you can perform a staining technique to better distinguish the bands.
   (a) Transfer the gel using a spatula to a rectangular piece of plastic film.
   (b) Place the film on a flat surface. Fill the wells with distilled water and cover the gel surface with Whatman #41 filter paper. Eliminate any air bubbles that may form between the gel surface and the filter paper. Then place a thick absorbent blotter (or lab towel) and a heavy book on the gel.
   (c) Press for 10 min.
   (d) Wash the pressed gel in 0.15 M NaCl for 15 min with agitation. Wash with fresh saline for an additional 15 min.
   (e) Wash for 15 min in distilled water.
   (f) Press for 10 min as described above.
   (g) Dry with a hair dryer for approximately 15 min with warm air or in a drying oven at 60°C.
   (h) Stain for 4 to 5 min with Coomassie blue (0.2% in methanol/acetic acid solution).
   (i) De-stain with methanol/acetic acid rinse solution. De-staining may require two to three changes of the rinse solution.
   (j) Dry with a hair dryer and observe bands.

Laboratory Report 73

Name ______________ Date ______________ Section ______________

Immunodiffusion-Precipitation

1. Indicate Additions to:

Center well	*T. vulgaris* antigen	
Well 1	antiserum to *T. vulgaris*	
Well 2	antiserum to *T. vulgaris*	
Well 3		
Well 4		
Well 5		If Greer template is used.
Well 6		

2. Draw the precipitin bands that are visible on the diagram.

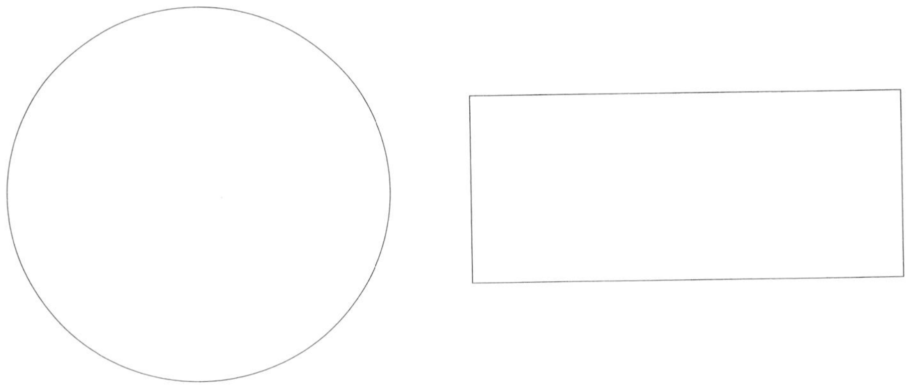

3. Does the staining procedure help to make the bands visible?

Review Questions

1. Why is azide added to the Noble agar in immunodiffusion techniques?

2. Describe how too much antibody can interfere with lattice formation in a precipitation reaction.

3. What is meant by a reaction of partial identity in an immunoprecipitation reaction? a reaction of non-identity?

4. How could it be determined whether the causative agents of hypersensitivity pneumonitis are in the air in a chicken processing plant?

LABORATORY

74 Enzyme-Linked Immunosorbent Assay (ELISA)*

Safety Issues

- Wear gloves when handling mice.

OBJECTIVES

1. Learn a protocol for an immunization procedure.
2. Distinguish antigen, antibody, and adjuvant.
3. Using the ELISA technique, detect antibody formation in mice in response to purified bovine immunoglobulin.

The enzyme-linked immunosorbent assay (**ELISA**) is a sensitive test that measures complexes of antigens and antibodies. It is a much more sensitive test than either the agglutination or precipitation test. Direct ELISA tests detect the presence of an antigen in a sample. Antibody that is specific for the antigen being sought is bound to wells in a plastic microtiter plate. A sample suspected to contain the antigen is placed in the wells, and if antigen is present, it will bind to the antibodies in the wells. Any unbound antigen is washed out of the wells. A second antibody, usually labeled with an enzyme such as horseradish peroxidase, is added to the wells. Excess secondary antibody is washed from the wells, and a substrate appropriate for the enzyme is then added. If a colored product is formed, then the enzyme coupled to the antibody must be present. If the antibody is present, the antigen specific for that antibody must have been present in the wells and in the original sample. The color is proportional to the amount of antigen present in the wells.

Indirect ELISAs are used to detect the presence of antibodies. In this test, antigen is bound to the microtiter wells, and serum suspected to contain antibodies is added to the wells. Excess antibody is washed from the wells, and if antibody is present in the serum, it will bind to the antigen in the wells. A secondary antibody, usually an anti-IgG antibody that has been conjugated with an enzyme, is then added to the wells. If any primary antibody has bound to antigen, then the secondary antibody will bind to it and, as a result, can be detected with the addition of an appropriate substrate.

This experiment will use an indirect ELISA procedure to detect the level of antibody formed in mice in response to an immunization with bovine IgG.

*This laboratory was contributed by Dr. Kenneth Klatt of the Biology Department at Denison University. His contribution of the lab for incorporation into this manual is appreciated.

Materials (per group)

1. Purified Bovine IgG (BIgG) (Sigma Chemical, catalog #I5506).
2. Sterile saline (0.85%)
3. Freund's complete adjuvant (Sigma Chemical, catalog #F5881 or F4258).
4. Anti-mouse IgG (secondary antibody) conjugated to horseradish peroxidase (BioRad Laboratories #172-1011).
5. Coupling buffer:
   (per liter distilled water)
   1.6 g sodium carbonate
   2.9 g sodium bicarbonate
   0.9 g sodium azide
6. Bovine serum albumin in PBS (1 mg/ml)
7. Phosphate buffered saline (PBS)
   0.02 M phosphate buffer, pH 7.2 in 0.85% sodium chloride
8. Washing solution:
   (in PBS)
   0.0015 M magnesium chloride
   0.002 M dithiothritol
   0.05% Tween 20
9. Color reagent (BioRad Laboratories #172-1064)
   BioRad solution A (9.0 ml)
   BioRad solution B (1.0 ml)
10. Sterile saline (0.85%)
11. Mice (any white outbred strain)
12. 3.0-ml luer-lock syringe with 26-gauge ½-inch needle
13. 96-well microtiter plates

Procedure

A. Immunization

1. Prepare a 4 mg/ml solution of Bovine IgG (BIgG) in 0.85% saline.
2. In a syringe, mix one volume of the BIgG solution with an equal volume of Freund's complete adjuvant until the mixture forms an emulsion.
3. Inject 0.1 ml of this emulsion intraperitoneally (i.p.) into a mouse. Mice bled two weeks after this injection will have a measurable titer for mouse IgG that is anti-BIgG (titer about 1:64 to 1:3000).
4. After two weeks, administer a booster dose of the antigen in Freund's complete adjuvant. After six weeks mice have titers in the 1:78,000 range.
5. Collect blood from the mouse by taking 0.1 ml via cardiac puncture.

B. Assay

1. The lab assistant or instructor has prepared a 10 mg/ml solution of the antigen Bovine IgG in coupling buffer. Wells of a microtiter plate have been filled with the antigen and kept overnight at 4°C.
2. Remove the antigen from microtiter plate by flicking the plate. Wash three times with the coupling buffer, flicking the plates after each wash.
3. Fill the wells of the plate with the solution of bovine serum albumin. Place the plate at 4°C for at least 1 hr. This bovine serum albumin will attach to all of the plastic binding sites not occupied by the antigen, BIgG.

4. While the plates are chilled, prepare dilutions of mouse serum. 1:5. 1:25, 1:125, 1:625, 1:3125, 1:15625, and 1:78125 dilutions should be prepared using PBS as the diluent.
5. After refrigeration, remove the solution from the plate and wash two times with PBS. Drain on absorbent tissue to remove any remaining PBS.
6. Add 0.05 ml of the sera to individual wells. As a control, add 0.05 ml PBS to one well.
7. Incubate the plate at room temperature for 1 hr. Remove the liquid from the plates and wash with the washing buffer four times. Remove the liquid between each washing.
8. Dilute out the anti-mouse IgG conjugated to horseradish peroxidase to 1:1000 to 1:3000.
9. Remove the liquid from the plates and wash four times with the washing buffer. Flick and drain after each wash.
10. Add 9.0 ml of the BioRad solution A to 1.0 ml of BioRad solution B and mix well. Add 0.1 ml of the reagent to each of the wells and incubate for 30 min.
11. Examine each of the wells for color production to determine the titer of antibody in the serum sample.

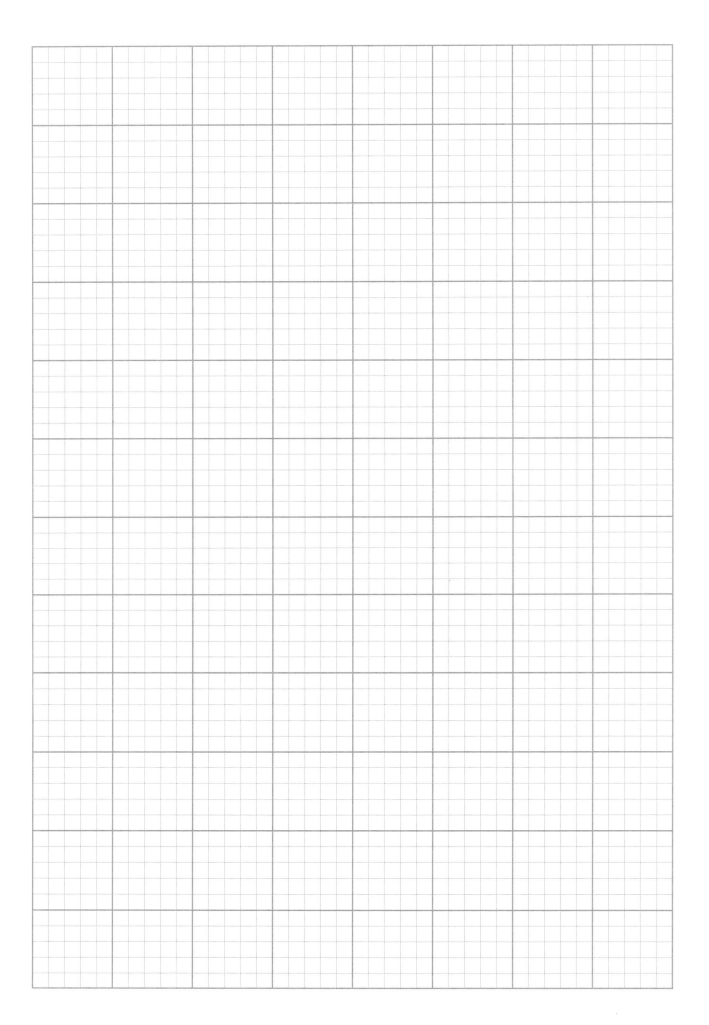

Laboratory Report 74

Name Date Section

Microtiter Plate Well Reactions

Well Number	Components	Color Reaction

Review Questions

1. What is the titer of antibody against the bovine IgG in the mouse serum?

2. How is ELISA used in determining whether a person is infected with HIV?

Appendix A

Stains and Reagents

Acid Alcohol

Concentrated HCl	8.0 g
95% ethyl alcohol	1000 ml

Mix slowly together.

Alcohol (Ethanol), 70%

95% alcohol	368 ml
Distilled water	132 ml

Basic Fuchsin

Basic fuchsin	0.3 g
95% ethanol	10 ml
Distilled water	95 ml

Dissolve fuchsin in alcohol and then add water.

Barritt's Reagent (for Voges-Proskauer Test)

Solution A:

Alpha-naphthol	6 g
95% alcohol	100 ml

Solution B:

Potassium hydroxide	6 g
Distilled water	100 ml

Bromocresol Green

Bromocresol green	0.05 g
Distilled water	100 ml
Sodium hydroxide (0.1 N)	0.7 ml

Mix the sodium hydroxide and water and dissolve the bromocresol green in the mixture.

Carbolfuchsin Stain

Solution A:

Basic fuchsin	0.3 g
95% ethyl alcohol	10.0 ml

Solution B:

Phenol	5.0 g
Distilled water	95.0 ml

Mix solutions A & B and filter before use.

Chloroform (Biotechnology grade)

Commercially available

Crystal Violet Stain

Crystal violet	2 g
95% ethanol	20 ml
Ammonium oxalate	0.8 g
Distilled water	80 ml

Dissolve the crystal violet in the ethanol and the ammonium oxalate in the distilled water. You may have to heat the ammonium oxalate to dissolve. Mix the two solutions together and filter before use.

Diphenylamine Reagent (for Nitrate test)

Diphenylamine	0.7 g
Concentrated sulfuric acid	60 ml
Distilled water	28.8 ml
Concentrated hydrochloric acid	11.3 ml

In a fume hood, dissolve the diphenylamine in the sulfuric acid, and then slowly add the water. Cool the mixture slowly and then add the hydrochloric acid. Allow to stand overnight at room temperature.

EDTA/NaCl (50 mM/0.1 M)

EDTA stock solution (see TE buffer)	5.0 ml
NaCl	0.58 g
Distilled water	100 ml

Ferric Chloride

$FeCl_3 \cdot 6\ H_2O$	10.0 g
Concentrated HCl	5.4 ml
Distilled water	94.6 ml

Add the concentrated HCl to the water and dissolve the ferric chloride.

Ferric Chloride Reagent (for API Test)

Ferric chloride	10 g
Distilled water	100 ml

Glycerol (30%)

Glycerin	30 ml
Distilled water	70 ml

Gram's Iodine (for Gram Stain and Detection of Starch)

Iodine	1 g
Potassium Iodide	2 g
Distilled water	300 ml

Dissolve the potassium iodide in the water and then add the iodine crystals.

Hydrogen Peroxide (for Detection of Catalase Activity)

Hydrogen Peroxide	3 ml
Distilled water	97 ml

3% hydrogen peroxide is available commercially and can be purchased and used directly. It is best kept refrigerated.

Kovac's Reagent (for Detection of Indole)

p-Dimethylaminobenzaldehyde	5 g
Amyl or butyl alcohol	75 ml
Concentrated hydrochloric acid	25 ml

Prepare the solution in a fume hood by dissolving the *p*-dimethylaminobenzaldehyde in the alcohol and then slowly adding the hydrochloric acid.

Loeffler's Methylene Blue

Methylene blue	0.3 g
95% ethanol	30 ml
Distilled water	100 ml

Dissolve the methylene blue in the ethanol. Add the distilled water and mix. Filter before use.

Malachite Green (for Spore Stain)

Malachite green	5.0 g
Distilled water	100 ml

McFarland Standards

Use the following amounts of 1% solutions of Barium Chloride and Sulfuric Acid to prepare the standards:

Tube 1	Barium Chloride (1%) (ml)	Sulfuric Acid (1%) (ml)	Cell Density ($\times 10^8$/ml)
1	0.1	9.9	3
2	0.2	9.8	6
3	0.3	9.7	9
4	0.4	9.6	12
5	0.5	9.5	15
6	0.6	9.4	18
7	0.7	9.3	21
8	0.8	9.2	24
9	0.9	9.1	27
10	1.0	9.0	30

Methylene Blue Stain

Methylene blue	0.3 g
Distilled water	100 ml

Methylene Blue (for Milk Analysis)

Methylene blue thiocyanate	25 mg
Distilled water	100 ml

Methyl Red

Methyl red	0.1 g
95% ethanol	300 ml

Dissolve the dye in the ethanol and add distilled water to make 500 ml of the final solution.

Naphthol (alpha)

Alpha-naphthol	5.0 g
95% ethanol	100 ml

Nigrosin Stain

Nigrosine	10.0 g
Distilled water	100.0 ml
Formaldehyde	0.5 ml

Boil 15 to 20 min until dissolved, and filter before dispersing into bottles.

Nitrite Test Solutions (Nitrate Test)

Solution A:

Sulfanilic acid	8.0 g
Acetic acid (5 N)	1000 ml

(100 ml glacial acetic acid to 250 ml distilled water)

Solution B:

Alpha-naphthylamine	5.0 g
Acetic acid (5 N)	1000 ml

Caution: Solution B may be carcinogenic. Avoid mouth pipetting and contact with skin.

Phenol (Equilibrated, Ultrapure), pH 8.0

Commercially available from AMRESCO or United States Biochemical (USB)

Phenol: Chloroform

Chloroform (Biotechnology grade)	48.0 ml
Phenol (Equilibrated and ultrapure)	48.0 ml
Isoamyl alcohol (Biotechnology grade)	4.0 ml

Phenolized Saline

NaCl	8.5 g
Phenol	5.0 g
Distilled water	1000 ml

Physiological Saline (Saline)

Sodium chloride	8.5 g
Distilled water	1000 ml

Potassium Hydroxide Creatine (for VP test)

Potassium hydroxide	40.0 g
Creatine	0.3 g
Distilled water	100 ml

Safranin (for Gram stain)

Safranin-O	2.5 g
95% ethanol	100 ml
Distilled water	900 ml

Safranin (for Spore Satin)

Safranin	1.0 g
Distilled water	100 ml

Sodium Acetate (3 M), pH 5.2

Sodium acetate (anhydrous)	12.3 g

Add to 35 ml distilled water and mix to dissolve. Adjust pH to 5.2 with glacial acetic acid. Make to 50 ml with distilled water. Autoclave.

Sodium Lauryl Sulfate (20%)

Sodium lauryl sulfate	20.0 g
Distilled water	80 ml

Heat to dissolve. Adjust pH 6 N HCl. Make to 100 ml with distilled water.

Caution: Avoid skin contact or inhalation of dry powder. Wear a mask when weighing.

TE Buffer [Tris (10 mM, ph 8.0; EDTA 1 mM)]

EDTA Stock Solution:

EDTA (disodium ethylene-diaminetetraacetate 2 · H_2O	7.4 g
Distilled water	20 ml

Mix and adjust pH to 8.0 with NaOH.

Tris Stock Solution:

Tris base	2.42 g
Distilled water	20 ml

Mix and adjust pH to 8.0 with concentrated HCl.

Use 0.1 ml EDTA stock and 1.0 ml Tris stock in 100 ml distilled water for buffer.

Toluidine Blue (for DNase test)

Toluidine blue	1.0 g
Distilled water	100 ml

APPENDIX B

Media

The media listed in this appendix are for experiments used in this manual. An excellent source for other media recipes is: *Handbook of Microbiological Media*, 2nd edition (1996), by Ronald Atlas and edited by Lawrence Parks, CRC Press, Inc., Boca Raton, Florida 33431.

There are many sources for the purchase of dehydrated media. Two main suppliers are Difco Laboratories, Detroit, Michigan 48232, and Baltimore Biological Laboratory (BBL), a division of Becton, Dickinson & Co., Cockeysville, Maryland 21030.

Unless specified, all chemical amounts are in grams (g) in a total volume of 1 L of distilled water.

*Indicates that the medium is commercially available.

Ames Test Minimal Medium

Glucose	20.0
Citric acid	3.0
$MgSO_4 \cdot 7\ H_2O$	0.3
K_2HPO_4	15.0
$Na(NH_4)HPO_4 \cdot 4\ H_2O$	5.2
Agar	15.0

Dissolve each component individually in the distilled water and autoclave for 15 min. Dispense into Petri plates.

Bacteriophage Broth (10×)

Peptone	100.0
Beef extract	30.0
Yeast extract	50.0
NaCl	25.0
KH_2PO_4	80.0

***Bile Esculin Agar**

Beef extract	3.0
Peptone	5.0
Oxgall	40.0
Esculin	1.0
Ferric citrate	0.5
Agar	15.0

Heat to dissolve and autoclave at 121°C for 15 min. Overheating may cause a darkening of the medium.

***Blood Agar Plates**

Unless a fresh supply of defibrinated blood is available, it is advisable to use commercially available prepared plates.

***Brain Heart Infusion Broth**

Infusion from calf brains	200.0
Infusion from beef hearts	250.0
Proteose peptone	10.0
Glucose	2.0
Sodium chloride	5.0
Na_2HPO_4	2.5

Brain Heart Infusion Broth with NaCl

Infusion from calf brains	200.0
Infusion from beef hearts	250.0
Proteose peptone	10.0
Glucose	2.0
Sodium chloride	5.0
Na_2HPO_4	2.5
NaCl	65.0

***Brain Heart Infusion Agar**

Infusion from calf brains	200.0
Infusion from beef hearts	250.0
Proteose peptone	10.0
Glucose	2.0
Sodium chloride	5.0
Na_2HPO_4	2.5
Agar	13.5

***Brilliant Green Lactose Bile Broth**

Peptone	10.0
Lactose	10.0
Oxgall	20.0
Brilliant green	0.0133

Add dehydrated powder to distilled water, mix, and heat gently to dissolve. Dispense into tubes with inverted Durham tubes in place. Autoclave for 15 min at 15 lb pressure (121°C). The final pH should be 7.2.

Casein Agar (for Caseinase) Skim Milk Agar

Nutrient broth	8.0
Glucose	1.0
Skim milk (sterilized)	150.0 ml
Agar	18.0

Sterilize the broth and the skim milk separately. Mix together after autoclaving and pour plates.

***Deoxyribonuclease (DNase) Test Agar**

Phytone	5.0
Trypticase	15.0
Sodium chloride	5.0
Agar	15.0

***Endo Broth (*m*-Endo Broth)**

Lactose	12.5
Peptone	10.0

NaCl	5.0
Pancreatic digest of casein	5.0
Peptic digest of animal tissue	5.0
K_2HPO_4	4.375
KH_2PO_4	1.375
Na_2SO_3	2.1
Yeast extract	1.5
Basic fuchsin	1.05
Sodium deoxycholate	0.1
Ethanol (95%)	20.0 ml

Prepare by dissolving 4.8 g of the dehydrated medium in 50 ml of distilled water to which has been added 2.0 ml of 95% ethanol. Dissolve completely and bring volume to 100 ml with distilled water. Heat in a boiling water bath until broth simmers. Cool and adjust the pH to 7.2. This medium should be prepared just prior to use.

***Eosin Methylene Blue Agar (Levine EMB)**

Peptone	10.0
Lactose	5.0
K_2HPO_4	2.0
Eosin Y	0.4
Methylene blue	0.065
Agar	13.5

***Glucose Minimal Salts Agar (Laboratory 63) (Minimal Agar, Davis)**

Glucose	1.0
Sodium citrate	0.5
K_2HPO_4	7.0
KH_2PO_4	2.0
$(NH_4)_2SO_4$	1.0
$MgSO_4 \cdot 7\ H_2O$	0.1
Agar	15.0

Glycerol Yeast Extract Agar

Glycerol	5.0 ml
Yeast extract	2.0
K_2HPO_4	1.0
Agar	15.0

***KF Streptococcus Agar**

Proteose peptone 3	10.0
Yeast extract	10.0
NaCl	5.0
Sodium glycerophosphate	10.0
Maltose	20.0
Lactose	1.0
Sodium azide	0.4
Bromocresol purple	0.015
Agar	20.0

***Lauryl Tryptose Broth**

Tryptose	20.0
Lactose	5.0
K_2HPO_4	2.75
KH_2PO_4	2.75
NaCl	5.0
Sodium lauryl sulfate	0.1

Rehydrate powder and dispense into tubes with inverted Durham tubes in place. Autoclave for 15 min at 15 lb pressure (121°C). The final pH should be 6.8.

***Litmus Milk**

Skim milk powder	100.0
Litmus	0.75

Dissolve and dispense into tubes. Autoclave for 15 min at 12 lb pressure.

****m*-FC Broth (Fecal Coliform Broth)**

Biosate peptone	10.0
Proteose peptone	5.0
Yeast extract	3.0
Lactose	12.5
NaCl	5.0
Bile salts	1.5
Aniline blue	0.1

Add 10 ml of rosolic acid (1% in a 0.2 N sodium hydroxide). Heat to boiling with gentle shaking. Do not autoclave.

****m*-HPC Agar (Formerly Called *m*-SPC Agar)**

Peptone	20.0
Gelatin	25.0
Glycerol	10.0 ml
Agar	15.0

Dissolve all ingredients except for glycerol in distilled water. Adjust pH to 7.1 with 1 N NaOH, heat to dissolve, add glycerol, and autoclave at 121°C for 5 min.

***MacConkey's Agar**

Peptone	17.0
Proteose peptone	3.0
Lactose	10.0
Bile salts	1.5
NaCl	5.0
Neutral red	0.03
Crystal violet	0.001
Agar	13.5

***Mannitol Salt Agar**

Beef extract	1.0
Peptone	10.0
NaCl	75.0
D-Mannitol	10.0
Phenol red	0.025

Minimal Medium (Laboratory 13)

NaCl	5.0
$MgSO_4$	0.2
K_2HPO_4	1.0
KH_2PO_4	1.0
$(NH_4)_2SO_4$	1.0
$FeSO_4$	0.05

Dissolve salts separately in distilled water and dispense.

Minimal Salts Medium (Laboratory 54)

KH_2PO_4	1.3
K_2HPO_4	3.5
$(NH_4)_2SO_4$	0.5
$MgSO_4 \cdot 7\ H_2O$	0.3
$CaCl_2 \cdot 2\ H_2O$	0.01
$FeSO_4 \cdot 7\ H_2O$	0.005
$MnSO_4 \cdot 7\ H_2O$	0.003

***MR-VP Medium**

Proteose peptone	7.0
Glucose (Dextrose)	5.0
K_2HPO_4	5.0

***Mueller-Hinton Agar**

Beef, infusion	300.0
Casamino acids	17.5
Starch	1.5
Agar	17.0

***Nitrate Broth**

Peptone	5.0
Beef extract	3.0
KNO_3	1.0

Noble Agar (for Immunodiffusion)

Noble agar	10.0
NaCl	8.5
Sodium azide	0.1

***Nutrient Agar**

Peptone	5.0
Beef extract	3.0
Agar	15.0

You may also use nutrient broth in place of the peptone and beef extract.

***Nutrient Broth**

Peptone	5.0
Beef extract	3.0

***Nutrient Gelatin**

Peptone	5.0

Beef extract	3.0
Gelatin	120.0

***Phenol Red Broth Base (for Carbohydrate Fermentation)**

Trypticase (proteose peptone)	10.0
Beef extract	1.0
NaCl	5.0
Phenol red	0.025
Sugar (glucose, fructose, lactose, or sucrose)	10.0

Autoclave at 12 lb pressure for 15 min and remove from the autoclave immediately after sterilizing. Addition of dilute base may be necessary before autoclaving to adjust phenol red indicator to appropriate neutral color.

***Phenylethanol Agar**

Trypticase	15.0
Phytane	5.0
NaCl	5.0
B-Phenylethyl alcohol	2.0
Agar	15.0

***Plate Count Agar**

Tryptone	5.0
Yeast extract	2.5
Glucose	1.0
Agar	15.0

***R2A Agar**

Yeast extract	0.5
Proteose peptone 3 or polypeptone	0.5
Casamino acids	0.5
Glucose	0.5
Soluble starch	0.5
K_2HPO_4	0.3
$MgSO_4 \cdot 7\ H_2O$	0.05
Sodium pyruvate	0.3
Agar	15.0

Dissolve powder in distilled water and sterilize at 121°C for 15 min.

***Sabouraud Dextrose Agar**

Peptone	10.0
Dextrose (Glucose)	40.0
Agar	15.0

******Salmonella-Shigella*** **Agar**

Beef extract	5.0
Proteose peptone	5.0
Lactose	10.0
Bile salts No. 3	8.5
Sodium citrate	8.5
Sodium thiosulfate	8.5
Ferric citrate	1.0

Brilliant green	0.00033
Neutral red	0.025
Agar	13.5

Suspend dehydrated medium in cold distilled water and heat to boiling to dissolve the medium completely. **Do not** sterilize by autoclaving.

***Simmon's Citrate Agar**

Sodium citrate	2.0
$MgSO_4$	0.2
$(NH_4)H_2PO_4$	1.0
K_2HPO_4	1.0
NaCl	5.0
Bromothymol blue	0.08
Agar	15.0

Skim Milk Agar (See Casein Agar)

***Snyder Test Agar**

Glucose	20.0
Tryptose	20.0
NaCl	5.0
Agar	20.0
Bromocresol Green	0.02

Soft (Top) Agar (for Bacteriophage)

Nutrient broth	8.0
Agar	7.5

Melt by boiling and distribute in 4 ml quantities to tubes and sterilize for 15 min.

***Spirit Blue Agar**

Spirit blue agar	35.0
Lipase reagent (Difco)	35.0 ml

Boil the agar in water to dissolve and sterilize for 15 min. Cool to 55°C and add the lipase reagent, mixing gently to evenly distribute. Pour into sterile Petri plates.

***Starch Agar**

Starch (soluble)	10.0
Beef extract	3.0
Peptone	5.0
Agar	15.0

***Sulfide Indole Motility (SIM) Agar**

Peptone	30.0
Beef extract	3.0
Peptonized iron	0.2
Sodium thiosulfate	0.025
Agar	3.0

Sulfate-Reducing Bacteria (SRB) Broth Medium

Medium A

KH_2PO_4	0.5 g

NH_4Cl	1.0
$MgSO_4 \cdot 7\ H_2O$	2.0
Sodium lactate (70% solution)	3.5 ml
$CaSO_4$	1.0
Yeast extract	1.0
Sodium thioglycolate	0.1
NaCl	23.4
Resazurin (1 mg/ml)	1.0 ml

Dissolve separately in 1 liter of distilled water and adjust pH to 7.4 to 7.6 with 2 M NaOH. Aliquot in 10 ml portions and autoclave at 121°C for 15 min.

Medium B ($FeSO_4$ Solution)

$FeSO_4 \cdot 7\ H_2O$	5.0

Add $FeSO_4 \cdot 7\ H_2O$ to 100 ml of distilled water. Mix thoroughly and filter sterilize using a 0.45 μm filter. Add 0.1 ml to 10 ml of Medium A.

Medium C (Sodium Ascorbate Solution)

Sodium ascorbate	1.0 g

Add sodium ascorbate to 100 ml of distilled water and filter sterilize using a 0.45 μm filter. Add 0.1 ml to 10 ml of Medium A.

Sulfate-Reducing Bacteria (SRB) Agar Medium

Medium A

KH_2PO_4	0.5 g
NH_4Cl	1.0
Na_2SO_4	2.9
$MgSO_4 \cdot 7\ H_2O$	2.0
Yeast extract	1.0
$CaSO_4$	1.0
Sodium thioglycollate	0.1
NaCl	23.4
Sodium lactate (70% solution)	3.5 ml
Resazurin (1 mg/ml)	1.0 ml
Agar	15.0

Adjust the medium to pH 7.6 with 2 M NaOH and autoclave at 121°C for 15 min.

Medium B ($FeSO_4$ Solution)

$FeSO_4 \cdot 7\ H_2O$	5.0

Add $FeSO_4 \cdot 7\ H_2O$ to 100 ml of distilled water. Mix thoroughly and filter sterilize. Add 10 ml to 1 liter of Medium A before pouring agar.

Medium C (Sodium Ascorbate Solution)

Sodium ascorbate	1.0

Add sodium ascorbate to 100 ml of distilled water and filter sterilize. Add 10 ml to 1 liter of autoclaved Medium A before pouring agar.

Top Agar for Ames Test

Solution A: Top Agar

Agar	6.0
NaCl	5.0

Solution B: Histidine/Biotin Stock

D-Biotin	123.6 mg
L-Histidine HCl	96.0

Heat the biotin and histidine to boiling to dissolve and then filter sterilize using a 0.22-μm filter. Store in a sterile container at 4°C. Just before use, add 10 ml of the biotin/histidine stock solution to 100 ml of the top agar and then aseptically transfer 3.0 ml to sterile tubes. Hold tubes at 45°C in a water bath.

Thiobacillus Medium

$(NH_4)_2SO_4$	0.3
$MgSO_4 \cdot 7\ H_2O$	0.5
KH_2PO_4	3.0
$CaCl_2$	0.1
$FeSO_4$	0.01

Add 0.5 g of elemental sulfur per 100 ml of medium. Adjust the pH of the medium to 4.5 to 5.0. Do not autoclave.

***Thioglycollate Broth**

Casein	15.0
Glucose	5.5
Yeast extract	5.0
NaCl	2.5
L-cystine	0.5
Sodium thioglycollate	0.5

Autoclave for 15 min at 15 psi pressure (121°C). Boil and cool the medium before use.

***Triple Sugar Iron Agar**

Beef extract	3.0
Yeast extract	3.0
Peptone	15.0
Proteose peptone	5.0
Lactose	10.0
Saccharose	10.0
Dextrose	1.0
Ferrous sulfate	0.2
Sodium chloride	5.0
Sodium thiosulfate	0.3
Phenol red	0.024
Agar	12.0

***Tryptic Soy Agar**

Tryticase (tryptone)	15.0
Phytone (Soytone)	5.0
Sodium chloride	5.0
Agar	15.0

Tryptic Soy Agar + Manganese (0.3%)

Tryptic soy agar	15.0
Manganous sulfate	3.0

Tryptic Soy Agar + Phenol (0.05%)

Tryptic soy agar	15.0
Phenol	0.5 ml

***Tryptic Soy Broth**

Tryptone	17.0
Soytone	3.0
Dextrose	2.5
Sodium chloride	5.0
K_2HPO_4	2.5

***Tryptone Broth**

Tryptone	10.0

***Urea Broth**

Peptone	1.0
Urea	10.0
NaCl	5.0
KH_2PO_4	2.0
Phenol red	0.012

Adjust pH to 6.8 to 6.9 and filter sterilize before dispensing to sterile tubes.

APPENDIX C

Microorganisms

Bacteria

Bacillus cereus
Bacillus stearothermophilus
Bacillus subtilis
Bacillus subtilis (ATCC #23059)
Bacillus subtilis (ATCC #23857)
Bacteroides fragilis (ATCC #23745)
Branhamella catarrhalis
Clostridium butyricum (ATCC #8260)
Desulfovibrio desulfuricans (ATCC #7757)
Enterobacter aerogenes (ATCC #13048)
Escherichia coli B
Escherichia coli (ATCC #11229)
Escherichia coli (ATCC #25257)
Escherichia coli (ATCC #25250)
Enterococcus faecalis
Lactobacillus casei
Micrococcus luteus
Mycobacterium phlei
Mycobacterium smegmatis
Peptostreptococcus magnus (ATCC #29328)
Proteus mirabilis
Proteus vulgaris
Pseudomonas aeruginosa
Pseudomonas fluorescens
Salmonella typhimurium (ATCC #29631)
Serratia marcescens
Shigella dysenteriae (ATCC #11456a)
Staphylococcus aureus
Staphylococcus aureus (ATCC #25923)
Staphylococcus epidermidis (ATCC #14990)
Streptococcus faecalis
Streptococcus mitis
Thiobacillus thiooxidans (ATCC #19377)

Molds

Acremonium
Alternia
Aspergillus
Chaetomium
Cladosporium
Fusarium
Gliocladium

Helminthosporium
Mucor
Nigrospora
Paecilomyces
Penicillium

Algae and Cyanobacteria

Anabaena
Chlamydomonas
Euglena
Gloeocapsa
Gonium
Lyngbya
Nostoc
Pandorina
Scytonema

Protozoa

Amoeba
Didinium
Paramecium
Stentor
Vorticella

Yeast

Saccharomyces cerevisiaie

APPENDIX D

The Use of a Spectrophotometer

A spectrophotometer is an instrument that measures the amount of light that is absorbed or transmitted when passed through a sample. When the absorbance of light is measured, the instrument is called an absorption spectrophotometer. Many biologically important molecules absorb specific types of radiant energy (light of differing wavelengths). In microbiological applications, a beam of light passes through a liquid suspension containing bacteria. The larger the population of bacteria the greater the amount of light scattering. As a result, less light passes through the sample.

The basic design of a spectrophotometer is shown in Figure D.1.

A source of white light is used as the light source. The light is focused on a dispersion device, usually a prism, to separate the light into individual bands of radiant energy. Each wavelength of light is focused through a narrow slit and then passes through the sample to be measured. The sample is usually held in a glass or plastic tube called a cuvette. After the light passes through the sample it strikes a light-detecting photoelectric cell. When the transmitted light strikes the photoelectric cell, an electric current is generated which is proportional to the transmitted light energy. The current is displayed on a meter or by digital readout. Most instruments display the amount of light transmitted (% transmission) and the amount of light absorbed.

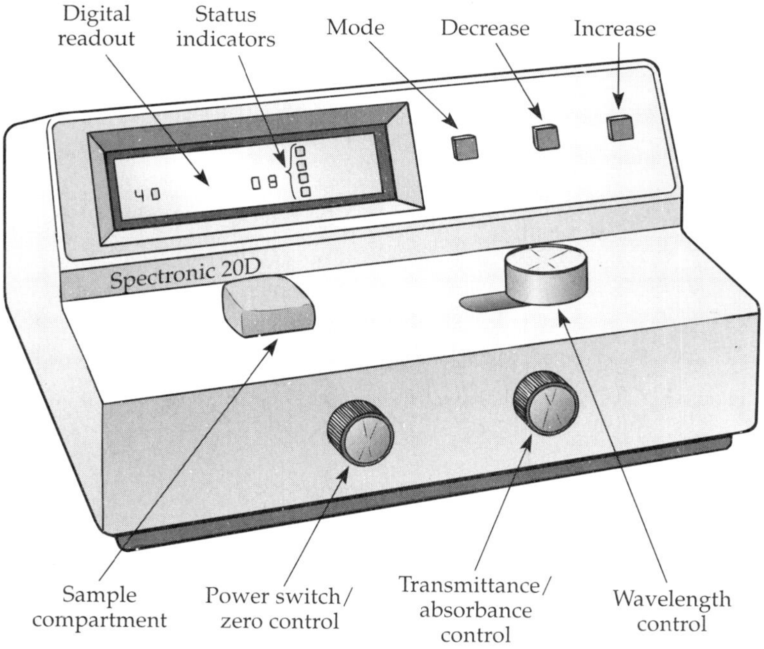

FIGURE D.1 Spectrophotometer

The light transmitted (T) is defined as

$$T = I/I_O \qquad (1)$$

and percentage of light transmitted as

$$\%T = T \times 100\% \qquad (2)$$

where

I = transmitted light beam that has passed through the sample

I_O = the incident light beam that passes through the sample to be measured.

Absorbance (A) is defined as:

$$A = \log(1/T) = -\log T \qquad (3)$$

The transmission of light is also influenced by the width of sample and the concentration of absorbing substance.

Procedure for Measuring Bacterial Turbidities

This procedure describes measurements taken with a Milton Roy Spectronic 20 or 21.

The wavelength control knob adjusts the angle of deflection of the light source in order to obtain a specific wavelength of light. The power switch also serves as the zero control knob and is used to turn the instrument on and off and to set the %T reading.

1. Turn on the spectrophotometer by turning the power switch knob clockwise. Allow the instrument to warm-up for 15 min prior to use.
2. Set the desired wavelength with the wavelength control knob.
3. On the Spectronic 20 adjust the meter to 0% T with the zero control knob. Notice that there is a mirror directly behind the needle. When taking a reading, be sure that the needle is directly in line with its reflection in the mirror. With the Spectronic 21, select the operating mode for either absorbance or % transmission (%T) and set the sensitivity switch to low.
4. Place a tube or cuvette containing the uninoculated culture medium into the sample compartment. This is a control that takes into account the light scattering or absorbance of the cuvette and the culture medium. We are interested in only determining the amount of light scattered by the bacteria present in the culture medium. Be sure that the cuvette or tube has been wiped clean to remove any liquid, dust, or fingerprints.
5. Adjust the meter to 100% T (0 Absorbance) with the transmittance/absorbance control.
6. Be sure that the inoculated culture tube or cuvette has been wiped clean and that it has been placed securely in the sample compartment. Close the lid and read the absorbance.
7. When all readings have been completed, turn off the instrument by turning the power switch counterclockwise until it clicks.

Procedural Hints

1. Inaccurate readings very often can be attributed to
   - (a) Improper zeroing of a blank solution. (Remember, the blank is the uninoculated tube of the particular culture medium used. If multiple media are used, each has to be zeroed separately.)
   - (b) Cuvette or tube not being placed securely in the sample compartment.
   - (c) Wavelength not being set properly. (When there are multiple users of a spectrophotometer, each user may be taking readings at different wavelengths.)
2. If any culture is spilled in the spectrophotometer, notify your lab instructor immediately.

APPENDIX E

The Use of Pipettes

Pipettes are used to accurately deliver specific volumes of fluids. They are made of glass or plastic and are graduated in milliliter units. The most commonly used pipettes are calibrated into 1.0-, 5.0-, or 10.0-ml units. Each pipette is marked to indicate the total volume and the smallest measurable unit. With a 1.0-ml pipette, the total volume and the smallest volume deliverable is indicated by the "1 in 1/100" designation. This indicates that the maximum deliverable volume is 1.0 ml and the smallest unit deliverable is 0.01 ml. Analyze the calibration for a 5.0-ml pipette. Take time to become familiar with the graduations on the pipettes. Two types of pipettes may be encountered in a microbiology laboratory: one is called a serological pipette and is labeled "TD" (To Deliver) and must be emptied completely to deliver the volume specified; a second type is not calibrated to the tip and delivers the volume specified by the graduations. Mistakes in pipetting commonly occur with this second type of pipette when fluids are allowed to run out of the pipette beyond the calibration point. Be careful to note the type of pipette being used before beginning work.

Aseptic technique must be employed when using pipettes in the microbiology laboratory. Pre-sterilized pipettes, either glass or plastic, may be available. They are packaged either individually or in bulk packages. Reusable glass pipettes must be placed with the delivery end at the bottom in metal or glass canisters and then steam sterilized. Canisters should be dried by using the drying cycle of the sterilizer or, alternatively, by placing them in a dry-heat oven at 150 to 170°C for 1 to 2 hr.

Never pipette any material by using your mouth. Use one of the pipetting devices described here. Two of the most commonly used are (a) a cylindrical, thumb wheel device, and (b) a bulb aspirator. When using the thumb wheel device, carefully attach the pipette to the end opposite the wheel. Be careful not to contaminate the sterilized pipette. Exercise caution in gently attaching the pipette, for excessive force may cause the pipette to break and cause injury. By turning the thumb wheel, liquid can be drawn into the pipette, and then by turning the wheel in the opposite direction, the fluid can be delivered. When using the bulb aspirator, first expel all of the air from the bulb by simultaneously squeezing the bulb and the A valve. Attach the pipette to the end of the bulb containing the S valve. Squeeze the S valve to draw fluid into the pipette. To release the fluid, squeeze the E valve. Be careful not to draw fluid into the bulb.

Extreme care must be exercised when using sterile pipettes. Remember all environmental laboratory surfaces, air, and skin harbor microorganisms which could be sources of contamination. Keep pipettes in canisters and remove one at a time as needed. With one hand, remove a pipette from the canister, being careful not to touch the pipette tip. Using your other hand, pick up the tube or flask to be sampled and remove the stopper with the little finger of the hand holding the pipette. Flame the lip of the tube or flask, fill the pipette, reflame the tube or flask, and replace the stopper. Repeat the process to deliver the sample to a new tube or flask. Be sure to keep the pipette in an upright position, for if inverted some of the fluid may run out into the pipetting device. Used pipettes must be placed in containers filled with disinfectant. If disposable pipettes are used, they should be discarded into the appropriate biohazard containers. **Never** place a used pipette on the lab bench or on your lab book or on any other surface. This pipetting technique involves some manual dexterity, and it is recommended that the technique be practiced using water and non-sterile pipettes in order to become more comfortable with the procedure.

Good judgment is called for in determining the size of pipette to use in a given situation. To accurately pipette volumes between 0.1 and 1.0 ml, use a 1.0-ml pipette. For volumes larger than 1.0 ml, use a 5.0- or 10.0-ml pipette. When volumes less than 0.1 ml need to be delivered, a micropipette should be used. Micropipettes are available in various ranges. They are expensive, carefully calibrated instruments and extreme care should be exercised when using them. Two of the most common micropipettes are (a) Eppendorf, and (b) Rainin Pipetman. Each operates in a slightly different way, so allow time to become familiar with the use of each.

To begin use, first press the micropipette on a disposable tip. Make sure that the tip sizes correspond to the range of the micropipette being used. With the Eppendorf micropipettes, the volume to be delivered is set by turning the ratchet mechanism to the desired volume. With the Rainin Pipetman, the volume is set by turning the thumb wheel near the top of the micropipette. **Never** attempt to set a volume greater than the calibrated range of the micropipette being used, as damage may occur to the micropipette.

The top button on the Eppendorf micropipette has three stops. In order to draw liquid into the tip, first depress the button to the first stop and then insert the tip in the liquid. Slowly release the button and liquid will be drawn into the tip. Be sure that no air has been drawn into the tip. To release the fluid, place the tip inside the receiving tube, touching the tip to the side wall of the tube, and press the button to the second stop. Pressing the button to the third stop will cause the tip to be ejected from the micropipette. The Rainin Pipetman operates in a similar way, except that the tip ejection is accomplished by pressing a separate eject button. All used tips should be discarded into a biohazard container.

APPENDIX F

Dilution Techniques

In the quantitation of the population size of microorganisms in water, soil, or food, it may be necessary to use a dilution scheme. The population size of microorganisms may be very large and, as a result, their direct enumeration on nutrient agars may be impossible. By diluting, that is, thinning down by mixing with another liquid, the population size can be reduced in order to make enumeration of individual colonies possible. Dilutions are made by pipetting samples into a diluting fluid (diluent), which is usually sterile distilled water or isotonic saline (0.85% sodium chloride).

The dilution factor is calculated by dividing the initial sample volume by the total volume obtained after mixing with the diluent.

$$\text{Dilution factor} = \frac{\text{volume of sample}}{\text{total volume (sample + diluent volume)}}$$

So, if 1.0 ml of a broth culture is pipetted into 9.0 ml of saline, a 1:10 (1/10, 10^{-1}) dilution of the original sample is prepared. Note that the dilution factor can be expressed in a number of different ways: either as a ratio, fraction, or exponent. Most often the dilution is represented in exponential notation form. If 1.0 ml of the 10^{-1} dilution is pipetted into another 9.0-ml saline tube, another 1/10 (10^{-1}) dilution is made.

What dilution is this of the original broth culture?

____

____

In order to calculate the final dilution made from the original, multiply each of the dilution factors. In this case:

$$10^{-1} \times 10^{-1} = 10^{-2} \quad \text{or } 1/100 \text{ dilution of the original sample}$$

(Remember that exponents are added in multiplication.) If 1.0 ml of the 10^{-2} dilution is added to 9.0 ml of saline, then another 10^{-1} dilution is made. To calculate the dilution of the original, multiply each of the individual dilution factors:

$$10^{-1} \times 10^{-1} \times 10^{-1} = 10^{-3} \quad \text{or } 1/1000 \text{ dilution of the original sample}$$

This type of dilution scheme is called a serial dilution technique since the same dilution is conducted repeatedly in a series of tubes.

A 10^{-1} dilution can also be made by changing either sample or diluent volumes. Pipetting 0.5 ml of sample into 4.5 ml of diluent creates a 10^{-1} dilution. Remember to calculate the dilution factor, divide the sample volume by the final volume created. In this case,

$$\text{Dilution factor} = \frac{\text{sample volume}}{\underset{\text{(diluent volume + sample volume)}}{\text{total volume}}} = \frac{0.5\text{ ml}}{4.5\text{ ml} + 0.5\text{ ml}} = \frac{0.5}{5.0} = \frac{1}{10}$$

Similarly, a 10^{-1} dilution can be made by using 0.1 ml of sample and 0.9 ml of diluent. Serial 10^{-2} dilutions may also be made. Here, 1.0 ml of sample is pipetted into 99.0 ml of diluent. By pipetting 1.0 ml of this 10^{-2} dilution into another 99.0 ml of diluent, a 10^{-4} dilution of the original sample is made.

The following example shows how dilutions are used to specifically quantitate the population size of bacteria in a turbid broth culture, using a spread-plate technique. Keep in mind that a slightly turbid culture has at least 10^6 microorganisms/ml. This information is helpful in making a decision about the number of dilutions required.

Procedure

1. In a rack, set up diluent tubes (9.0 ml) and label 10^{-1} to 10^{-4}.
2. With aseptic technique, withdraw 1.0 ml of the turbid broth culture with a pipetting device and transfer to the tube labeled 10^{-1}. Place the pipette either in the disinfectant solution or in a biohazard container.
3. Mix the tube by gentle shaking or use a vortex mixer.
4. Using a fresh sterile pipette, transfer 1.0 ml from the 10^{-1} tube to the 9.0-ml diluent tube marked 10^{-2}. Discard the pipette and mix the tube contents well.
5. Continue in a similar fashion until all dilutions through 10^{-4} have been prepared.
6. Label a duplicate series of eight plates of nutrient agar 10^{-1} to 10^{-4}.

At this point, decisions must be made concerning the dilutions and the volumes to be plated. Usually the maximum volume plated in a spread-plate technique is 0.1 ml. A number of dilutions are usually sampled to ensure that nutrient agar plates with countable colonies are obtained. In this example all dilutions are plated.

7. Using a sterile pipette, transfer 0.1 ml of the 10^{-4} dilution to the agar plate marked 10^{-5}. In a similar way, using the same pipette, transfer 0.1 ml to the second 10^{-5} plate.
8. Using the same pipette, transfer two 0.1-ml aliquots of the 10^{-3} dilution to two nutrient agar plates labeled 10^{-4}.
9. Continue in a similar manner until all dilutions have been transferred to the appropriately labeled plates.

Separate pipettes could have been used to do the platings, but to conserve pipettes only one is used. Explain why this is possible. The same pipette cannot be used if the platings are done in the order from 10^{-1} to 10^{-4}. Why not?

____

____

____

____

10. Spread the plates and incubate.
11. Report the number of colonies on the plates, counting only those plates containing less than 300 colonies. Plates that have greater than 300 colonies are scored **TNTC** (Too Numerous To Count).

The following hypothetical results are used to demonstrate how to calculate the population size of microorganisms present in the original broth culture. The concentration of microorganisms is referred to as the **titer** and is reported as the number of colony-forming units per milliliter (cfu/ml) of sample.

RESULTS

Dilution Plate	Number of Colonies
10^{-1}	TNTC
10^{-1}	TNTC
10^{-2}	TNTC
10^{-2}	TNTC
10^{-3}	TNTC
10^{-3}	TNTC
10^{-4}	150
10^{-4}	190

Calculate the average number of colonies for each dilution. In this example the mean that can be calculated is for the 10^{-4} dilution, and that is 170. To determine the cfu/ml in the original broth culture use the following equation:

$$\text{Number of colony-forming units/ml of original sample} = (\text{number of colonies}) \times \{1/(\text{dilution factor})\}$$

$$\text{cfu/ml} = (170) \times (10^4)$$
$$= 1.7 \times 10^6$$

While ten-fold dilutions are most popular, other dilutions can be used.

Dilution	Volume of Sample (ml)	Volume of Diluent (ml)
1:4	1.0	3.0
1:5	1.0	4.0
1:20	0.5	9.5
1:25	1.0	24.0
1:25	0.5	12.0

Fill in the appropriate dilution factors and final dilutions in the following protocol:

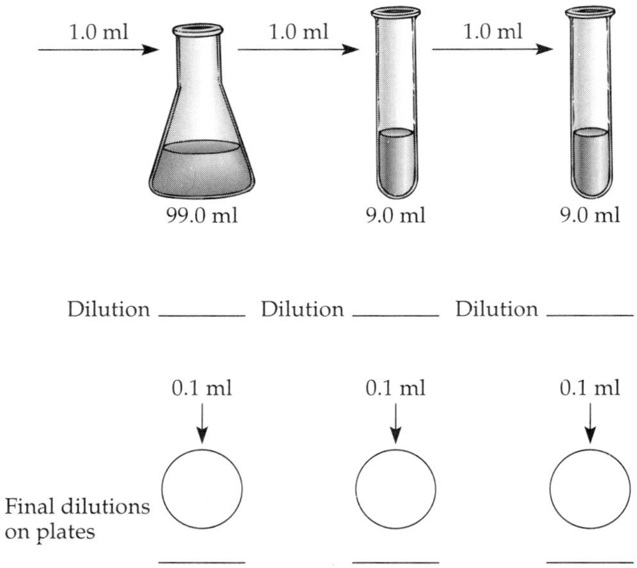

APPENDIX G

Bacterial Identification Charts

ID Scheme for Gram-Positive Rods

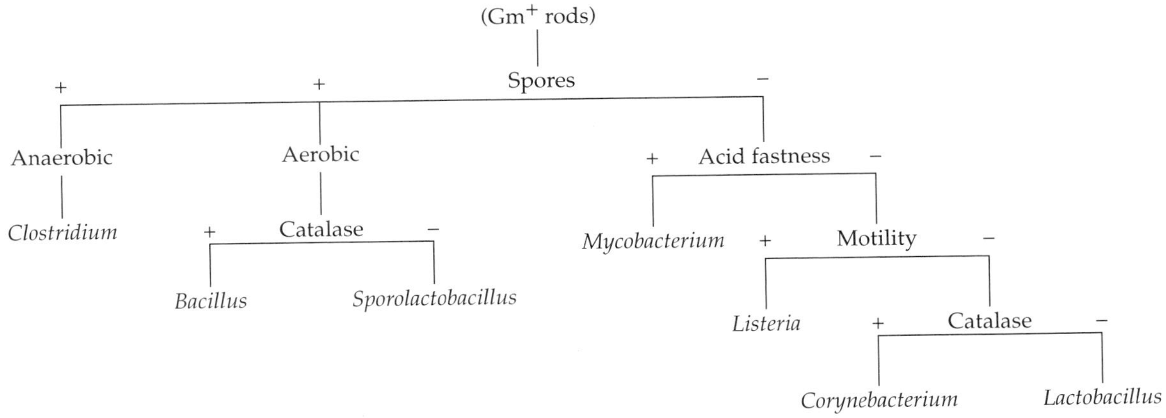

ID Scheme for Gram-Negative Rods

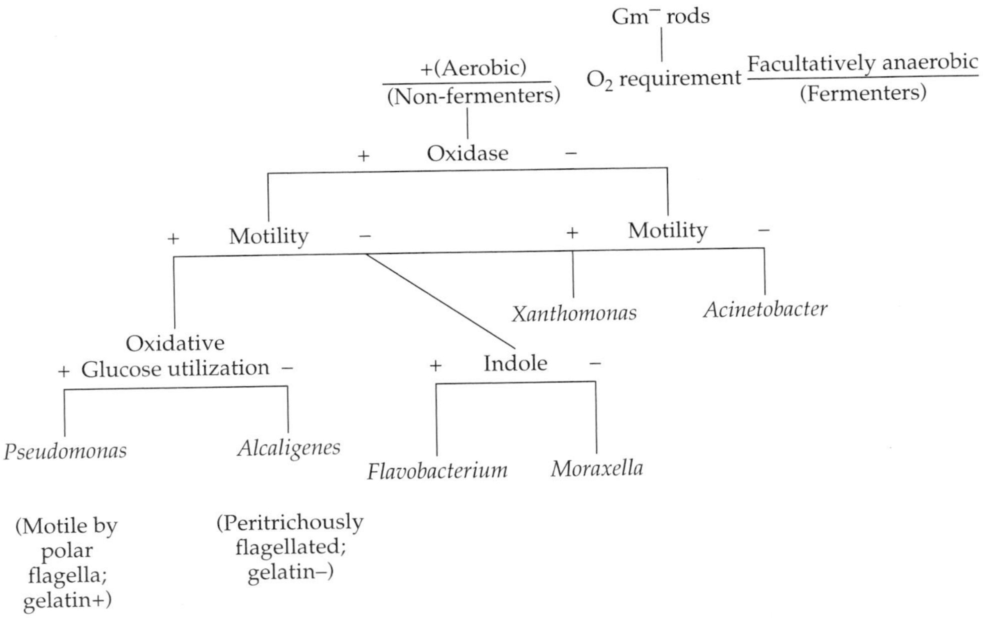

ID Scheme for Gram-Negative Glucose Fermenters

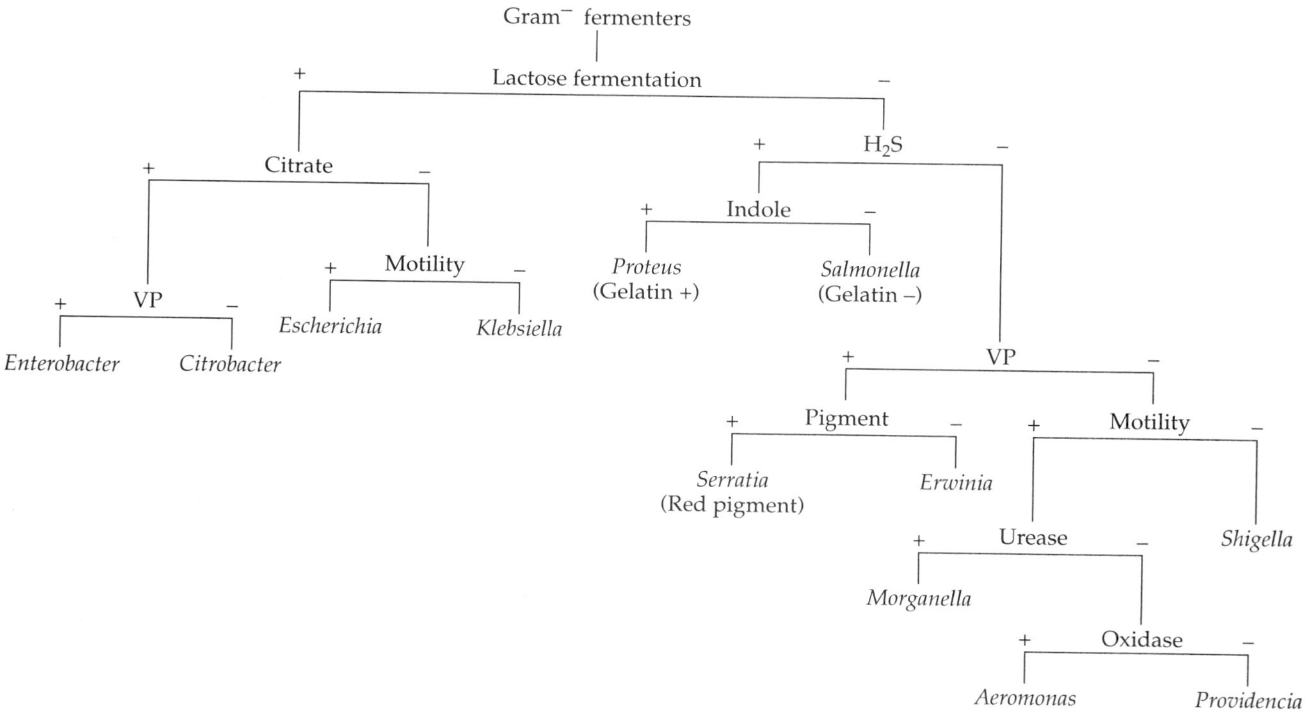

ID Scheme for Gram-Negative Cocci

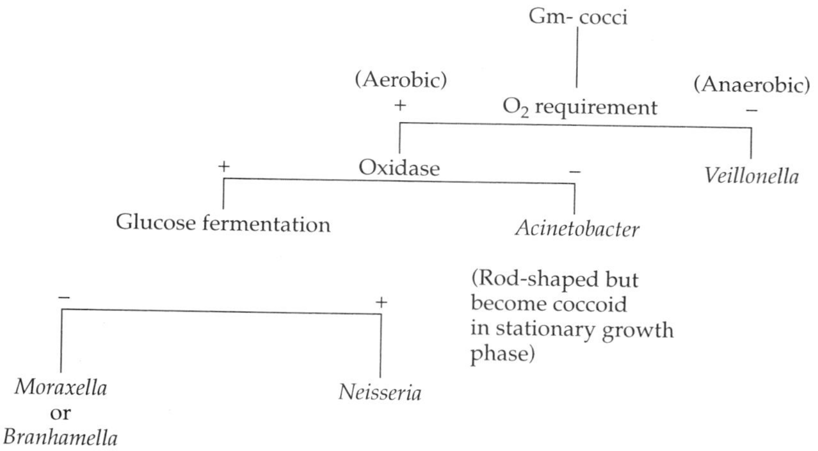

ID Scheme for Gram-Positive Cocci

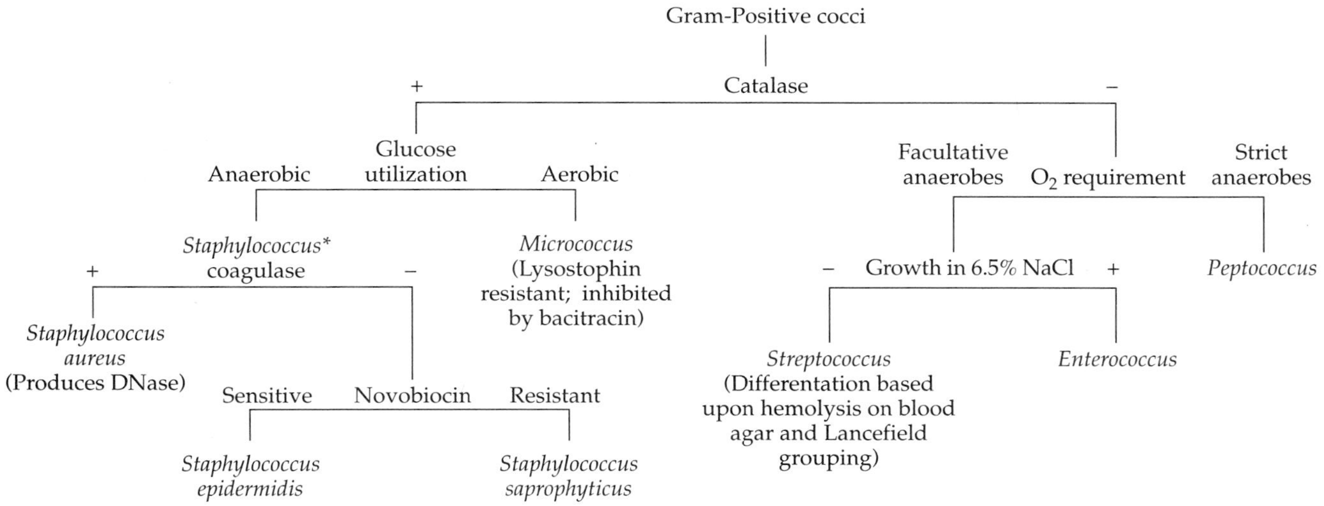

* Staphylococci can be identified by using commercially available kits such as Staph-IDENT or Staph-TRAC.

APPENDIX H

Instructions to Authors*

How to Submit Manuscripts

Submit manuscripts directly to: Journals Department, American Society for Microbiology, 1325 Massachusetts Ave., N.W., Washington, DC 20005-4171. *Since all submissions must be processed through this office, alternate routings, such as to an editor, will delay initiation of the review process.* The manuscript must be accompanied by a cover letter stating the following: the journal to which the manuscript is being submitted; the most appropriate section of the journal; the complete mailing address (including the street), e-mail address (if available), and telephone and fax numbers of the corresponding author; and the former ASM manuscript number and year if it is a resubmission. It is expected that the author will include written assurance that permission to cite unpublished data or personal communications has been granted.

Authors may suggest an appropriate editor for new submissions. If we are unable to comply with such a request, the corresponding author will be notified before the manuscript is assigned to another editor. To expedite the review process, authors may recommend at least two or three reviewers who are not members of their institution(s) and have never been associated with them or their laboratory(ies). Please provide the name, address, phone and fax numbers, and area of expertise for each. Note that reviewers so recommended will be used at the discretion of the editor.

Submit three complete copies of each manuscript, including figures and tables. Type every portion of the manuscript **double spaced** (a minimum of 6 mm between lines), including figure legends, table footnotes, and References, and number all pages in sequence, including the abstract, figure legends, and tables. Place the last two items after the References section. Manuscript pages must have margins of at least 1 inch on all four sides and **should have line numbers if possible**. It is recommended that the following sets of characters be easily distinguishable in the manuscript: the numeral zero (0) and the letter "oh" (O); the numeral one (1), the letter "el" (l), and the letter "eye" (I); and a multiplication sign (×) and the letter "ex" (x). If such distinctions cannot be made, please mark these items at the first occurrence for cell lines, strain and genetic designations, viruses, etc., on the modified manuscript so that they may be identified properly for the printer by the copy editor. See p. vi for detailed instructions about illustrations.

Copies of in-press and submitted manuscripts that are important for judgment of the present manuscript should be enclosed to facilitate the review. Three copies of each such manuscript should be provided.

*From Applied and Environmental Microbiology, Jan. 1996.

Authors who are unsure of proper English usage should have their manuscripts checked by someone proficient in the English language. Manuscripts may be rejected on the basis of poor English or lack of conformity to accepted standards of style.

Editorial Policy

Manuscripts submitted to the journal must represent reports of original research. **All authors of a manuscript must have agreed to its submission and are responsible for its content**, including appropriate citations and acknowledgments, and must also have agreed that the corresponding author has the authority to act on their behalf on all matters pertaining to publication of the manuscript. By submission of a manuscript to the journal, the authors guarantee that the manuscript, or one with substantially the same content, was not published previously, is not being considered or published elsewhere, and was not rejected on scientific grounds by another ASM journal.

By publishing in the journal, the authors agree that any plasmids, viruses, and living materials such as microbial strains and cell lines newly described in the article are available from a national collection or will be made available in a timely fashion and at reasonable cost to members of the scientific community for noncommercial purposes.

Failure to comply with the above-mentioned policies may result in a suspension of publishing privileges in ASM journals and notification of the authors' institutions.

Primary Publication

The American Society for Microbiology accepts the definition of primary publication as defined in *How to Write and Publish a Scientific Paper*, 4th ed., by Robert A. Day, to wit: "… (1) the first publication of original research results, (2) in a form whereby peers of the author can repeat the experiments and test the conclusions, and (3) in a journal *or other source document* [emphasis added] readily available within the scientific community."

A scientific paper or its substance published in a conference report, symposium proceeding, or technical bulletin, posted on a host computer to which there is access via Internet, or made available through any other retrievable source, including CD-ROM and other electronic forms, is unacceptable for submission to an ASM journal on grounds of *prior publication*. A preliminary disclosure of research findings published in abstract form as an adjunct to a meeting, e.g., part of a program, is not considered prior publication because it does not meet the criteria for a scientific paper.

It is incumbent upon the author to acknowledge any prior publication of the data contained in a manuscript submitted to an ASM journal. A copy of the relevant work should accompany the paper.

Permissions

The corresponding author is responsible for obtaining permissions from both the original publisher and the original author [i.e., the copyright owner(s)] to reproduce figures, tables, or text (in whole or in part) from previous publications. The signed permissions must be submitted to ASM, and each should be identified as to the relevant item in the ASM manuscript (e.g., "permissions for Fig. 1 in AEM 123-96").

Authorship

An author is one who made a substantial contribution to the "overall design and execution of the experiments"; therefore, ASM considers all authors responsible for the entire paper. Individuals who provided assistance, e.g., supplied strains or reagents or critiqued the paper, need not be listed as authors but may be recognized in the Acknowledgment section.

All authors must agree to the order in which their names are listed in the byline. Footnotes regarding attribution of work (e.g., X. Jones and Y. Smith contributed equally to . . .) are not permitted. If necessary, such statements may be included in the Acknowledgment section.

Page Charges

It is anticipated that page charges, currently $50 per page for the first five pages and $70 per page for each page in excess of five (prices subject to change), will be paid by authors whose research was supported by grants, special funds, or contracts (departmental, governmental, institutional, etc.) or whose research was done as part of their official duties. A bill for page charges is sent with the page proofs and reprint order form.

If the research was not supported by any of the means described above, a request to waive the charges may be sent to the Journals Department, American Society for Microbiology, 1325 Massachusetts Ave., N.W., Washington, DC 20005-4171, with the submitted manuscript. This request, which must be separate from the cover letter, must indicate how the work was supported and should be accompanied by a copy of the Acknowledgement section.

Minireviews and Letters to the Editor (see p. vi) are not subject to page charges.

Copyright

To maintain and protect the Society's ownership and rights and to protect the original authors from misappropriations of their work, ASM requires the corresponding author to sign a copyright transfer agreement on behalf of all the authors. This agreement is sent to the corresponding author when the manuscript is accepted and scheduled for publication. Unless this agreement is executed (without changes and/or addenda), ASM will not publish the manuscript.

If *all* authors were employed by the U.S. government when the work was performed, the corresponding author should not sign the copyright transfer agreement but should, instead, attach to the agreement a statement attesting that the manuscript was prepared as a part of their official duties and, as such, is a work of the U.S. government not subject to copyright.

If *some* of the authors were employed by the U.S. government when the work was performed but the others were not, the corresponding author should sign the copyright transfer agreement as it applies to that portion performed by the non-government employee authors.

Scope

Applied and Environmental Microbiology (AEM) publishes descriptions of all aspects of applied research as well as research of a genetic and molecular nature that focuses on topics of practical value and basic research on microbial ecology. Topics that are considered include microbiology in relation to foods, agriculture, industry, biotechnology, public health, plants, invertebrates, and basic biological properties of bacteria, fungi, protozoa, and other simple eucaryotic organisms as related to microbial ecology. Submitted manuscripts should primarily emphasize microbiology-based research. The plant microbiology section will consider manuscripts dealing with all aspects of plant-microorganism interactions, including symbiotic and rhizosphere bacteria and phytopathogenic microorganisms. Manuscripts submitted to the invertebrate microbiology section should address interactions between invertebrates and microorganisms, ranging from commensalism and mutualism to parasitism and pathogenicity. Manuscripts describing work dealing with the metabolites or toxins from animal, plant, or insect cells or the physiology of such cells are not suitable for AEM unless it affects a microbial community or individual microorganisms.

ASM publishes a number of different journals covering various aspects of the field of microbiology. Each journal has a prescribed scope which must be considered in determining the most appropriate journal for each manuscript. The following guidelines may be of assistance.

(i) AEM will consider manuscripts describing properties of enzymes and proteins that are produced by either wild-type or genetically engineered microorganisms and that are significant or have potential significance in industrial or environmental settings. Studies dealing with basic biological phenomena of enzymes or proteins or in which enzymes have been used in investigations of basic biological functions are more appropriate for the *Journal of Bacteriology*.

(ii) AEM will consider papers which describe the use of antimicrobial or anticancer agents as tools for elucidating aspects of applied and environmental microbiology. Other papers dealing with antimicrobial or anticancer agents, including manuscripts dealing with the biosynthesis and metabolism of such agents, are more appropriate for *Antimicrobial Agents and Chemotherapy*.

(iii) Papers on the biology of bacteriophages and other viruses are more appropriate for the *Journal of Virology* or the *Journal of Bacteriology*. AEM does, however, consider manuscripts dealing with viruses in relation to environmental, public health, or industrial microbiology.

(iv) Manuscripts dealing with the immune system or with topics of basic medical interest or oral microbiology are more appropriate for *Infection and Immunity*. Reports of clinical investigations and environmental biology applied to hospitals should be submitted to the *Journal of Clinical Microbiology*.

(v) Papers that include mainly taxonomic material (e.g., descriptions of new taxa) should be submitted to the *International Journal of Systematic Bacteriology*, which is published by ASM for the International Union of Microbiological Societies.

(vi) In most cases, AEM will not consider reports that emphasize nucleotide sequence data alone (without experimental documentation of the functional and evolutionary significance of the sequence).

Questions about these guidelines may be directed to the editor in chief of the journal being considered.

If transfer to another ASM journal is recommended by an editor, the corresponding author will be contacted.

Note that a manuscript rejected by one ASM journal on scientific grounds or on the basis of its general suitability for publication is considered rejected by all other ASM journals.

Culture Deposition

AEM encourages authors to deposit important strains in publicly accessible culture collections and to refer to the collections and strain numbers in the text. Since the authenticity of subcultures of culture collection specimens that are distributed by individuals cannot be ensured, authors should indicate laboratory strain designations and donor source as well as original culture collection identification numbers.

Nucleotide Sequences

It is expected that GenBank/EMBL accession numbers for primary nucleotide and/or amino acid sequence data will be included in the original manuscript or be inserted when the manuscript is modified. (The accession number should be included in a separate paragraph at the end of the Materials and Methods section for full-length papers or at the end of the text of Notes.)

If conclusions in a manuscript are based on the analysis of sequences and a GenBank/EMBL accession number is not provided at the time of the review, authors may be required to provide the sequence data as a file on a floppy disk.

GenBank may be contacted at GenBank Submissions, National Center for Biotechnology Information, Bldg. 38A, Rm. 8N-803, 8600 Rockville Pike, Bethesda, MD 20894; e-mail (new submissions): gb-sub@ncbi.nlm.nih.gov; e-mail (updates): update@ncbi.nlm.nih.gov. The EMBL Data Library may be contacted at EMBL Data Library Submissions, Postfach 10.2209, Meyerhofstrasse 1, 69012 Heidelberg, Germany; telephone: 49 (6221) 387258; fax: 49 (6221) 387306; e-mail (data submissions): datasubs@embl.bitnet.

See p. viii for nucleic acid sequence formatting instructions.

Editorial Style

The editorial style of ASM journals conforms to the *ASM Style Manual for Journals and Books* (American Society for Microbiology, 1991) and Robert A. Day's *How to Write and Publish a Scientific Paper*, 4th ed. (Oryx Press, Phoenix, Ariz., 1994), as interpreted and modified by the editors and the ASM Journals Department. The editors and the Journals Department reserve the privilege of editing manuscripts to conform with the stylistic conventions set forth in the aforesaid publications and in these instructions.

Review Process

All manuscripts are reviewed by the editors, members of the editorial board, or qualified ad hoc reviewers. When a manuscript is submitted to the journal, it is given a number (e.g., AEM 47-96) and assigned to one of the editors. All coauthors are notified of this number and the editor to whom the manuscript has been assigned. (Always refer to this number in communications with the editor and Journals Department.) It is the responsibility of the corresponding author to inform the coauthors of the manuscript's status throughout the review and publication processes. The reviewers operate under strict guidelines set forth in "Guidelines for Reviewers" and are expected to complete their reviews within 3 weeks after receiving the manuscript. The corresponding author is notified, an average of 8 weeks after submission, of the editor's decision to accept, reject, or require modification. When a manuscript is returned to the corresponding author for modification, it should be returned to the editor within 2 months; otherwise it may be considered withdrawn. A point-for-point response to the reviews must be included with the revised manuscript; an extra copy of the revised manuscript (without figures) should have the changes highlighted. Manuscripts that have been rejected, or withdrawn after being returned for modification, may be resubmitted if the major criticisms have been addressed. As with initial submissions, resubmitted manuscripts should be sent to the Journals Department, *not to the editor*, and should be accompanied by a cover letter stating that the manuscript is a resubmission. A point-for-point response to the original reviews, as well as a copy of the resubmitted manuscript with the changes highlighted, should be included. Resubmitted manuscripts are normally handled by the original editor. Manuscripts cannot be resubmitted more than once unless permission has been obtained from the original editor or from the editor in chief.

Notification of Acceptance

When an editor has decided that a manuscript is acceptable for publication on the basis of scientific merit, it is sent to the Journals Department, where it is checked by the production editor. If the manuscript has been prepared according to the criteria set forth in these instructions, it is scheduled for the next available issue and an acceptance letter that indicates the month of publication, approximate page proof dates, and section is mailed to the corresponding author. The editorial staff of the ASM Journals Department completes the editing of the manuscript to bring it into conformity with prescribed style.

Page Proofs

The printer sends page proofs, the copyedited manuscript, and the page charge/reprint order form to the corresponding author. As soon as the page proofs are corrected and signed by the person who proofread them (within 48 h), they should be mailed to the ASM Journals Department.

The proof stage is not the time to make extensive corrections, additions, or deletions. Important new information that has become available between acceptance of the manuscript and receipt of the proofs may be inserted as an addendum in proof with the permission of the editor. If references to unpublished data or personal communications are added, it is expected that written assurance granting permission for the citation will be included. Limit changes to correction of spelling errors, incorrect data, and grammatical errors and updated information for references to submitted and in-press articles.

Questions about late proofs and problems in the proofs should be directed to the ASM Journals Department (telephone, 202 942-9219).

Reprints

Reprints (in multiples of 100) may be purchased by all coauthors. An order form that includes a table showing the cost of reprints is sent with the proofs to the corresponding author.

Organization and Format

Regular Papers

Regular full-length papers should include the elements described in this section.

Title, running title, and byline. Each manuscript should present the results of an independent, cohesive study; thus, numbered series titles are not permitted. Exercise care in composing a main title. Avoid the main title/subtitle arrangement, complete sentences, and unnecessary articles. On the title page, include the title, running title (not to exceed 54 characters and spaces), name of each author, address(es) of the institutions(s) at which the work was performed, each author's affiliation, and a footnote indicating the present address of any author no longer at the institution where the work was performed. Place an asterisk after the name of the author to whom inquiries regarding the paper should be directed, and **give that author's telephone and fax numbers**.

Correspondent footnote. The complete mailing address, telephone number, fax number, and e-mail address of the corresponding author will be published as a footnote if so desired by the author. This information should be included in the lower left corner of the manuscript title page and must be labeled "Correspondent Footnote." **Such footnotes will not be added at the proof stage.**

Abstract. Limit the abstract to **250 words or fewer** and concisely summarize the basic content of the paper without presenting extensive experimental details. Avoid abbreviations and do not include diagrams. When it is essential to include a reference, use the References entry but omit the article title. Because the abstract will be published separately by abstracting services, it must be complete and understandable without reference to the text.

Introduction. The introduction should supply sufficient background information to allow the reader to understand and evaluate the results of the present study without referring to previous publications on the topic. The introduction should also provide the rationale for the present study. Use only those references required to provide the most salient background rather than an exhaustive review of the topic.

Materials and Methods. The Materials and Methods section should include sufficient technical information to allow the experiments to be repeated. When centrifugation conditions are critical, give enough information to enable another investigator to repeat the procedure: make of centrifuge, model of rotor, temperature, time at maximum speed, and centrifugal force ($\times\, g$ rather than revolutions per minute). For commonly used materials and methods (e.g., media and protein concentration determinations), a simple reference is sufficient. If several alternative methods are commonly used, it is helpful to identify the method briefly as well as to cite the reference. For example, it is preferable to state, "cells were broken by ultrasonic treatment as previously described (9)," rather than to state, "cells were broken as previously described (9)." The reader should be allowed to assess the method without constant reference to previous publications. Describe new methods completely, and give sources of unusual chemicals, equipment, or microbial strains. When large numbers of microbial strains or mutants are used in a study, include tables identifying the sources and properties of the strains, mutants, bacteriophages, plasmids, etc.

A method, strain, etc., used in only one of several experiments reported in the paper may be described in the Results section or very briefly (one or two sentences) in a table footnote or figure legend.

Results. In the Results section, include only the results of the experiments; reserve extensive interpretation of the results for the Discussion section. Present the results as concisely as possible in **one** of the following: text, table(s), or figure(s). Avoid extensive use of graphs to present data that might be more concisely presented in the text or tables. For example, except in unusual cases, double-reciprocal plots used to determine apparent K_m values should not be presented as graphs; instead, the values should be stated in the text. Similarly, graphs illustrating other methods commonly used to derive kinetic or physical constants (e.g., reduced viscosity plots and plots used to determine sedimentation velocity) need not be shown except in unusual circumstances. Limit photographs (particularly photomicrographs and electron micrographs) to those that are absolutely necessary to show the experimental findings. Number figures and tables in the order in which they are cited in the text, and be sure to cite all figures and tables.

Discussion. The Discussion should provide an interpretation of the results in relation to previously published work and to the experimental system at hand and should not contain extensive repetition of the Results section or reiteration of the introduction. In short papers, the Results and Discussion sections may be combined.

Acknowledgments. The source of any financial support received for the work being published must be indicated in the Acknowledgments section. (It will be assumed that the absence of such an acknowledgment is a statement by the authors that no support was received.) The usual format is as follows: "This work was supported by Public Health Service grant CA-01234 from the National Cancer Institute."

Recognition of personal assistance should be given as a separate paragraph.

Appendixes. Appendixes, which contain supplementary material to aid the reader, are permitted. Titles, authors, and References sections that are distinct from those of the primary article are not allowed. If it is not feasible to list the author(s) of the appendix in the byline or the Acknowledgment section of the primary article, rewrite the appendix so that it can be considered for publication as an independent article, either full length or Note style. Equations, tables, and figures should be labeled with the letter "A" preceding the numeral to distinguish them from those cited in the main body of the text.

References. The References section must include **all** relevant sources, and all listed references **must** by cited in the text. Arrange the citations in **alphabetical order**, by first author, and **number consecutively**. Abbreviate journal names according to *Serial Sources for the BIOSIS Previews Data Base* (BioSciences Information Service, Philadelphia, 1995). Cite each listed reference by number in the text.

Follow the styles shown in the examples below.

1. **Armstrong, J. E., and J. A. Calder.** 1978. Inhibition of light-induced pH increase and O_2 evolution of marine microalgae by water-soluble components of crude and refined oils. Appl. Environ. Microbiol. **35**:858-862.

2. **Barton, B., G. Harding, and A. Zuccarelli.** 1994. A general method for detecting and sizing large plasmids, abstr. H-249, p. 244. *In* Abstracts of the 94th General Meeting of the American Society for Microbiology 1994. American Society for Microbiology, Washington, D.C.
3. **Berry, L. J., R. N. Moore, K.J. Goodrum, and R.E. Couch, Jr.** 1977. Cellular requirements for enzyme inhibition by endotoxin in mice, p. 321-325. *In* D. Schlessinger (ed.), Microbiology—1977. American Society for Microbiology, Washington, D.C.
4. **Cox, C. S., B. R. Brown, and J. C. Smith.** J. Gen. Genet., in press.* [*Article title is optional.*]
5. **Finegold, S. M., W. E. Shepherd, and E. H. Spaulding.** 1977. Cumitech 5, Practical anaerobic bacteriology. Coordinating ed., W. E. Shepherd. American Society for Microbiology, Washington, D.C.
6. **Fitzgerald, G., and D. Shaw.** *In* A.E. Waters (ed.), Clinical microbiology, in press. EFH Publishing Co., Boston. [*Chapter title is optional.*]
7. **Gill, T. J., III.** 1976. Principles of radioimmunoassay, p. 169-171. *In* N. R. Rose and H. Friedman (ed.), Manual of clinical immunology. American Society for Microbiology, Washington, D.C.
8. **Gustlethwaite, F. P.** 1985. Letter. Lancet **ii**:327.
9. **Jacoby, J., R. Grimm, J. Bostic, V. Dean, and G. Starke.** Submitted for publication. [*Article title is optional.*]
10. **Jensen, C., and D. S. Schumacher.** Unpublished data. [*Date is optional.*]
11. **Jones, A.** Personal communication. [*Date is optional.*]
12. **Leadbetter, E. R.** 1974. Order II. *Cytophagales* nomen novum, p. 99. *In* R. E. Buchanan and N. E. Gibbons (ed.), Bergey's manual of determinative bacteriology, 8th ed. The Williams & Wilkins Co., Baltimore.
13. **Sacks, L. E.** 1972. Influence of intra- and extracellular cations on the germination of bacterial spores, p. 437-442. *In* H. O. Halvorson, R. Hanson, and L. L. Campbell (ed.), Spores V. American Society for Microbiology, Washington, D.C.
14. **Sigma Chemical Co.** 1989. Sigma manual. Sigma Chemical Co., St. Louis, Mo.
15. **Smith, J. C.** April 1970. U.S. patent 484,363,770.
16. **Smyth, D. R.** 1972. Ph.D. thesis. University of California, Los Angeles. [*Title is optional.*]
17. **Yagupsky, P., and M. A. Menegus.** 1989. Intraluminal colonization as a source of catheter-related infection. Antimicrob. Agents Chemother. **33**:2025. (Letter.)

*Note that a reference to an in-press ASM publication should state the control number (e.g., AEM 576-96) if it is a journal article or the name of the publication if it is a book.

Notes

Submit Notes in the same way as full-length papers. They receive the same review, they are not published more rapidly than full-length papers, and they are not considered preliminary communications. The Note format is intended for the presentation of brief observations that do not warrant full-length papers.

Each Note must have an **abstract of no more than 50 words**. Do not use section headings in the body of the Note; report methods, results, and discussion in a single section. Paragraph lead-ins are permissible. The text should be kept to a minimum and if possible **should not exceed 1,000 words**;

the number of figures and tables should also be kept to a minimum. **Materials and methods should be described in the text, not in figure legends or table footnotes**. Present acknowledgments as in full-length papers, but do not use a heading. The References section is identical to that of full-length papers.

Minireviews

Minireviews are brief summaries (**limit of 6 printed pages exclusive of references**) of developments in fast-moving areas. They must be based on published articles; they may address any subject within the scope of AEM. Minireviews may be either solicited or proffered by authors responding to a recognized need. Irrespective of origin, minireviews are subject to editorial review. Three double-spaced copies must be provided.

Letters to the Editor

Letters to the Editor must include data to support the writer's argument and are intended only for comments on articles published previously in the journal. They may **be no more than 500 words long**. Send three copies to the Journals Department. The letter will be processed and sent to the editor who handled the article in question. If the editor believes that publication is warranted, he will solicit a reply from the corresponding author of the article and make a recommendation to the editor in chief. Final approval for publication rests with the editor in chief. All letters intended for publication must be **typed double spaced**.

Errata

The Erratum section provides a means of correcting errors that occurred during the writing, typing, editing, or printing (e.g., a misspelling, a dropped word or line, mislabeling in a figure, etc.) of a published article. Send errata directly to the Journals Department.

Author's Corrections

The Author's Correction section provides a means of correcting errors of omission (e.g., author names or citations) and errors of a scientific nature that do not alter the overall basic results or conclusions of a published article.

For omission of an author's name, the authors of the article and the author whose name was inadvertently omitted must agree to publication of the correction. Letters from both parties must accompany the correction and be sent directly to the Journals Department.

Corrections of a scientific nature (e.g., an incorrect unit of measurement or order of magnitude used throughout; contamination of one of numerous cultures; misidentification of a mutant or strain, causing erroneous data for only a portion [noncritical] of the study; etc.) must be sent directly to the editor who handled the article. If the editor believes that publication is warranted, he will send the correction to the Journals Department for publication. Note that the addition of new data is not permitted.

Retractions

Retractions are reserved for major errors or breaches of ethics that, for example, may call into question the source of the data or the validity of the results and conclusions of an article. Send a retraction and an accompanying explanatory letter signed by all of the authors directly to the editor in chief of the journal. The editor who handled the paper and the chairman of the

ASM Publications Board will be consulted. If all parties agree to the publication and content of the retraction, it will be sent to the Journals Department for publication.

Disclaimers

Statements disclaiming governmental or any other type of endorsement or approval will be deleted by the Journals Department.

Illustrations and Tables

The figure number and authors' names should be written on all figures, either in the margin or on the back (marked lightly with a soft pencil). For micrographs especially, the top should be indicated as well.

Do not clasp figures to each other or to the manuscript with paper clips. Insert small figures in an envelope. To avoid damage in transit, do not submit illustrations larger than 8½ by 11 inches.

Illustrations in published articles will not be returned to authors.

Continuous-Tone and Composite Photographs

When submitting continuous-tone photographs (e.g., polyacrylamide gels), keep in mind the journal page width: 3⁵⁄₁₆ inches for a single column and 6⅞ inches for a double column (maximum). Include only the significant portion of an illustration. Photos must be of sufficient contrast to withstand the inevitable loss of contrast and detail inherent in the printing process. **Submit one photograph of each continuous-tone figure for each copy of the manuscript; photocopies are not acceptable.** If possible, the figures submitted should be the size they will appear when published so that no reduction is necessary. If they must be reduced, make sure that *all* elements, including labeling, can withstand reduction and remain legible.

If a figure is a composite of a continuous-tone photograph and a drawing or labeling, the **original composite** (i.e., not a photograph of the composite) **must be provided** for the printer. This original, labeled "printer's copy," may be sent with the modified manuscript to the editor. Composites should be mounted on lightweight flexible backing, not on heavy cardboard.

Electron and light micrographs must be direct copies of the original negative. Indicate the magnification with a scale marker on each micrograph.

Color Photographs and Illustrations

If it is necessary to include color photographs or illustrations in an article, include an extra copy (in addition to the three required for review) at the time of manuscript submission so that a cost estimate for printing may be obtained. The cost of printing in color must be borne by the author. Adherence to the following guidelines will help to minimize costs and to ensure color reproduction that is as accurate as possible.

Keep in mind the journal page width (3⁵⁄₁₆ inches for a single column and 6⅞ inches for a double column) and height (9¹⁄₁₆ inches) and submit figures the size they should appear when published so that no reduction is necessary. Include only the significant portions of illustrations so that the number of printed pages containing color figures is minimized. Make sure that all edges are straight and corners are square. If a figure is a composite "plate" comprising a number of individual parts, add any necessary labels and tooling (i.e., thin white lines between the parts) that is of even width. Composites should be mounted on lightweight flexible backing so that they can be wrapped around a scanner drum. (If a composite is mounted on a heavy board and cannot be wrapped on the drum, a transparency will have to be produced, at additional cost.)

For optimal color reproduction, plates should comprise parts containing similar colors of similar lightness or darkness. If necessary, separate unlike photos on a single plate into two separate plates; this will increase the cost somewhat, but the color rendition will be more accurate since the two plates will be scanned separately.

Computer-Generated Images

ASM has discontinued accepting computer-generated images submitted on floppy disks until further notice. At the present time, the highest-quality, simplest, least-expensive reproduction of gels and similar illustrations continues to be scanning of author-supplied continuous-tone photographs or dye sublimation prints by the printer.

If continuous-tone photographs or dye sublimation prints cannot be supplied, computer-generated hard-copy images may be submitted. However, be advised that the quality of reproduction may not be as good as that for continuous-tone photos. The images produced by desktop systems are digitized and printed as patterns of dots. This presents problems for reproduction of such prints in the journal because the printing process also requires that the images be broken up into a dot pattern. Performing this process twice may result in some degradation of image quality and resolution. It is possible to use prints produced by desktop systems, but they may have to be scanned slightly out of focus to avoid interference of the dot patterns, and thus ASM cannot guarantee the quality of their reproduction.

Since the contents of computer-generated images can be manipulated for better clarity, the Publications Board at its May 1992 meeting indicated that a description of the software/hardware used should be included in the figure legend(s).

Drawings

Submit graphs, charts, complicated chemical or mathematical formulas, diagrams, and other drawings as glossy photographs made from finished drawings not requiring additional artwork or typesetting. Computer-generated graphics produced on high-quality laser printers are also usually acceptable. No part of the graph or drawing should be handwritten. Both axes of graphs must be labeled. Most graphs will be reduced to one-column width ($3\frac{5}{16}$ inches), and *all* elements in the drawing should be large enough to withstand this reduction. Avoid heavy letters, which tend to close up when reduced, and unusual symbols, which the printer may not be able to reproduce in the legend.

In figure ordinate and abscissa scales (as well as table column headings), avoid ambiguous use of numbers with exponents. Usually, it is preferable to use the Système International d'Unités (SI) (μ for 10^{-6}, m for 10^{-3}, k for 10^3, M for 10^6, etc.). A complete listing of SI symbols can be found in the International Union of Pure and Applied Chemistry (IUPAC) "Manual of Symbols and Terminology for Physicochemical Quantities and Units" (Pure Appl. Chem. **21**:3–44, 1970). Thus, a representation of 20,000 dpm on a figure ordinate is to be made by the number 20 accompanied by the label kdpm.

When powers of 10 must be used, the journal requires that the exponent power be associated with the number shown. In representing 20,000 cells per ml, the numeral on the ordinate would be "2" and the label would be "10^4 cells per ml" (not "cells per ml × 10^{-4}"). Likewise, an enzyme activity of 0.06 U/ml would be shown as 6, accompanied by the label 10^{-2} U/ml. The preferred designation would be 60 mU/ml (milliunits per milliliter).

Presentation of Nucleic Acid Sequences

Nucleic acid sequences of limited length which are the primary subject of a study may be presented freestyle in the most effective format. Longer nucleic acid sequences must be presented in the following format to conserve space. Submit the sequence as camera-ready copy with dimensions of 8½ by 11 inches (or slightly less) in standard (portrait) orientation. Print the sequence in lines of 100 bases, each in a nonproportional (monospace) font which is easily legible when published at 100 bases/6 inches. Uppercase and lowercase letters may be used to designate the exon-intron structure, transcribed regions, etc., if the lowercase letters remain legible at 100 bases/6 inches. Number the sequence line by line; place numerals, representing the first base of each line, to the left of the lines. **Minimize spacing between lines of sequence, leaving room only for annotation of the sequence.** Annotation may include boldface, underlining, brackets, boxes, etc. Encoded amino acid sequences may be presented, if necessary, immediately above or below the first nucleotide of each codon, by using the single-letter amino acid symbols. Comparisons of multiple nucleic acid sequences should conform as nearly as possible to the same format.

Figure Legends

Legends should provide enough information so that the figure is understandable without frequent reference to the text. However, detailed experimental methods must be described in the Materials and Methods section, not in a figure legend. A method that is unique to one of several experiments may be reported in a legend only if the discussion is very brief (one or two sentences). Define all symbols used in the figure and define all abbreviations that are not used in the text.

Tables

Type each table on a separate page. Arrange the data so that **columns of like material read down, not across**. The headings should be sufficiently clear so that the meaning of the data will be understandable without reference to the

TABLE 1. DISTRIBUTION OF PROTEIN AND ATPASE IN FRACTIONS OF DIALYZED MEMBRANES[a]

		ATPase	
Membranes	Fraction	U/mg of protein	Total U
Control	Depleted membrane	0.036	2.3
	Concentrated supernatant	0.134	4.82
E1 treated	Depleted membrane	0.034	1.98
	Concentrated supernatant	0.11	4.6

[a]Specific activities of ATPase of nondepleted membranes from control and treated bacteria were 0.21 and 0.20, respectively.

text. See "Abbreviations" in these instructions for those that should be used in tables. Explanatory footnotes are acceptable, but more extensive table "legends" are not. Footnotes should not include detailed descriptions of the experiment. Tables must include enough information to warrant table format; those with fewer than six pieces of data will be incorporated into the text by the copy editor. A well-constructed table is shown above.

Tables that can be photographically reproduced for publication without further typesetting or artwork are referred to as "camera ready." They should not be hand lettered and must be carefully prepared to conform with the style of the journal. The advantage of submitting camera-ready copy is that the material will appear exactly as envisioned by the author, and no second proof-reading is necessary. This is particularly advantageous when there are long, complicated tables and when the division of material and spacing are important.

Nomenclature

Chemical and Biochemical Nomenclature

The recognized authority for the names of chemical compounds is *Chemical Abstracts* (Chemical Abstracts Service, Ohio State University, Columbus) and its indexes. *The Merck Index*, 11th ed. (Merck & Co., Inc., Rahway, N.J., 1989), is also an excellent source. For biochemical terminology, including abbreviations and symbols, consult *Biochemical Nomenclature and Related Documents* (1978; reprinted for The Biochemical Society, London) and the instructions to authors of the *Journal of Biological Chemistry* and the *Archives of Biochemistry and Biophysics* (first issues of each year).

Do not express molecular weights in daltons; molecular weight is a unitless ratio. Molecular mass is expressed in daltons.

For enzymes, use the recommended (trivial) name assigned by the Nomenclature Committee of the International Union of Biochemistry (IUB) as described in *Enzyme Nomenclature* (Academic Press, Inc., New York, 1992). If a nonrecommended name is used, place the proper (trivial) name in parentheses at first use in the abstract and text. Use the EC number when one has been assigned, and express enzyme activity either in katals (preferred) or in the older system of micromoles per minute.

Nomenclature of Microorganisms

Binary names, consisting of a generic name and a specific epithet (e.g., *Escherichia coli*), must be used for all microorganisms. Names of categories above the genus level may be used alone, but specific and subspecific epithets may not. A specific epithet must be preceded by a generic name, written out in full the first time it is used in a paper. Thereafter, the generic name should be abbreviated to the initial capital letter (e.g., *E. coli*), provided there can be no confusion with other genera used in the paper. Names of all taxa (phyla, classes, orders, families, genera, species, subspecies) are printed in italics and should be underlined (or italicized) in the manuscript; strain designations and numbers are not.

The spelling of names should follow the *Approved Lists of Bacterial Names* (amended edition) (V.B.D. Skerman, V. McGowan, and P.H.A. Sneath, ed.) and the *Index of the Bacterial and Yeast Nomenclatural Changes Published in the International Journal of Systematic Bacteriology since the 1980 Approved Lists of Bacterial Names (1 January 1980 to 1 January 1989)* (W. E. C. Moore and L. V.

H. Moore, ed.), both published by the American Society for Microbiology in 1989, and the validation lists and articles published in the *International Journal of Systematic Bacteriology* since 1 January 1989. If there is reason to use a name that does not have standing in nomenclature, the name should be enclosed in quotation marks and an appropriate statement concerning the nomenclatural status of the name should be made in the text (for an example, see Int. J. Syst. Bacteriol. **30**: 547-556, 1980).

It is recommended that a strain be deposited in a recognized culture collection when that strain is necessary for the description of a new taxon (see *Bacteriological Code*, 1990 Revision, American Society for Microbiology, 1992).

Since the classification of fungi is not complete, it is the responsibility of the author to determine the accepted binomial for a given organism. Some sources for these names include *The Yeasts: a Taxonomic Study*, 3rd ed. (N. J. W. Kreger-van Rij, ed., Elsevier Science Publishers B.V., Amsterdam, 1984), and *Ainsworth and Bisby's Dictionary of the Fungi, Including the Lichens*, 7th ed. (Commonwealth Mycological Institute, Kew, Surrey, England, 1983).

Names used for viruses should be those approved by the International Committee on Taxonomy of Viruses (ICTV) and published in the 4th Report of the ICTV, Classification and Nomenclature of Viruses (Intervirology **17**:23–199, 1982), with the modifications contained in the 5th Report of the ICTV (Arch, Virol., Suppl. 2, 1991). If desired, synonyms may be added parenthetically when the name is first mentioned. Approved generic (or group) and family names may also be used.

Microorganisms, viruses, and plasmids should be given designations consisting of letters and serial numbers. It is generally advisable to include a worker's initials or a descriptive symbol of locale, laboratory, etc., in the designation. Each new strain, mutant, isolate, or derivative should be given a new (serial) designation. This designation should be distinct from those of the genotype and phenotype, and genotypic and phenotypic symbols should not be included.

Genetic Nomenclature

Bacteria. The genetic properties of bacteria are described in terms of phenotypes and genotypes. The phenotype describes the observable properties of an organism. The genotype refers to the genetic constitution of an organism, usually in reference to some standard wild type. Use the recommendations of Demerec et al. (Genetics **54**:61-76, 1966) as a guide to the use of these terms.

(i) Phenotypic designations must be used when mutant loci have not been identified or mapped. They can also be used to identify the protein product of a gene, e.g., the OmpA protein. Phenotypic designations generally consist of three-letter symbols; these are *not* italicized, and the first letter of the symbol is capitalized. It is preferable to use roman or arabic numerals (instead of letters) to identify a series of related phenotypes. Thus, a series of nucleic acid polymerase mutants might be designated Pol1, Pol2, Pol3, etc. Wild-type characteristics can be designated with a superscript plus (Pol^+), and, when necessary for clarity, negative superscripts (Pol^-) can be used to designate mutant characteristics. Lowercase superscript letters may be used to further delineate phenotypes (e.g., Str^s for streptomycin sensitivity). Phenotypic designations should be defined.

(ii) Genotypic designations are similarly indicated by three-letter locus symbols. In contrast to phenotypic designations, these are lowercase italic (e.g., *ara his rps*). If several loci govern related functions, these are distinguished by italicized capital letters following the locus symbol (e.g., *araA araB araC*). Promoter, terminator, and operator sites should be indicated as described by Bachmann and Low (Microbiol. Rev. **44**:1-56, 1980), e.g., *lacZp*, *lacAt*, and *lacZo*.

(iii) Wild-type alleles are indicated with a superscript plus (ara^+ his^+). A superscript minus is not used to indicate a mutant locus; thus, one refers to an *ara* mutant rather than an ara^- strain.

(iv) Mutation sites are designated by placing serial isolation numbers (allele numbers) after the locus symbol (e.g., *araA1 araA2*). If it is not known in which of several related loci the mutation has occurred, a hyphen is used instead of the capital letter (e.g., *ara-23*). It is essential in papers reporting the isolation of new mutants that allele numbers be given to the mutations. For *Escherichia coli*, there is a registry of such numbers: *E coli* Genetic Stock Center, Department of Biology, Yale University, New Haven, CT 06511-5188. For the genus *Salmonella*, the registry is *Salmonella* Genetic Stock Center, Department of Biology, University of Calgary, Calgary, Alberta, T2N 1N4 Canada. For the genus *Bacillus*, the registry is *Bacillus* Genetic Stock Center, Ohio State University, Columbus, OH 43210. A registry of allele numbers and insertion elements (omega [Ω] numbers) for chromosomal mutations and chromosomal insertions of transposons and other insertion elements has been established in conjunction with the ISP collection of *Staphylococcus aureus* at Iowa State University. Blocks of allele numbers and Ω numbers are assigned to laboratories on request. Blocks of numbers and additional information can be obtained from Peter A. Pattee, Department of Microbiology, Iowa State University, Ames, IA 50011. A registry of plasmid designations is maintained by E. Lederberg, Plasmid Reference Center, Department of Medical Microbiology and Immunology, 5402, Stanford University School of Medicine, Stanford, CA 94305-2499.

(v) The use of superscripts with genotypes (other than + to indicate wild-type alleles) should be avoided. Designations indicating amber mutations (Am), temperature-sensitive mutations (Ts), constitutive mutations (Con), cold-sensitive mutations (Cs), production of a hybrid protein (Hyb), and other important phenotypic properties should follow the allele number [e.g., *araA230* (Am) *hisD21*(Ts)]. All other such designations of phenotype must be defined at the first occurrence. If superscripts *must* be used, they must be approved by the editor and they must be defined at the first occurrence.

Subscripts may be used in two situations. Subscripts may be used to distinguish between genes (having the same name) from different organisms or strains, e.g., $his_{E.\ coli}$ or $his_{K\text{-}12}$ for the *his* genes of *E. coli* or strain K-12 in another species or strain, respectively. An abbreviation may also be used if it is explained. Similarly, a subscript is also used to distinguish between genetic elements that have the same name. For example, the promoters of the *gln* operon can be designated $glnAp_1$ and $glnAp_2$. This form departs slightly from that recommended by Bachmann and Low (e.g., *desC1p*).

(vi) Deletions are indicated by the symbol Δ placed before the deleted gene or region, e.g., Δ*trpA432*, Δ(*aroP-aceE*)419, or Δ*his*(*dhuA hisJ hisQ*)*1256*. Similarly, other symbols can be used (with appropriate definition). Thus, a fusion of the *ara* and *lac* operons can be shown as Φ(*ara-lac*)*95*. Likewise, Φ(*araB′-lacZ*$^+$)*96* indicates that the fusion results in a truncated *araB* gene fused to an intact *lacZ*, and Φ(*malE-lacZ*)*97*(Hyb) shows that a hybrid protein is synthesized. An inversion is shown as IN(*rrnD-rrnE*)*1*. An insertion of an *E. coli his* gene into plasmid pSC101 at zero kilobases (0 kb) is shown as pSC101 Ω(0kb::K-12*hisB*)*4*. An alternative designation of an insertion can be used in simple cases, e.g., *galT236*::Tn*5*. The number *236* refers to the locus of the insertion, and if the strain carries an additional *gal* mutation, it is listed separately. Additional examples, which utilize a slightly different format, can be found in the papers by Campbell et al. and Novick et al. cited below. It is important in reporting the construction of strains in which a mobile element was inserted and subsequently deleted that this latter fact be noted in the strain table. This can be done by listing the genotype of the strain used as an intermediate, in a table footnote or by a direct or parenthetical remark in the genotype. e.g., (F^-), ΔMu *c*ts, *mal*::ΔMu *c*ts::*lac*. In setting parenthetical remarks within the genotype or dividing the genotype into constituent elements, parentheses and square brackets are used without special meaning; square brackets are used outside parentheses. To indicate the presence of an episome, parentheses (or brackets) are used (λ, F^+). Reference to an integrated episome is indicated as described above for inserted elements, and an exogenote is shown as, for example, W3110/F′8(*gal*$^+$).

Any deviations from standard genetic nomenclature should be explained in Materials and Methods or in a table of strains. For information about the symbols in current use, consult Bachmann (B. J. Bachmann, p. 807–876, *in* J. L. Ingraham, K. B. Low, B. Magasanik, M. Schaechter, and H. E. Umbarger, ed., *Escherichia coli and Salmonella typhimurium*: *Cellular and Molecular Biology*, American Society for Microbiology, Washington, D.C., 1987) for *E. coli* K-12, Sanderson and Roth (Microbiol. Rev. **52:**485-532, 1988) for *Salmonella typhimurium*, Holloway et al. (Microbiol. Rev. **43:**73-102, 1979) for the genus *Pseudomonas*, Piggot and Hoch (Microbiol. Rev. **49:**158-179, 1985) for *Bacillus subtilis*, Perkins et al. (Microbiol. Rev. **46:**426-570, 1982) for *Neurospora crassa*, and Mortimer and Schild (Microbiol. Rev. **49:**181-213, 1985) for *Saccharomyces cerevisiae*. For yeasts, *Chlamydomanas* spp., and several fungal species, symbols such as those given in the *Handbook of Microbiology* (A. I. Laskin and H. A. Lechevalier, ed., CRC Press, Inc., 1974) should be used.

Conventions for Naming Genes. It is recommended that (entirely) new genes be given names that are mnemonics of their function, avoiding names that are already assigned and earlier or alternative gene names, irrespective of the bacterium for which such assignments have been made. Similarly, it is recommended that, whenever possible, homologous genes present in different organisms receive the same name. When homology is not apparent or the function of a new gene has not been established, a provisional name may be given by one of the following methods. (i) The gene may be named on the basis of its map location in the style *yaaA*, analogous to the style used for recording transposon insertions (*zef*) as discussed below. A registry of such names in use for *E. coli* is maintained by K. Rudd at the National Center for Biotechnology Information (fax, 301 480-9241; e-mail address, rudd@ncbi.nlm.nih.gov) and should be consulted. (ii) A provisional name may be given in the style described by Demerec et al. (e.g., *usg*, gene upstream of *folC*). Such names should be unique, and names such as *orf* or *genX* should not be used. For reference, the *E. coli* Genetic Stock Center's database includes an updated listing of *E. coli* gene names and gene products. It is accessible on Internet by Gopher (cgsc.biology.yale.edu) or Mosaic World Wide

Web (http://cgsc.biology.yale.edu/cgsc.html). The Center's relational database can also be searched via Telnet; for access, send a request to berlyn@cgsc.biology.yale.edu. A list can also be found in the work of Riley (Microbiol. Rev. **57**:862–952, 1993). For the genes of other bacteria, consult the references given above.

"Mutant" vs. "Mutation." Keep in mind the distinction between a *mutation* (an alteration of the primary sequence of the genetic material) and a *mutant* (a strain carrying one or more mutations). One may speak about the mapping of a mutation, but one cannot map a mutant. Likewise, a mutant has no genetic locus, only a phenotype.

Strain Designations. Do not use a genotype as a name (e.g., "subsequent use of *leuC6* for transduction"). If a strain designation has not been chosen, select an appropriate word combination (e.g., "another strain containing the *leuC6* mutation").

Viruses. The genetic nomenclature for viruses differs from that for bacteria. In most instances, viruses have no phenotype, since they have no metabolism outside host cells. Therefore, distinctions between phenotype and genotype cannot be made. Superscripts are used to indicate hybrid genomes. Genetic symbols may be one, two, or three letters. For example, a mutant strain of λ might be designated as λ *A*am11 *int*2 *red*114 *c*I857; this strain carries mutations in genes *c*I. *int*, and *red* and an amber-suppressible (am) mutation in gene *A*. A strain designated λ att^{434} imm^{21} would represent a hybrid of phage λ which carries the immunity region (*imm*) of phage 21 and the attachment (*att*) region of phage 434. Host DNA insertions into viruses should be delineated by square brackets, and the genetic symbols and designations for such inserted DNA should conform to those used for the host genome. Genetic symbols for phage λ can be found in Szybalski and Szybalski (Gene **7**:217–270, 1979) and in Echols and Murialdo (Microbiol. Rev. **42**:577–591, 1978).

Transposable Elements, Plasmids, and Restriction Enzymes. Nomenclature of transposable elements (insertion sequences, transposons, phage Mu, etc.) should follow the recommendations of Campbell et al. (Gene **5**:197–206, 1979), with the modifications given in section vi. The system of designating transposon insertions at sites where there are no known loci, e.g., *zef123*::Tn5, has been described by Chumley et al. (Genetics **91**:639–655, 1979). The nomenclature recommendations of Novick et al. (Bacteriol. Rev. **40**:168–189, 1976) for plasmids and plasmid-specified activities, of Low (Bacteriol. Rev. **36**:587–607, 1972) for F-prime factors, and of Roberts (Nucleic Acids Res. **17**:r347–r387, 1989) for restriction enzymes and their isoschizomers should be used when possible. The nomenclature for recombinant DNA molecules constructed in vitro follows the nomenclature for insertions in general. DNA inserted into recombinant DNA molecules should be described by using the gene symbols and conventions for the organism from which the DNA was obtained. The Plasmid Reference Center (E. Lederberg, Plasmid Reference Center, Department of Medical Microbiology and Immunology, 5402, Stanford University School of Medicine, Stanford, CA 94305-2499) assigns Tn and IS numbers to avoid conflicting and repetitive use and also clears nonconflicting plasmid prefix designations.

Tetracycline Resistance Determinants. The nomenclature for tetracycline resistance determinants is based on the proposal of Levy et al. (Antimicrob. Agents Chemother. **33:**1373–1374, 1989). The style for such determinants is, e.g., Tet B; the space helps distinguish the determinant designation from that for phenotypes and proteins (TetB). Table 2 of the above-referenced article shows the correct format for genes, proteins, and determinants in this family.

Abbreviations and Conventions

Verb Tense

ASM strongly recommends that for clarity you use the **past** tense to narrate particular events in the past, including the procedures, observations, and data of the study that you are reporting. Use the present tense for your own general conclusions, the conclusions of previous researchers, and generally accepted facts. Thus, most of the abstract, Materials and Methods, and Results will be in the past tense, and most of the introduction and some of the Discussion will be in the present tense.

Be aware that it may be necessary to vary the tense in a single sentence. For example, it is correct to say "White (30) demonstrat*ed* that XYZ cells *grow* at pH 6.8," "Figure 2 shows that ABC cells fail*ed* to grow at room temperature," and "Air *was* removed from the chamber and the mice *died*, which *proves* that mice *require* air." In reporting statistics and calculations, it is correct to say "The values for the ABC cells *are* statistically significant, indicating that the drug inhibit*ed* . . ."

For an in-depth discussion of tense in scientific writing, see p. 164–166 in *How to Write and Publish a Scientific Paper*, 4th ed.

Abbreviations

General. Abbreviations should be used as an aid to the reader rather than as a convenience to the author, and therefore their **use should be limited**. Abbreviations other than those recommended by the IUPAC-IUB (*Biochemical Nomenclature and Related Documents*, 1978) should be used only when a case can be made for necessity, such as in tables and figures.

It is often possible to use pronouns or to paraphrase a long word after its first use (e.g., "the drug," "the substrate"). Standard chemical symbols and trivial names or their symbols (folate, Ala, Leu, etc.) may also be used.

It is strongly recommended that all abbreviations except those listed below be introduced in the first paragraph in Materials and Methods. Alternatively, define each abbreviation and introduce it in parentheses the first time it is used; e.g., "cultures were grown in Eagle minimal essential medium (MEM)." Generally, eliminate abbreviations that are not used at least five times in the text (including tables and figure legends).

Not Requiring Introduction. In addition to abbreviations for Système International d'Unités (SI) units of measurement, other common units (e.g., bp, kb, and Da), and chemical symbols for the elements, the following should be used without definition in the title, abstract, text, figure legends, and tables: DNA (deoxyribonucleic acid); cDNA (complementary DNA); RNA (ribonucleic acid); cRNA (complementary RNA); RNase (ribonuclease); DNase (deoxyribonuclease); rRNA (ribosomal RNA); mRNA (messenger RNA);

tRNA (transfer RNA); AMP, ADP, ATP, dAMP, ddATP, GTP, etc. (for the respective 5′ phosphates of adenosine and other nucleosides) (add 2′-, 3′-, or 5′- when needed for contrast); ATPase, dGTPase, etc. (adenosine triphosphatase, deoxyguanosine triphosphatase, etc.); NAD (nicotinamide adenine dinucleotide); NAD^+ (nicotinamide adenine dinucleotide, oxidized); NADH (nicotinamide adenine dinucleotide, reduced); NADP (nicotinamide adenine dinucleotide phosphate); NADPH (nicotinamide adenine dinucleotide phosphate, reduced); $NADP^+$ (nicotinamide adenine dinucleotide phosphate, oxidized); poly(A), poly(dT), etc. (polyadenylic acid, polydeoxythymidylic acid, etc.); oligo(dT), etc. (oligodeoxythymidylic acid, etc.); P_i (orthophosphate); PP_i (pyrophosphate); UV (ultraviolet); PFU (plaque-forming units); CFU (colony-forming units); MIC (minimal inhibitory concentration); MBC (minimal bactericidal concentration); Tris [tris(hydroxymethyl) aminomethane]; DEAE (diethylaminoethyl); A_{260} (absorbance at 260 nm); EDTA (ethylenediaminetetraacetic acid); PCR (polymerase chain reaction); and AIDS (acquired immunodeficiency syndrome). Abbreviations for cell lines (e.g., HeLa) also need not be defined.

The following abbreviations should be used without definition in tables:

amt (amount)
approx (approximately)
avg (average)
concn (concentration)
diam (diameter)
expt (experiment)
exptl (experimental)
ht (height)
mo (month)
mol wt (molecular weight)
no. (number)
prepn (preparation)
SD (standard deviation)
SE (standard error)
SEM (standard error of the mean)
sp act (specific activity)
sp gr (specific gravity)
temp (temperature)
tr (trace)
vol (volume)
vs (versus)
wk (week)
wt (weight)
yr (year)

Reporting Numerical Data

Standard metric units are used for reporting length, weight, and volume. For these units and for molarity, use the prefixes m, μ, n, and p for 10^{-3}, 10^{-6}, 10^{-9}, and 10^{-12}, respectively. Likewise, use the prefix k for 10^3. Avoid compound prefixes such as mμ or $\mu\mu$. Parts per million (ppm) may be used when that is the common measure for the science in that field. Units of temperature are presented as follows: 37ºC or 324 K.

When fractions are used to express such units as enzymatic activities, it is preferable to use whole units, such as g or min, in the denominator instead of fractional or multiple units, such as μg or 10 min. For example, "pmol/min" is preferable to "nmol/10 min," and "μmol/g" is preferable to "nmol/μg." It is also preferable that an unambiguous form, such as exponential notation, be used; for example, "μmol g^{-1} min^{-1}" is preferable to "μmol/g/min." Always report numerical data in the applicable SI units.

Representation of data as accurate to more than two significant figures must be justified by presentation of appropriate statistical analyses.

Statistics

If biological variation within a treatment (coefficient of variation, the standard deviation divided by the mean) is small (less than 10%) and the difference among treatment means is large (greater than 3 standard deviations), it is not necessary to report statistics. If the data do not meet these criteria, however, the authors must include an appropriate statistical analysis (e.g., Student's *t* test, analysis of variance, Tukey's test, etc.). Statistics should represent the variation among biological units (e.g., replicate incubations) and not just the variation due to method of analysis.

Equations

In mathematical equations, indicate the order of operations clearly by enclosing operations in parentheses, brackets, and braces, in that order: $(a + b) \times c$ or $a + (b \times c)$, $100 \times \{[(a/b) \times c] + d\}$ or $100 \times \{a/[(b \times c) + d]\}$. Italicize (by underlining) variables and constants (but not numerals), and use roman type for designations: E_0, E_h, M_r, K_m, K_s, $a + 2b = 1.2$ mM, $\mathrm{Ca}^{2+}V_\mathrm{max} = \exp(1.5x + y)$, $\mathrm{BOD} = 2.7x^2$.

Isotopically Labeled Compounds

For simple molecules, isotopic labeling is indicated in the chemical formula (e.g., $^{14}CO_2$, $^{3}H_2$, $H^{35}SO_4$). Brackets are not used when the isotopic symbol is attached to the name of a compound that in its natural state does not contain the element (e.g., ^{32}S-ATP) or to a word that is not a specific chemical name (e.g., ^{131}I-labeled protein, ^{14}C-amino acids, ^{3}H-ligands, etc.).

For specific chemicals, the symbol for the isotope introduced is placed in brackets directly preceding the part of the name that describes the labeled entity. Note that configuration symbols and modifiers precede the isotopic symbol. The following examples illustrate correct usage.

[^{14}C]urea	[γ-^{32}P]ATP
L-[*methyl*-^{14}C]methionine	UDP-[U-^{14}C]glucose
[2,3-^{3}H]serine	*E. coli* [^{32}P]DNA
[α-^{14}C]lysine	fructose 1,6-[1-^{32}P]bisphosphate

AEM follows the same conventions for isotopic labeling as the *Journal of Biological Chemistry*, and more detailed information can be found in the instructions to authors of that journal (first issue of each year).

For your convenience, these Instructions are also available on the ASM HomePage through the Internet at the following address: http://www.asmusa.org.

APPENDIX I

pH Indicators Commonly Used in Microbiology

Indicator	Acid Color	Alkaline Color	pH range
Brilliant Green	Yellow	Green	0.0–2.6
Thymol Blue	Red	Yellow	1.2–2.8
Bromophenol Blue	Yellow	Blue	3.0–4.6
Bromocresol Green	Yellow	Blue	4.4–6.4
Methyl Red	Red	Yellow	4.4–6.4
Litmus	Red	Blue	4.5–8.3
Bromocresol Purple	Yellow	Purple	5.2–6.8
Bromothymol Blue	Yellow	Blue	6.0–7.6
Neutral Red	Red	Amber	6.8–8.4
Phenol Red	Yellow	Red	6.8–8.4
Cresol Red	Yellow	Red	7.2–8.8
Thymol Blue	Yellow	Blue	8.0–9.6
Phenolphthalein	Colorless	Red	8.3–10.0
Alizarin	Yellow	Red	10.1–12.0

APPENDIX J

API-20E Information*

SUMMARY OF RESULTS—18–24-HOUR PROCEDURE

Tube	Interpretation of Reactions		
		Positive	**Negative**
ONPG		Yellow	Colorless
ADH	Incubation 18–24 hr	Red or orange	Yellow
	36–48 hr	Red	Yellow or orange
LDC	18–24 hr	Red or orange	Yellow
	36–48 hr	Red	Yellow or orange
ODC	18–24 hr	Red or orange	Yellow
	36–48 hr	Red	Yellow or orange
CIT		Turquoise or dark blue	Light green or yellow
H_2S		Black deposit	No black deposit
URE	18–24 hr	Red or orange	Yellow
	36–48 hr	Red	Yellow or orange
TDA		Add 1 drop 10% Ferric chloride	
		Brown-red	Yellow
IND		Add 1 drop Kovacs' reagent	
		Red ring	Yellow
VP		Add 1 drop of 40% potassium hydroxide, then 1 drop of alpha-naphthol	
		Red	Colorless
GEL		Diffusion of the pigment	No diffusion
GLU		Yellow or gray	Blue or blue-green
MAN INO SOR RHA SAC MEL AMY ARA		Yellow	Blue or blue-green
GLU		After reading GLU reaction, add 2 drops 0.8% sulfanilic acid and 2 drops 0.5% N, N-dimethyl-alpha-naphthoylamine	
Nitrate reduction	NO_2 N_2 gas	Red Bubbles: yellow after reagents and zinc	Yellow orange after reagents and zinc
MAN INO SOR Catalase		After reading carbohydrate reaction, add 1 drop 1.5% H_2O_2	
		Bubbles	No bubbles

*Courtesy of bioMérieux Vitek.

Summary of Results—18–24 Hour Procedure (continued)

Interpretation of Reactions
Comments
(1) Any shade of yellow is a positive reaction. (2) VP tube, before the addition of reagents, can be used as a negative control.
Orange reactions occurring at 36–48 hr should be interpreted as negative.
Any shade of orange within 18–24 hr is a positive reaction. At 36–48 hr, orange decarboxylase reactions should be interpreted as negative.
Orange reactions occurring at 36–48 hr should be interpreted as negative.
(1) Both the tube and cupule should be filled. (2) Reaction is read in the aerobic (cupule) area.
(1) H_2S production may range from a heavy black deposit to a very thin black line around the tube bottom. Carefully examine the bottom of the tube before considering the reaction negative. (2) A "browning" of the medium is a negative reaction unless a black deposit is present. "Browning" occurs with TDA positive organisms.
A method of lower sensitivity has been chosen. *Klebsiella*, *Proteus* and *Yersinia*, routinely give positive reactions.
(1) Immediate reaction. (2) Indole positive organisms may produce a golden orange color due to indole production. This is a negative reaction.
(1) The reaction should be read within 2 min after the addition of the Kovacs' reagent and the results recorded. (2) After several minutes, the HCI present in Kovacs' reagent may react with the plastic of the cupule resulting in a change from negative (yellow) color to a brownish-red. This is a negative reaction.
(1) Wait 10 min before considering the reaction negative. (2) A pale pink color (after 10 min) should be interpreted as negative. A pale pink color which appears immediately after the addition of reagents but which turns dark pink or red after 10 min should be interpreted as positive.
Motility may be observed by hanging drop or wet mount preparation.
(1) The solid gelatin particles may spread throughout the tube after inoculation. Unless diffusion occurs, the reaction is negative. (2) Any degree of diffusion is a positive reaction.
Comments for All Carbohydrates **Fermentation** (*Enterobacteriaceae, Aeromonas, Vibrio*) (1) Fermentation of the carbohydrates begins in the most anaerobic portion (bottom) of the tube. Therefore, these reactions should be read from the bottom of the tube to the top. (2) A yellow color at the bottom of the tube only indicates a weak or delayed positive reaction. **Oxidation** (Other Gram-negatives) (1) Oxidative utilization of the carbohydrates begins in the most aerobic portion (top) of the tube. Therefore, these reactions should be read from the top to the bottom of the tube. (2) A yellow color in the upper portion of the tube and a blue in the bottom of the tube indicates oxidative utilization of the sugar. This reaction should be considered positive **only** for non-*Enterobacteriaceae* gram-negative rods. This is a negative reaction for fermentative organisms such as *Enterobacteriaceae*.
(1) Before addition of reagents, observe GLU tube (positive or negative) for bubbles. Bubbles are indicative of reduction of nitrate to the nitrogenous (N_2) state. (2) A positive reaction may take 2–3 min for the red color to appear. (3) Confirm a negative test by adding zinc dust or 20 mesh granular zinc. A pink-orange color after 10 min confirms a negative reaction. A yellow color indicates reduction of nitrates to the nitrogenous (N_2) state.
(1) Bubbles may take 1–2 min to appear. (2) Best results will be obtained if the test is run in tubes which have no gas from fermentation.

SUMMARY OF CHEMICAL AND PHYSICAL PRINCIPLES OF THE TESTS ON THE API-20E

Tube	Chemical/Physical Principles	Components		Ref.
		Reactive Ingredients	Quantity	
ONPG	Hydrolysis of ONPG by beta-galactosidase releases yellow orthonitrophenol from the colorless ONPG; IPTG (isopropylthiogalactopyranoside) is used as inducer.	ONPG IPTG	0.2 mg 8.0 μg	12 13 14
ADH	Arginine dihydrolase transforms arginine into ornithine, ammonia and carbon dioxide. This causes a pH rise in the acid-buffered system and a change in the indicator from yellow to red.	Arginine	2.0 mg	15
LDC	Lysine decarboxylase transforms lysine into a basic primary amine, cadaverine. This amine causes a pH rise in the acid-buffered system and a change in the indicator from yellow to red.	Lysine	2.0 mg	15
ODC	Ornithine decarboxylase transforms ornithine into a basic primary amine, putrescine. This amine causes a pH rise in the acid-buffered system and a change in the indicator from yellow to red.	Ornithine	2.0 mg	15
CIT	Citrate is the sole carbon source. Citrate utilization results in a pH rise and a change in the indicator from green to blue.	Sodium citrate	0.8 mg	21
H_2S	Hydrogen sulfide is produced from thiosulfate. The hydrogen sulfide reacts with iron salts to produce a black precipitate.	Sodium thiosulfate	80.0 μg	6
URE	Urease releases ammonia from urea: ammonia causes the pH to rise and changes the indicator from yellow to red.	Urea	0.8 mg	7
TDA	Tryptophane deaminase forms indolepyruvic acid from tryptophane. Indolepyruvic acid produces a brownish-red color in the presence of ferric chloride.	Tryptophane	0.4 mg	22
IND	Metabolism of tryptophane results in the formation of indole. Kovacs' reagent forms a colored complex (pink to red) with indole.	Tryptophane	0.2 mg	10
VP	Acetoin, an intermediary glucose metabolite, is produced from sodium pyruvate and indicated by the formation of a colored complex. Conventional VP tests may take up to 4 days, but by using sodium pyruvate, API has shortened the required test time. Creatine intensifies the color when tests are positive.	Sodium pyruvate Creatine	2.0 mg 0.9 mg	3
GEL	Liquefaction of gelatin by proteolytic enzymes releases a black pigment which diffuses throughout the tube.	Kohn charcoal gelatin	0.6 mg	9
GLU MAN INO SOR RHA SAC MEL AMY ARA	Utilization of the carbohydrate results in acid formation and a consequent pH drop. The indicator changes from blue to yellow.	Glucose Mannitol Inositol Sorbitol Rhamnose Sucrose Melibiose Amygdalin (L^+) Arabinose	2.0 mg 2.0 mg 2.0 mg 2.0 mg 2.0 mg 2.0 mg 2.0 mg 2.0 mg 2.0 mg	5 6 12
GLU Nitrate reduction	Nitrites form a red complex with sulfanilic acid and N, N-dimethyl-alpha-naphthylamine. In case of negative reaction, addition of zinc confirms the presence of unreduced nitrates by reducing them to nitrites (pink-orange color). If there is no color change after the addition of zinc, this is indicative of the complete reduction of nitrates through nitrites to nitrogen gas or to an anaerogenic amine.	Potassium nitrate	80.0 μg	6
MAN INO SOR Catalase	Catalase releases oxygen gas from hydrogen peroxide.			24

APPENDIX K

API-NFT Information*

Recommended Methodology

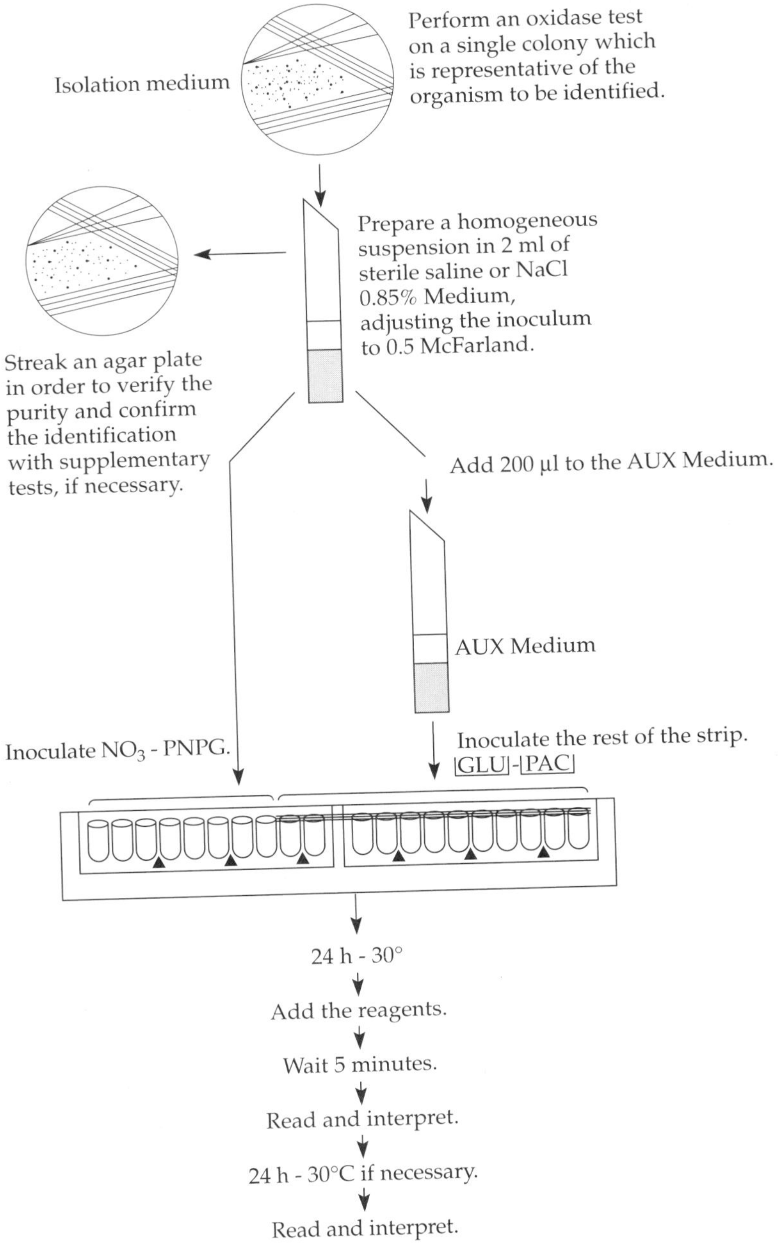

*Courtesy of bioMérieux Vitek.

Interpretation of Reactions

Biochemical Tests	Reagent(s) to be Added	Interpretation: Positive	Interpretation: Negative	Physio-Chemical Principles	Reactive Ingredient	Quantity
NO_3 (Nitrate reduction)	1 drop each of NIT 1 and NIT 2 reagents. Read reaction within 5 min.	Pale pink to dark red	Colorless	Reduction of NO_3 to NO_2. A negative result should be confirmed by adding zinc powder to the cupule.	KNO_3	0.12 mg
	Zinc dust. Read reaction within 5 min.	Colorless	Pink	Reduction of NO_3 to N_2. A pink coloration indicates that no nitrogen gas is being produced.		
TRP (Tryptophanase)	1 drop of JAMES reagent. Immediate reaction.	Pink	Colorless pale green to yellow	The metabolism of tryptophane results in the formation of indole by tryptophanase, which forms a colored complex (pink) with JAMES reagent.	Tryptophane	0.4 mg
<u>GLU</u> (Glucose fermentation)		Yellow	Green to blue	Carbohydrate fermentation forms acids which lower the pH and bring about a color change in the indicator (brom thymol blue).	Glucose	2.0 mg
<u>ADH</u> (Arginine dihydrolase)		Orange to pink-red*	Yellow	Arginine is transformed into ornithine, ammonia and carbon dioxide. An increase in the pH causes a color change in the indicator (phenol red).	Arginine	2.0 mg
<u>URE</u> (Urease)		Orange to pink-red*	Yellow	Urease releases ammonia from urea, causing a change in color of the pH indicator (phenol red).	Urea	0.8 mg
ESC (Esculin hydrolysis)		Grey, brown, black	Yellow	Esculin hydrolized by beta-glucosidase into glucose and esculetin which reacts with an iron salt to give a black coloration.	Esculin	0.56 mg
GEL (Gelatinase)		Diffusion of the pigment	No diffusion	Proteolytic enzymes cause liquefaction of the gelatin and allow charcoal to diffuse.	Kohn Charcoal Gelatin	0.6 mg
PNPG (Beta-galactosidase)		Yellow	Colorless	Hydrolysis of PNPG (p-nitro-phenyl-beta-galactopyranoside) by beta-galactosidase release yellow para-nitrophenol. IPTG (isopropylthiogalacto-pyranoside) is used as an inducer.	PNPG IPTG	0.2 mg 8.0 μg

*(Weak orange occurring after 48 hours should be interpreted as negative)

Interpretation of Reactions

Assimilation Tests	Reagent(s) to be Added	Interpretation Positive	Interpretation Negative	Physio-Chemical Principles	Reactive Ingredient	Quantity
GLU D-Glucose		Growth	No growth	Growth depends upon the assimilation of a single carbon source in the presence of ammonium sulfate in the AUX medium. Assimilation is observed as opacity in the open portion (Fig. 1) of the cupule.	D-Glucose	1.5 mg
ARA L-Arabinose		Growth	No growth		L-Arabinose	1.4 mg
MNE D-Mannose		Growth	No growth		D-Mannose	1.4 mg
MAN D-Mannitol		Growth	No growth		D-Mannitol	1.4 mg
NAG N-Acetyl-D glucosamine		Growth	No growth		N-Acetyl-D glucosamine	1.3 mg
MAL Maltose		Growth	No growth		Maltose	1.4 mg
GNT D-Gluconate		Growth	No growth		D-Gluconic acid	1.8 mg
CAP Caprate		Growth	No growth		Capric acid	0.8 mg
ADI Adipate		Growth	No growth		Adipic acid	1.1 mg
MLT L-Malate		Growth	No growth		L-Malic	1.5 mg
CIT Citrate		Growth	No growth		Citric acid	2.3 mg
PAC Phenyl-acetate		Growth	No growth		Phenylacetic acid	0.8 mg

APPENDIX L

Enterotube II Information*

Reactions			Remarks
	Negative	**Positive**	**Compartment 1**
Glucose	Red	Yellow	**Glucose (GLU):** The end products of bacterial fermentation of glucose are either acid, or acid and gas. The shift in pH due to the production of acid is indicated by change in the color of the indicator in the medium from red (alkaline) to yellow (acidic). Any degree of yellow should be interpreted as a positive reaction; orange should be considered negative.
Gas production	Wax not lifted	Wax lifted	**Gas production (GAS):** This is evidenced by definite and complete separation of the wax overlay from the surface of the glucose medium, but not by bubbles in the medium. Since the amount of gas produced by different bacteria varies, the amount of separation between medium and overlay will also vary with the strain being tested.
Lysine	Yellow	Purple	**Compartment 2** **Lysine decarboxylase (LYS):** Bacterial decarboxylation of lysine, which results in the formation of the alkaline end product cadaverine, is indicated by a change in the color of the indicator in the medium from pale yellow (acidic) to purple (alkaline). Any degree of purple should be interpreted as a positive reaction. The medium remains yellow if decarboxylation of lysine does not occur.
Ornithine	Yellow	Purple	**Compartment 3** **Ornithine decarboxylase (ORN):** Bacterial decarboxylation of ornithine, which results in the formation of the alkaline end product putrescine, is indicated by a change in the color of the indicator in the medium from pale yellow (acidic) to purple (alkaline). Any degree of purple should be interpreted as a positive reaction. The medium remains yellow if decarboxylation of ornithine does not occur.
H_2S	Beige	Black	**Compartment 4** **H_2S production (H_2S):** Hydrogen sulfide is produced by bacteria capable of reducing sulfur-containing compounds, such as peptones and sodium thiosulfate present in the medium. The hydrogen sulfide reacts with the iron salts also present in the medium to form a black precipitate of ferric sulfide usually along the line of inoculation. Some *Proteus* and *Providencia* strains may produce a diffuse brown coloration in this medium, however, this should not be confused with true H_2S production, i.e., presence of black color. **NOTE:** The black precipitate may fade or revert back to negative if **Enterotube® II** is read after 24 hr incubation.
Indole formation	Colorless	Pink-red	**Indole formation (IND):** The production of indole from the metabolism of tryptophan by the bacterial enzyme tryptophanase is detected by the development of a pink to red color after the addition of Kovacs' indole reagent, which is injected into the compartment after 18 to 24 hr incubation of the tube. (See final step in "Procedure" section.)
Adonitol	Red	Yellow	**Compartment 5** **Adonitol (ADON):** Bacterial fermentation of adonitol, which results in the formation of acidic end products, is indicated by a change in color of the indicator present in the medium from red (alkaline) to yellow (acidic). Any sign of yellow should be interpreted as a positive reaction; orange should be considered negative.

*Courtesy of Becton Dickinson and Company.

Reactions			Remarks
	Negative	**Positive**	**Compartment 6**
Lactose	Red	Yellow	**Lactose (LAC):** Bacterial fermentation of lactose, which results in the formation of acidic end products, is indicated by a change in color of the indicator present in the medium from red (alkaline) to yellow (acidic). Any sign of yellow should be interpreted as a positive reaction; orange should be considered negative. This test is useful to confirm the lactose reaction of colonies taken from various enteric differential media, e.g., *Salmonella-Shigella* (SS), MacConkey (MAC), Eosin Methylene Blue (EMB).
Arabinose	Red	Yellow	**Compartment 7** **Arabinose (ARAB):** Bacterial fermentation of arabinose, which results in the formation of acidic end products, is indicated by a change in color of the indicator present in the medium from red (alkaline) to yellow (acidic). Any sign of yellow should be interpreted as a positive reaction; orange should be considered negative.
Sorbitol	Red	Yellow	**Compartment 8** **Sorbitol (SORB):** Bacterial fermentation of sorbitol, which results in the formation of acidic end products, is indicated by a change in color of indicator present in the medium from red (alkaline) to yellow (acidic). Any sign of yellow should be interpreted as a positive reaction; orange should be considered negative.
Voges-Proskauer	Colorless	Red	**Compartment 9** **Voges-Proskauer (VP):** Acetylmethylcarbinol (acetoin) is an intermediate in the production of butylene glycol from glucose fermentation. The production of acetoin is detected by the addition of two drops of a 20% w/v aqueous solution of potassium hydroxide containing 0.3% w/v of creatine and three drops of a 5% w/v solution of alphanaphthol in absolute ethyl alcohol. The presence of acetoin is indicated by the development of a red color within 20 min. However, most positive reactions are evident within 10 min.
Dulcitol	Green	Yellow	**Compartment 10** **Dulcitol (DUL):** Bacterial fermentation of dulcitol, which results in the formation of acidic end products, is indicated by a change in color of the indicator present in the medium from green (alkaline) to yellow or pale yellow (acidic).
PA	Green	Black-smoky-gray	**Phenylalanine deaminase (PA):** This test detects the formation of pyruvic acid from the deamination of phenylalanine. The pyruvic acid formed reacts with a ferric salt present in the medium to produce a characteristic black to smoky gray color. Ferric chloride need not be added since the medium already contains an iron salt.
Urea	Beige	Red-purple	**Compartment 11** **Urea (UREA):** Urease, an enzyme possessed by various microorganisms, hydrolyzes urea to ammonia causing the color of the indicator in the medium to shift from yellow (acidic) to red-purple (alkaline). The urease test is strongly positive for *Proteus* species and may be evident as early as 4 to 6 hr after incubation; it is weakly positive (light pink color) after 18 to 24 hr incubation for *Klebsiella* and *Enterobacter* species.

Reactions			Remarks
	Negative	**Positive**	**Compartment 12**
Citrate	Green	Blue	**Citrate (CIT):** This test detects those organisms which are capable of utilizing citrate, in the form of its sodium salt, as the sole source of carbon. Organisms capable of utilizing citrate produce alkaline metabolites which change the color of the indicator from green (acidic) to deep blue (alkaline). Any degree of blue should be considered positive. **NOTE:** Certain microorganisms will not always produce the ideal "strong" positive color change. Lighter shades of the same basic color should also be considered positive.

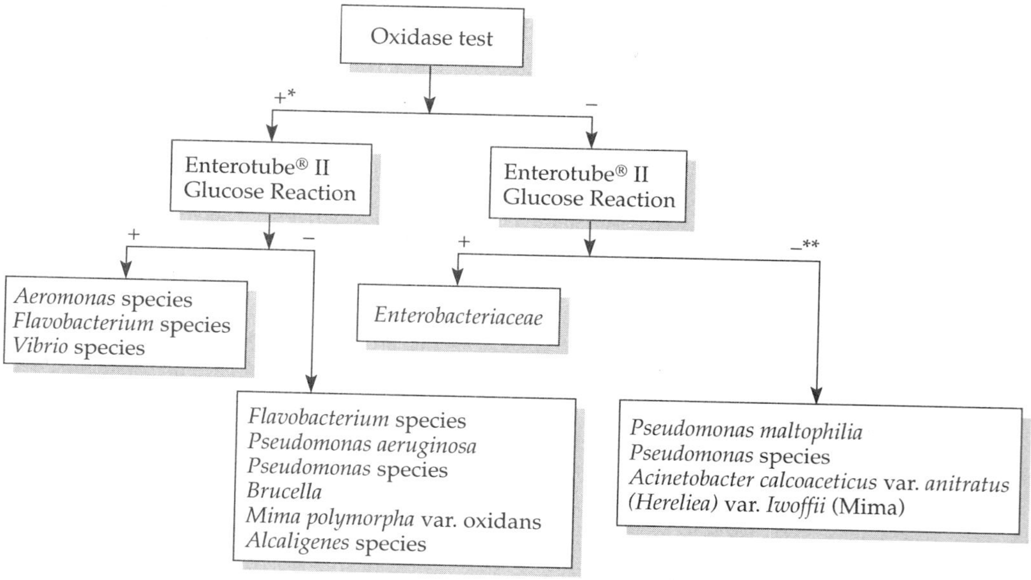

* This side of the flow diagram may be found useful if an oxidase-positive is inadvertently inocculated into the Enterotube® II. It is suggested that Oxi/Ferm® Tube be used for their identification.
** Common biotypes of this group of organisms may be identified with Enterotube® II. Check section entitled "Oxidase-negative-non-fermenters, biocodes," located in the CCIS.

APPENDIX M

Semester-Length Laboratory Projects

An endospore-forming thermophilic bacteria.
Brock, T.D. 1978. *Thermophilic Microorganisms and Life at High Temperatures.* New York: Springer-Verlag.

A Member of the genus *Nitrobacter* or *Nitrosomonas.*
Belser. 1979. "Nitrifying Bacteria", in Starr, *Annual Review of Microbiology 33*: 310–329.

A member of the genus *Caulobacter.*
Poindexter, J.S. 1981. "The Caulobacters: Ubiquitous Unusual Bacteria." *Microbiological Reviews 45*: 123–179.

A member of the genus *Thiobacillus.*
Hutchins, Davidson, Brierley, and Brierley. 1986. "Microorganisms in Reclamation of Metals." *Annual Review of Microbiology 40*: 311–336.

An alginate-degrading bacterium.
Kitamikado, M., et al. 1990. "Method Designed to Detect Alginate-Degrading Bacteria." *Applied and Environmental Microbiology 56*: 2939–2940.

A bacterial virus.
Farrah, Samuel R. 1987. "Ecology of Phage in Freshwater Environments," *in* Goyal, S.M. et al., *Phage Ecology*. New York: John Wiley and Sons, pp. 125–136.

An oligotrophic microorganism.
Fry, J.C. 1990. "Oligotrophs," *in* Edwards, Clive (ed.), *Microbiology of Extreme Environments*. New York: McGraw-Hill, pp. 93–116.

Bacteria from an invertebrate.
West, P.A. 1988. "Bacteria of Aquatic Invertebrates," *in* Austin, B. (ed.), *Methods in Aquatic Bacteriology*. New York: John Wiley and Sons, pp. 143–170.

A cyanobacterium.
Daft, M.J. 1988. "Cyanobacteria: Isolation, Interactions and Ecology," in Austin, B. (ed.), *Methods in Aquatic Bacteriology*. New York: John Wiley and Sons, pp. 241–268.

A microbial contaminant in a pharmaceutical, cosmetic, or toiletry product.
Baird, R.M. 1988. "Microbiological Contamination of Manufactured Products: Official and Unofficial Limits," in Bloomfield, S.F. et al. (ed.), *Microbial Quality Assurance in Pharmaceuticals, Cosmetics and Toiletries.* New York: Halsted Press, pp. 61–76.

Microoganism in food.
Smith, H.R. et al. 1991 "Examination of Retail Chickens and Sausages in Britain for Vero Cytotoxin-Producing *Escherichia coli." Applied and Environmental Microbiology 57*: 2091–2093.

Formic acid degrading organism.
Pronk, J.T. et al. 1991. "Growth of *Thiobacillus ferrooxidans* on Formic Acid." *Applied and Environmental Microbiology 57*: 2057–2062.

A sulfate-reducing bacterium.

Battersby, N.S. "Sulfate-Reducing Bacteria," *in* Austin, B. (ed.), *Methods in Aquatic Bacteriology*. New York: John Wiley and Sons, pp. 269–299.

A cellulose decomposer.

Schneider, J. and G. Rheinheimer. 1988. "Isolation Methods," in Austin, B. (ed.), *Methods in Aquatic Bacteriology*. New York: John Wiley and Sons, pp. 73–94.

A magnetotactic organism.

Mann, Stephen, et al. 1990. "Magnetotactic Bacteria: Microbiology, Biomineralization, Palaemagnetism and Biotechnology," *in* Rose, A.H. and D.W. Tempest (eds.), *Advances in Microbial Physiology 31*: 125–181.

Bacteriophage recovered from food.

Kennedy, James E. and Gabriel Bitton. 1987. "Bacteriophages in Foods," in Goyal, S.M., et al. (eds.), *Phage Ecology*. New York: John Wiley and Sons, pp. 289–316.

An antibiotic-producing organism.

Parkinson, D., T.R.G. Gray and S. T. Williams (eds.). 1971. *Methods for Studying the Ecology of Soil Microorganisms*. Oxford, England: Blackwell Scientific Publications.

A thermophile.

Lacey, J. 1990. "Isolation of Thermophilic Microorganisms," *in* Labeda, David P., *Isolation of Biotechnological Organisms from Nature*, New York: McGraw-Hill, pp. 141–181.

A member of the genus *Lactobacillus*.

Swain, Roger. 1979. "The Cultured Cabbage." *Horticulture 57*: 16–18.

Pediococcus.

Whiting, Michael, et al. 1992. "Detection of *Pediococcus sp.* in Brewing Yeast by a Rapid Immunoassay." *Applied and Environmental Microbiology 58*: 713–716.

Organism capable of degrading EDTA.

Nortemann, Bernd. 1992. "Total Degradation of EDTA by Mixed Cultures and a Bacterial Isolate." *Applied and Environmental Microbiology 58*: 671–676.

A psychrophile.

Wiebe, W.J. et al. 1992. "Bacterial Growth in the Cold: Evidence for an Enhanced Substrate Requirement." *Applied and Environmental Microbiology 58*: 359–364.

A methane-producing organism.

Zeikus, J.G. 1977. "The Biology of Methanogenic Bacteria." *Bacteriological Reviews 41*: 514–541.

A bacteriophage from the environment.

Asghari, Abdolkarim et al. 1992. "Use of Hydrogen Peroxide Treatment and Crystal Violet Agar Plates for Selective Recovery of Bacteriophages from Natural Environments." *Applied and Environmental Microbiology 58*: 1159–1163.

A naphthalene-utilizing organism.

Wheelis, M.L. 1975. "The Genetics of Dissimilarity Pathways in *Pseudomonas*." *Annual Review of Microbiology 29*: 505–524.

Fecal coliform.

Curis, Thomas P. et al. 1992. "Influence of pH, Oxygen, and Humic Substances on Ability of Sunlight to Damage Fecal Coliforms in Waste Stabilization Pond Water." *Applied and Environmental Microbiology 58*: 1335–1343.

A naturally occurring auxotroph.
Gerhardt, Phillip. 1981. *Manual of Methods for General Microbiology*. Washington, DC: American Society for Microbiology.

Organism smaller than 0.45 micrometers.
Shirley, James J. and Gary K. Bissonnette. 1991. "Detection and Identification of Groundwater Bacteria Capable of Escaping Entrapment on 0.45 micrometer-Pore Size Membrane Filters." *Applied and Environmental Microbiology 57*: 2251–2254.

A member of the genus *Azotobacter*.
Yao-Tsing, Tehan and New. 1984. Genus 1 : *Azotobacter*. *in* Krieg and Holt, *Bergey's Manual of Systematic Bacteriology*, Vol. 1. Baltimore, MD: Williams and Wilkins, p. 223.

An acetic-acid–producing organism.
Asai, T. 1968. *Acetic Acid Bacteria*. Baltimore MD: University Park Press.

Polyethylene glycol degrader.
Obradors, N. and J. Aguilar. 1991. "Efficient Biodegradation of High-Molecular-Weight Polyethylene Glycols by Pure Cultures of *Pseudomonas stutzeri*." *Applied and Environmental Microbiology 57*: 2383–2388.

Oleic acid degrader.
El-Sharkawy, Saleh H. 1992. "Microbial Oxidation of Oleic Acid." *Applied and Environmental Microbiology 58*: 2116–2122.

An obligate anaerobe.
Gerhardt, Phillip. 1981. *Manual of Methods for General Microbiology*. American Society for Microbiology Washington, DC.

A member of the genus *Legionella*.
Barker, John. 1992. "Relationship Between *Legionella pneumophila* and *Acanthamoeba polyphaga*: Physiological Status and Susceptibility to Chemical Inactivation." *Applied and Environmental Microbiology 58*: 2420–2425.

An acidophile.
Johnson, D. Barrie, M.A. Ghauri, and M.F. Said. 1992. "Isolation and Characterization of an Acidophilic, Heterotrophic Bacterium Capable of Oxidizing Ferrous Iron." *Applied and Environmental Microbiology 58*: 1423–1428.

A sulfur-oxidizing organism.
Southam, G. and T. J. Beveridge. 1992. "Enumeration of *Thiobacilli* within pH-Neutral and Acidic Mine Tailings and Their Role in the Development of Secondary Mineral Soil." *Applied and Environmental Microbiology 58*: 1904–1912.

An organism capable of producing hydrogen sulfide gas.
Moats, W.A. and J.A. Kinner. 1976. "Observations on Brilliant Green Agar with an H_2S Indicator. *Applied and Environmental Microbiology 31*: 380–384.

A representative of the genus *Sphaerotilus* or *Leptothrix*.
VanVeen, W.L., E.G. Mulder, and M.H. Deinema. 1978. "The *Sphaerotilus-Leptothrix* Group of Bacteria." *Microbiolological Reviews 42*: 329–356.

An obligate osmophile.
Munitis, M.T., E. Cabera, and A. Rodriguez-Navarro. 1976. "An Obligate Osmophilic Yeast From Honey." *Applied and Environmental Microbiology 32*: 320–323.

A representative of the genus *Zymomonas*.
Swings, J. and J. DeLey. 1977. "The Biology of *Zymomonas*." *Bacteriological Reviews 41*: 1–46.

A representative of the genus *Beggiatoa.*
Burton, S.D. and J.E. Lee. 1978. "Improved Enrichment and Isolation Procedures for Obtaining Pure Cultures of *Beggiatoa.*" *Applied and Environmental Microbiology 35*: 614–617.

A microorganism capable of metabolizing polyaromatic hydrocarbons.
Kiyohara, H., K. Nagao, and K. Yana. 1982. "Rapid Screen For Bacteria Degrading Water-Insoluble Solid Hydrocarbon on Agar Plates." *Applied and Environmental Microbiology 43*: 454–457.

A representative of the genus *Hyphomicrobium.*
Mooren, R.L. and K.C. Marshall. 1981. "Attachment and Rosette Formation by Hyphomicrobia." *Applied and Environmental Microbiology 42*: 752–757.

A microorganism capable of using petroleum as its sole carbon source.
Atlas, R.M. 1981. "Microbial Degradation of Petroleum Hydrocarbons: An Environmental Perspective." *Microbiological Reviews 45*: 180–209.

A chemoheterotrophic *Spirilla*.
Krieg, N.R. 1976. "Biology of the Chemoheterotrophic *Spirilla*." *Bacteriological Reviews 40*: 55–115.

A representative of the genus *Xenorhabdus.*
Akhurst, R. 1980. "Morphology and Functional Dimorphism in *Xenorhabdus sp*. Bacteria Symbiotically Associated with the Insect Pathogenic Nematodes *Neoaplectana* and *Heterorhabditis*." *Journal of General Microbiology 121*: 303–309.

Index

Page numbers in **boldface** indicate a figure; page numbers in *italics* indicate tabular material.